Advance Praise for

Making Our World

"*Making Our World* offers an expansive view of the continued evolution of discourses on hacking and making. Readers will appreciate the revitalizing commentary from various geographies and viewpoints that trouble taken-for-granted associations of making and hacking in the contemporary global economy and culture. This book is good reading for those seeking to understand not just the ideals and values that continue to be attached to making and hacking but also the variegated situated practices that accompany their reproduction and repurposes in various sociopolitical contexts."

—Seyram Avle, Assistant Professor at the University of Massachusetts Amherst

"This brilliant book analyses hacking and making: the leading spirits of our technological age. *Making Our World* is essential reading for anyone trying to understand how information technologies are being created and are creating our society and the strategies being used by those who make such technologies outside of corporations and governments. The book brings together world-leading experts on some of the most recent technological innovations, and these authors deliver a powerful analysis of the global meaning of both making and hacking technologies. Importantly, the book has a genuinely international reach because it refuses to take the 'global' to be some generic all-encompassing idea and instead analyses in specific contexts around the world the different ways hacking and making create and are being created in society."

—Tim Jordan, Professor of Digital Cultures at the University of Sussex

"*Making Our World* draws important and under-realized connections between often disparate strands of scholarship around the promise of hacking and making. From critical perspectives to hopeful exemplars, the authors pose new forms of world-building as central sites to illuminate the often mystifying influence of technology cultures. Hacking is not only ordinary, the authors show, but also tied up in the production of ordinariness—of the everyday concerns, identities, and collectives through which transformations of civil society unfold. Whether in makerspaces, start-up lofts, hackathons, or protests—online or out on the streets—these grounded studies interrogate and, in some cases, recover the often precarious terms of neoliberal educational and economic reform."

—Daniela Rosner, Assistant Professor at the University of Washington

Making Our World

Digital Formations

Steve Jones
General Editor

Vol. 120

The Digital Formations series is part of the Peter Lang Media and Communication list.
Every volume is peer reviewed and meets
the highest quality standards for content and production.

PETER LANG
New York • Bern • Berlin
Brussels • Vienna • Oxford • Warsaw

Making Our World

The Hacker and Maker Movements in Context

Edited by Jeremy Hunsinger
& Andrew Schrock

PETER LANG
New York • Bern • Berlin
Brussels • Vienna • Oxford • Warsaw

Library of Congress Cataloging-in-Publication Data

Names: Hunsinger, Jeremy, editor. | Schrock, Andrew, editor.
Title: Making our world: the hacker and maker movements in context /
edited by Jeremy Hunsinger and Andrew Schrock.
Description: New York: Peter Lang, 2019.
Series: Digital formations, vol. 120 | ISSN 1526-3169
Includes bibliographical references.
Identifiers: LCCN 2018039551 | ISBN 978-1-4331-6001-1 (hardback: alk. paper)
ISBN 978-1-4331-6000-4 (paperback: alk. paper)
ISBN 978-1-4331-6002-8 (ebook pdf) | ISBN 978-1-4331-6003-5 (epub)
ISBN 978-1-4331-6004-2 (mobi)
Subjects: LCSH: Social movements—Technological innovations.
Internet and activism. | Digital media. | Makerspaces.
Classification: LCC HM881 .M338 2019 | DDC 303.48/4—dc23
LC record available at https://lccn.loc.gov/2018039551
DOI 10.3726/b14469

Bibliographic information published by **Die Deutsche Nationalbibliothek**.
Die Deutsche Nationalbibliothek lists this publication in the "Deutsche Nationalbibliografie"; detailed bibliographic data are available on the Internet at http://dnb.d-nb.de/.

29 Broadway, 18th floor, New York, NY 10006
www.peterlang.com

Table of Contents

Introduction vii
JEREMY HUNSINGER AND ANDREW R. SCHROCK

Section I: Histories Introduction 1
ANDREW R. SCHROCK

1. Learning by Doing: The Tenuous Alliance of the "Maker Movement" and Education Reform 3
T. PHILIP NICHOLS AND DEBORA LUI

2. Kevin Mitnick, The New York Times, *and the Media's Conception of the Hacker* 21
MOLLY R. SAUTER

3. Making Civic Media in the Post-Fukushima Japanese Media Ecology 37
YASUHITO ABE

4. Project Chanology and the Formation of Anonymous as an Activist Movement 55
RHEA VICHOT

Section II: Politics Introduction 75
ANDREW R. SCHROCK

5. Conscientious Hacking and the Weak Collective 79
NATHANAEL BASSETT

6. Policy Hacking: Opening Up the Code of Media and Communications Regulation 97
ARNE HINTZ

7. Hacking Administration—A Report From Los Angeles 115
MORGAN CURRIE

8. Why Locality and Presence (Still) Matter for Political Activism 133
SEBASTIAN KUBITSCHKO

Section: III: Organizing Introduction 151
JEREMY HUNSINGER

9. Basteln, Tinkering, and Bricolage: A Cultural History of Hacking 153
ALEXANDER VON LÜNEN

10. Women's Hacking of the Poison Gift of Free/Libre/Open Source Software 171
JENNIFER MAHER

11. Making Space for a Revolution: Occupy Wall Street as a Maker Movement 197
ALISON E. VOGELAAR AND CHARLOTTE M. MCKERNAN

12. The Détente Model of Managing Divergent Values in the Maker-Sphere 217
ANN LIGHT

Section IV: Case Studies Introduction 235
JEREMY HUNSINGER

13. Hacker Agency and the Raspberry Pi: Informal Education and Social Innovation in a Belfast Makerspace 237
PIP SHEA

14. Hacking as a Way of Life: "Makers" at the Margins of Global Digital Culture 255
NICHOLAS BALAISIS

15. The Paradox of Maker Movement in China 271
XIN GU

16. Our Community Hacks: Exploring Hive Toronto's Open Infrastructures 293
KAREN LOUISE SMITH

Afterword: Hackers and Makers are Ordinary 313
ANDREW R. SCHROCK

Introduction

JEREMY HUNSINGER
Wilfrid Laurier University

ANDREW R. SCHROCK
Chapman University

In this volume, we seek to provide an introduction and an overview of important projects in the worlds of hacking and making. In performing those tasks, we had to be inclusive and engaged in the debate about how these movements have been interpreted across disciplines. Communication studies, media studies, political science, policy studies, information science, computer science, and other fields have all engaged in studies of hacking and making from a variety of perspectives and methods. While this volume recognizes the scholarship as it exists and uses it, it attempts to provide another foundation from which to build. We use examples and cases as the primary location of analysis of hacking and making. This provides our readers conceptual tools to research future hacking and making, and examples to think through the present.

The book is an engagement with many tangents that are coming together to define a set of intersections which construct a mutating, coalescing, and diverging space. For this reason, we have exercised a light editorial influence, only organizing papers across four sections: history, politics, organizing, and cases. These sections illuminate different aspects of the ongoing development of the world of hacking and making. Each section has a selection of materials based on its title and an introduction that goes into greater detail. As such, we invite you to read these chapters either as stand-alone pieces, or in conversation with each section's introduction.

This broader introduction adds some conceptualizations and contexts to ongoing debates beyond this text. It is our hope that larger projects of hacking and making might exist, such as democracy and the politics of hope it fosters. Between the possibilities for mass action and individual agency exists the possibilities of group action; action to change the world such as hacking and making. Hacking and making are frequently found in both online and offline hackerspaces/makerspaces. These spaces enable group actions to change the world through the empowerment of both individuals and groups. It is in these spaces that there must be an active politics beyond hacking and making. It is a politics of activities and practices that disseminate, reproduce, and introduces an assemblage of knowledges and related materialities. Like most institutions around knowledge such as schools and universities, democratic aspirations exist in tension with institutional and societal forces.

Hackerspaces are new versions of old cultural institutions oriented toward hacking and making (Hunsinger, 2017). Hacking can be thought of as a creative activity centered on the use of technology to "scratch an itch" or solve a problem (Hunsinger, 2017; Raymond, 2001). Making tends to include the broader categories found in DIY creativity that aren't necessarily resolving personal goals, but might also be contributing to formal and informal economies in other ways. Hacking and making materialize possibilities in that they tend to require space, which is occupied by an assemblage of people and things that come together to reproduce and/or generate objects. In these spaces, they produce social, cultural and economic relations. It is in these possibilities that people might find hope, or at least a new politics of hope might arise (Fenton, 2012; Lewis, 2009; Salovaara, 2015).

Following the 2016 *New Media & Society* issue on the Democratization of Hacking and Making, there has been a shift away from the politics of hope, and a shift from the democratic faith (Deneen, 2009; Dewey, 1944). The hope for the demos and thus for the masses to become hackers or makers and to use technologies in a way that makes them more equal and free is not lost. That is, the ideologies of freedom and equality still exist within the movement, as they do in most hacker and technological progress oriented movements. The existence of the ideologically oriented teleologies, however, does not entail that there is any substantive change being made that will achieve those ends. While this saddens us, we have to admit that the belief that these ends could come to fruition does comfort us and likely others. While democratic hope still exists, the driving force of these communities seems to be internally-directed. In other words, hackers and makers, while occasionally interconnected with others, have a problem that most of hackerspaces are independent entities and tend toward maintaining that independence. There have been larger organizations of several hackerspaces in some countries, but rarely have they had any large political effect. Most of their effect in the social sphere has been commercial, and arguably that effect is primarily as consumers instead of producers (Day, 2016).

The tendency of independence and internal-directedness is not entirely surprising as the mythos of the independent creator is still a significant ideological construct in information technology. This tendency is also the cause of some of the problems with unionization of information technology workers, which is deeply needed but deeply resisted by the workers themselves though the work toward organizing them is increasing. These ideological constructs and others are slowly transforming this solidly middle-class occupation into a worker class precariat (Gregg, 2015; Neilson & Rossiter, 2008; Wark, 2013). Hacking and making are developing in relation to the needs of that precariat.

This growing precarity is also one of the reasons why hacking and making and thus hackerspaces exist (Lindtner, 2017). They provide access to the tools, infrastructures, and knowledges necessary to "innovate". The mythogenesis of "innovation" as an economic driver has been tied to technology and its development for some time. The imagination of hacking and making as "innovation" is tied to various historical examples that may not be actual progenitors of the movement (Levy, 2001). However, even if the progenitors of hacking and making were sites of innovation, times and situations have changed enough that the continued imagination and mythologizing is less warranted. The mythological garages where several companies (HP, Apple) were launched are less possible in today's economic and ecological system of innovation than they were in the past.

Tools and tooling are one of many costly expenses of "innovation," which hacking and making mythologized and attempted to transfer to the hackers and makers as hobbies, interests and passions. Hackers and makers are encouraged to pursue these activities as "innovative," but they also can be considered exploitative. That is, they transfer the costs of innovation in time and labor from institutions to individuals and communities while allowing capital to accumulate in the institutions exploiting these community's "innovations". Some communities and some companies have invested in hacking and making to drive the system of innovation to capture or if not capture then localize the innovation.

As argued elsewhere, the practices and locations of hacking and making are not centers of innovation. Rather, they are centers of reproduction and dissemination of knowledge (Hunsinger, 2009, 2016). Mostly, hackers and makers don't create new things. Instead, they reproduce things that are built upon ideas that already exist from parts that are closer to advanced LEGOs or Erector sets. They buy the parts from any number of local or online suppliers, with instructions for use. Sometimes they build basic and advanced kits with minor deviations, much like much of the world does with birdhouses. In this case, the birdhouses are electronics and the boards are electronic bits and the nails are wires, solder, and other interconnections. We should not be surprised that the majority of "innovation" is based on consumer electronics, since stores such as Radio Shack have long existed. The "do-it-yourself" (DIY) ethos has always been a rich area for market exploitation (Hunsinger, 2016).

International communities organizing around hacking and making resist diversity (Dunbar-Hester, 2016), excepting those with feminist principles built into their founding. For example, hackerspaces tend to be overwhelmingly male, and parallel the hegemonic ethnicity of their locality. This results in hacking and making being less diverse than the communities that they inhabit. The tendency is not intentional, but the result of social forces and individual choices within the DIY milieu in which they exist. The international space in which they occupy is notably interconnected by the internet, which these groups of makers use extensively for research, discussion, dissemination, planning, and organizational matters. The interconnectedness provides the opportunity and possibility of collaboration, which they have done in the past. There have been national and international conferences for, and there is a significantly commercialized set of conferences for the maker movement. These opportunities should allow for them to reach out to each other and to reach out further into their local community, but given that the growth is limited as described above, this tends not to happen as dynamically we might hope.

This book discusses the myriad of institutions that ally with hackers and makers, from libraries to colleges, to universities, to governments, to museums, to businesses, to community organizations, to radical groups, and onward. The hacking and making movements have long been seen either as the development of hobbies in the era of precarity or as systems of innovations attached to other organizations. However, both of these imaginations of hacking and making are born of ideologies of hope, resisting the knowledge that books like this one provide in favor of imagination of possibility for positive social change and new possibilities for institutions in communities. Sometimes this has worked well, especially in regards to library hackerspaces where knowledge is a good in its own right. However, frequently hackerspaces fail to be justified in other systems, such as profit-seeking corporations. The structures of legitimation in which FIY movements such as hackerspaces and maker movement flourish requires a vibrant ecosystem that appreciates knowledge over profit, though inarguably they rely on profit within their ecosystem.

As our authors show, there is a plurality of forms of hacking and making, and with that, there is a plurality of academic disciplines researching them. There have been some great studies of hacking and making in the last few years such as David Gauntlett (2011), Sarah Davies (2017) and many of the contributors of this volume. This work has sought to bring forth the topic, to stabilize the space of debate and the definitions; our volume is not engaged significantly in the definition debates around hacking and making. It is an engagement with many tangents that are coming together to define a set of intersections which construct a space of possibility.

The possibility that this book creates is a possibility of meaning and perhaps the development of the knowledges around hacking and making. The direction of research represented are diverse and transversally connected in numerous ways. It is in the interconnection that possibilities arise for future research and learning about hacker and makers.

References

Davies, S. R. (2017). *Hackerspaces: making the maker movement.* London, UK: Polity

Day, A. (2016). *DIY Utopia: Cultural imagination and the remaking of the possible.* Lanham, MD: Lexington Books.

Deneen, P. (2009). *Democratic faith.* Princeton, NJ: Princeton University Press.

Dewey, J. (1944). The democratic faith and education. *The Antioch Review, 4*(2), 274.

Dunbar-Hester, C. (2016). "Freedom from Jobs" or learning to love to labor? Diversity advocacy and working imaginaries in Open Technology Projects. *Revista Teknokultura, 13*(2), 541–566.

Fenton, N. (2012). The internet and radical politics. In J. Curran, D. Freedman, & N. Fenton (Eds.), *Misunderstanding the internet* (pp. 149–176). London; New York, NY: Routledge.

Gauntlett, D. (2011). *Making is connecting: The social meaning of creativity from DIY and knitting to YouTube and Web 2.0.* Cambridge; Malden, MA: Polity Press.

Gregg, M. (2015). Hack for good: Speculative labour, app development and the burden of austerity. *The Fibreculture Journal, 25.*

Hunsinger, J. (2009). Knowledge and cultural production in the context of contemporary capitalism: A response to wittkower. *Fast Capitalism, 4*(1). Retrieved from http://www.uta.edu/huma/agger/fastcapitalism/4_1/hunsinger.html

Hunsinger, J. (2016). Our knowledge is our market: Consuming the DIY world. In A. Day (Ed.), *DIY Utopia: Cultural imagination and the remaking of the possible.* Lanham, MD: Lexington Books.

Hunsinger, J. (2017). Hacking together globally. *Digital Culture & Society, 3*(1), 95–108.

Levy, S. (2001). The hacker ethic, hackers: Heroes of the computer revolution. *Penguin-USA, Cap, 2*, 32–41.

Lewis, J. (2009). Digitopians: Transculturalism, computers and the politics of hope. *International Journal of Cultural Studies, 1*, 373–389.

Lindtner, S. (2017). Laboratory of the precarious: Prototyping entrepreneurial living in Shenzhen. *WSQ: Women's Studies Quarterly, 45*(3), 287–305.

Neilson, B., & Rossiter, N. (2008). Precarity as a political concept, or, Fordism as exception. *Theory, Culture & Society, 25*(7–8), 51–72.

Raymond, E. S. (2001). *The cathedral and the bazaar.* Sebastopol, CA: O'Reilly Media, Inc.

Salovaara, I. (2015). Spaces of emotion: Technology, media and affective activism. In T. Miller (Ed.), *The Routledge companion to global popular culture* (pp. 471–480). New York, NY: Routledge, Taylor & Francis Group.

Wark, M. (2013). *Considerations on a hacker manifesto.* London: Routledge.

Section I. Histories Introduction

Andrew R. Schrock
Chapman University

"Studying the past," suggested John Durham Peters, is "a problem *of* communication." Hacking and making present just such a challenge. They have had ephemeral histories, passed down through word-of-mouth, textfiles, and guides. Hacker culture has often prided itself on technological skills and obscurity. Researching histories of hackers often entails unearthing communications of a particular group or publication. The maker movement, although relatively recent, has seen more significant financial support and investment. Both hacking and making presents particular, although quite different, challenges for understanding how identities, practices, and media are developed. We have set aside a section for histories to better understand the historical contours of these movements.

The Maker Movement has prided itself on its ability to reform education and improve work conditions through hands-on education. Philip Nichols and Debora Lui suggest that this claim tenuously relies on reconciling two vastly different historical perspectives on education. They describe how Maker Media, Inc. situates making as an "ethos" that echoes previous waves of experiential learning, particularly that of John Dewey. The Maker Movement embraces quite different—even conflicting—outcomes: experiential education, STEM-based learning, and entry into the world of entrepreneurship. Making is thus revealed to be an extension of historical tensions around education and economics that may even work counter to their emancipatory goals. "Without an intentional parsing of these conflicts," write Nichols and Lui, "making risks becoming another means for reproducing the same inequities that have followed previous educational movements."

The media has often circumscribed deviant behavior and technological anxieties by defining hackers as evildoers. For example, Helen Nissenbaum has suggested that law and the media essentially criminalized hackers for doing what they always have—exploring technological systems for the thrill of discovery. For this reason, Kevin Mitnick has been a pivotal figure in hacker cultures. His 1995 arrest and imprisonment drew protests, "Free Kevin" t-shirts, and articles in both underground and mainstream publications. In this section, Molly Sauter analyzes articles on Micknick during this period published in the *New York Times.* They suggest the media initially referred to Mitnick as a criminal hacker, "one of the prime social enemies of the technological age," both deviant and "superuser." Eventually, the hacker persona had to be jettisoned to complete his symbolic rehabilitation and return to society. Mitnick was increasingly referred to as a reformed hacker to clarify his motives as related to monetary gain and sanitize his identity. They conclude that the metaphors used to describe hacker operate as a kind of morality play to talk about technological anxieties and crime: "the hacker folk devil is still present in modern tech journalism and media."

At other historical moments hackers and makers have seized power by becoming media creators and appropriators on the world stage. A massive 9.0 earthquake struck Japan in 2011, resulting in the Fukushima nuclear reactor suffering a near meltdown and began leaking radiation. The Japanese government provided scant information on the disaster, and news organizations largely followed suit. Citizens were left with little way to find out the extent the radiation had spread. Yasuhito Abe traces the actions of an unlikely combination of hackerspace members and environmentally aware techies that banded together to collect and share radiation data. He argues that Safecast, the organization that led this effort, can be considered as creating civic media. The "press club system" for news distribution unique to Japan made it easy for vital information to be suppressed, even during such a sudden event. The presence of Safecast—and its unfiltered statistics and maps—was particularly vital for citizens' "need to know" during moments of crisis.

Of course, history will never be a singular narrative. Rhea Vichot argues that attention has been paid to Anonymous' activities came at the expense of understanding its emergence from online communities. She traces how the political tactics and awareness emerged from 2chan and 4chan, anonymous Japanese message boards. A conflict between the Church of Scientology and Anonymous precipitated their evolution into a "protest organization" around what was dubbed "Project Chanology." The group started to explore offline protests to discredit, provoke and disrupt the Church of Scientology. In the process, they created guidelines for conduct and began to appropriate popular media for their messages. The sophisticated combinations of détournement, tactical media, and culture jamming began to mark this shift as significant in terms of media practices.

1. *Learning by Doing: The Tenuous Alliance of the "Maker Movement" and Education Reform*

T. Philip Nichols
University of Pennsylvania

Debora Lui
University of Pennsylvania

Over the last decade, the promises of the Maker Movement—a growing public interested in do-it-yourself designing, remixing, and building using physical and digital tools—have found resonance in the field of education, sparking discussion among researchers, policymakers, and practitioners. In many of these conversations, *making* has been positioned as a catalyst for educational change—an intervention to transform the rote activities and bureaucratic structures that often characterize schooling in the popular imagination. In policy, for example, President Obama's *Educate to Innovate* campaign inaugurated a National Week of Making, where he called on students to be not just passive consumers of information and products, but to become a "nation of *makers*" (White House, 2014). In research, the *Harvard Educational Review* devoted a full symposium to "The Maker Movement in Education," featuring scholarly accounts of youth *making* practices and their possibilities for reimagining school instruction (Halverson & Sheridan, 2014). And among teachers, the National Writing Project—an organization focused on improving writing in K-12 contexts—has offered workshops and professional development on "Writing as Making," integrating tenets of the movement into classroom literacy instruction (National Writing Project, 2013). In each of these forums, *making* has been framed as an alternative to the routines and rituals of "traditional" schooling—offering, instead, a different vision for education: learning through active, hands-on problem-solving, or, "learning by doing."

However, this narrative of *making* as a counterpoint to more conventional models of instruction elides a more complex history of 20th century U.S. education reform and the Maker Movement's place therein. While it is certainly true that many classrooms have foregrounded routine, efficiency, and standardization—what Tyack and Tobin (1994) call "the grammar of schooling"—there also exists a rich parallel tradition of "progressive" school reform, rooted in a "learning by doing" ethos similar to that invoked by the present-day Maker Movement (Cremin, 1961; Dewey, 1938). These progressive approaches have not always enjoyed unanimous public support, but as education historians have shown, their central interest in experiential learning has left a considerable mark on research, teacher preparation, and education policy (e.g. Zilversmit, 1993). For example, in the last decade, as publications associated with *making* have advocated for greater emphasis on "learning by doing," many similar conversations were already taking place across formal and informal educational spheres: from increasing attention to "hands-on" training in science and technology (National Research Council, 2012); to renewing interest in creativity and innovation across curricula (Bronson & Merryman, 2010; White House, 2014); to broadening conceptions of literacy that extend to new and non-traditional media (Kress, 2010; Lankshear & Knobel, 2009). This is not to suggest, of course, that all schools fully embody or practice these "progressive" ideals—only that such ideals have congealed as recognizable goals toward which multiple and diverse educational stakeholders are striving (cf. Egan, 2004). Given the alignment of these aims with those of the larger Maker Movement, then, we can begin to understand the current resonance of *making* not as a break or intervention in existing practice, but rather, as a recent extension of a familiar mode of progressive critique.

In this chapter, we aim to situate *making* within this larger landscape of contemporary education reform, analyzing the claims of researchers and policymakers alongside those of the Maker Movement itself. We do so in order to better understand the alliance these groups have found with one another—and to consider some of its implications as the Maker Movement continues to find resonance in education policy, pedagogy, and practice. In discussing "the Maker Movement," we limit our attention to Maker Media, Inc. and its associated figures, publications, and activities rather than all informal, grassroots activities that might be included under the umbrella of *making*. We do so not because the latter is less important, but because of the substantive role the former has already played in shaping the discourse of *making* in education. While the stated pedagogical aims of Maker Media Inc. are ostensibly consistent with those of other educational stakeholders, we argue that they are internally conflicted. Though the Maker Movement maintains rhetorical alignment with the "learning by doing" ethos of progressive education, it is equally committed to cultivating social mobility through independent entrepreneurship and nurturing STEM labor for global economic competition—three educational outcomes that are not easily reconciled with one another. In this way, even as *making* and education have been positioned as unproblematic allies in the effort to reform schools, we argue that, without critical inquiry on the part of teachers and researchers, the competing visions latent in each have potential to upend progress toward effective and equitable teaching and learning.

Making and Education

At the outset, it is worth stating directly: education is political. Beneath the activities, assignments, experiences, and assessments—it is a process of molding individuals, of producing certain kinds of human subjects (and, by extension, of not producing others). Media theorist and progressive educator Neil Postman (1970) once said,

> In the broadest sense, a political ideology is a conglomerate of systems for promoting certain modes of thinking and behavior. And there is no system that more directly tries to do this than schools. There is not one thing that is done to, for, with, or against a student in school that is not rooted in a political ideology, bias, or notion. (p. 244)

As such, while it is common to think about education as a matter of what a student ought to *know*, the practice of education is often rooted in unstated assumptions about what a student ought to *be*. Schools are not only concerned with instilling knowledge, but also with cultivating practices, dispositions, and ways of being. This phenomenon has been explored by a range of social theorists. Foucault (1977), for example, delineates the disciplinary techniques used in school contexts to produce desirable forms of obedience and docility. Bourdieu (1990), likewise, examines how these mechanisms contribute to the creation, sustenance, and reproduction of class identities and structures of inequality. With this in mind, the recent resonance of *making* among educators raises important questions for those who study and participate in the movement: What kinds of subjects does *making* aim to produce? And for what purpose? In the section that follows we explore this first question in the context of education—analyzing how Maker Media,

Inc., the commercial arm of the Maker Movement, positions *making* not as learned content or skills, but as a learned "ethos." We then turn to the second question, examining the different ways this ethos is put to work for diverse educational stakeholders.

Making as "Ethos"

To understand the kinds of student subjects that *making* is intended to produce, it is necessary to attend to the vocabulary of the movement itself. As David Lang, author of Maker Media, Inc.'s *Zero to Maker* (2013) says, "Learning the *maker* lingo is as important to understanding the *maker* culture as speaking Spanish is to understanding Mexico" (p. 22). For those unaccustomed to the lexicon of *making*, one of its most striking idiosyncrasies is its tendency to ascribe uncommon meanings to common words. Most notable among these is the term "make" itself. In common usage, "making" is a transitive verb, paired with a particular object that receives the verb's action—for instance, a chef can make *dinner*; or a carpenter can make *a birdhouse*. However, Lang (2013) defines the term as the process of "creating or exploring new possibilities through building and experimenting with tools, technology, and materials" (p. 22). In such a framing, "making" becomes intransitive—and the object being made is subordinated to "making" as a standalone *process*. In other words, *making* becomes less an act of creation and more of a stance, or "ethos."

A similar subordination occurs in the use of "learning"—which, like "making," also commonly takes a transitive form, emphasizing some particular skill, practice, or subject-area that is being grasped or mastered. However, in the literature of the movement, this too becomes intransitive—characterized, once again, as a non-specific process by which cognition and personal growth occur irrespective of the content or activity involved. Dale Dougherty, the founder of Maker Media Inc. and the first to coin the phrase "Maker Movement," anticipates the skepticism such a vague definition might elicit from educators. In his forward to Curt Gabrielson's *Tinkering: Kids Learn by Making Stuff* (2013), he raises the hypothetical question, "But what is actually being *learned*?" then offers his response:

> [Students] are learning something and it may be the most valuable thing they've learned all week, and it may raise all sorts of questions in their minds that inspire them to learn more about what they're tinkering with, and it may start them on a path to a satisfying career, not to mention good fun on their own time, and it may put them in the driver's seat of their education by realizing their competence and ability to learn through tinkering, and they may begin to demand more of just this sort of learning opportunity. (p. xii)

In other words, the substance that is learned through *making* is not disciplinary content or skills, so much as it is an orientation: the *maker ethos*. AnnMarie Thomas, co-founder of Maker Media Inc.'s education initiative, reinforces this when she outlines the traits, values, and practices she sees as most necessary in formal and informal learning environments: curiosity, playfulness, willingness to take risks, persistence, collaboration, and optimism (2014, p. 5). Together, these traits constitute a model disposition toward learning that Dougherty (2013) refers to as "the maker mindset"—a phrase that has found traction among educators and researchers who are, likewise, interested in fostering creative impulses toward design, play, and innovation among students (e.g. Honey & Kanter, 2013).

As this "maker mindset" and the attendant terminology of *making* are taken up in discussions of teaching and learning, it is worth emphasizing that this perspective—one that situates learning as an orientation rather than the absorbing of discrete facts and figures—has a longer history in the field of education. While Maker Media publications play up the subversiveness of the "maker ethos" as a foil to the mindlessness of discipline-specific, teacher-centered instruction in traditional schools, calls for similar orientations abound. From "computational thinking" (Papert, 1980, 1996) and the "scientific mindset" (Kalman & Aulls, 2003) to the ethos of "new literacies" that animates youth practices in emerging digital environments (Lankshear & Knobel, 2009)—each of these recognizes learning as a social process rooted in action, and by extension, sees education as a means of cultivating orientations that allow students to effectively navigate this process. Importantly, each of these "mindsets" bolster their respective claims by situating themselves in the larger historiography of "progressive education"—a pedagogical approach popularized by John Dewey that foregrounds the experience of the learner over the content being learned. With this in mind, we can begin to see the Maker Movement not as representing a hygienic break from the education of the past, but as an outgrowth and extension of experiential learning and progressive reform.

All of this speaks to the political ideology of the Maker Movement in education and the work it does to shape particular kinds of student subjects. Positioning "making" and "learning" as orientations rather than activities linked to specific content illuminates the constellation of qualities that the movement aims to produce in an ideal *maker*: creativity, spontaneity, curiosity, resourcefulness. Further, *makers* are to be self-motivated in following their desires and interests, and flexible in navigating the structural, organizational, and logistical challenges that might stand in their way. The irony, of course, is that while the literature of the movement positions these traits as subversive, there are few (if any) educational stakeholders who would object to students exhibiting such qualities. Indeed, even some of the most rigid federal reforms of the post-war era were couched in language that foregrounded similar dispositions (Rudolph, 2002; Urban, 2010). The "subversiveness" of *making*, then, appears to be less about the tangible consequences it yields for education reform, than it is about the discursive work that it does when juxtaposed with the popular image of schooling as joyless, routinized, and teacher-centered. Whether these descriptors are accurate representations of how most classrooms operate is beside the point—it is a narrative that people believe, and as a result, helps fuel the need for the counterweight of "progressive" reforms, of which *making* is the latest iteration (cf. Labaree, 2004). With that said, the continuity between *making* and earlier iterations of progressive reform does not mean there are no important differences in the ways these alternatives position students or theorize schooling. In the section that follows we examine how different educational stakeholders leverage this *making* ethos for diverse, and at times, conflicting purposes.

Aims of Education in the Maker Movement

While it is clear there is an overlap in the language and aims of the Maker Movement and certain historical progressive reform efforts, it remains to be see why the ethos of *making* has found the resonance it has across educational stakeholders, including those not typically aligned ideologically. Throughout the United States, researchers, policy makers, principals, and teachers have called for maker-based programs and initiatives in both formal and informal learning contexts—from library makerspaces and after school programs (Vossoughi, Hooper, & Escudé, 2016) to school curricula explicitly designed around principles of *making* (Stornaiuolo & Nichols, 2018). A major reason for this broad appeal, we argue, is that the concept of *making* is malleable—able to be leveraged, at different times and by different actors, to fulfill a plurality of competing goals that hold weight in the current political, economic, and cultural landscape. In what follows, we elaborate on three such goals: experiential education, STEM learning, and individual entrepreneurship.

Experiential Education

As we have already suggested, the assumptions underpinning *making* in education are not entirely novel—but rather, an extension of earlier pedagogical modes that stress the importance of "learning by doing" (cf. Blikstein, 2013; Martinez & Stager, 2013). Such approaches are often referred to as progressive, or experiential education. As the name suggests, experiential education emphasizes the experience of the student as the principal driver and motivator of teaching and learning. Its proponents argue that learning has value for its own sake—not just to the degree that it can be used for some larger instrumental outcome—and they often foreground tactile and material encounters as opportunities to cultivate these learning experiences.

While these methods were popularized in the work of John Dewey and the Progressive Education Society (Cremin, 1961)—their history in education reform stretches back even further. In the mid-19th century, for example, Edward Sheldon drew on the writings of Rousseau and Froebel to develop an instructional method he called "object teaching." This strategy aimed to put students in direct contact with material artifacts—wood, sugar, glass, leather—in order to help them learn about their properties and uses firsthand, through observation and experimentation (Kohlstedt, 2010). At the turn of the century, advocates of the Nature-Study Movement worked to expose students to hands-on, experiential inquiry through the building and maintenance of outdoor gardens (Armitage, 2009). In 1918, William Kilpatrick—a colleague of Dewey's at Teachers College, Columbia University—published his "Project Method," which argued that for learning to be "wholehearted and purposeful," it needed to connect students to real problems and activities (Schneider, 2014). In such models, we can see threads of the sort of experiential education that today's Maker Movement looks to cultivate.

While progressive, experiential education was, and remains, far from a monolith, Dewey's articulation—which has had the most decisive impact on education research and practice in the last century (e.g. Labaree, 2004)—combined these experiential traditions with emerging psychological and political theory. Using studies of cognitive development and evolutionary biology, he argued that students learn best when their curiosity is permitted to move them from the concrete and familiar to the abstract and foreign—thereby adding the weight and prestige of scientific discovery to "learning by doing" methods sometimes criticized for their lack of rigor (Cremin, 1961; Kohlstedt, 2010). This was the basis for his Laboratory School at the University of Chicago, which supplanted books and lectures with hands-on, problem-based activities (Egan, 2004). Importantly, Dewey also imagined a political valence to this work: progressive pedagogy was intended as a social process to develop in students a sense of civic virtue and a desire for democratic participation. For this reason, Dewey was critical of those who used schooling or the name of "progressive education" to advocate for narrow, utilitarian aims like job preparation. Such approaches, according to Dewey (1916), not only reduced education to an extension of the "industrial regime," but also "perpetuated social divisions with their counterpart intellectual and moral dualisms" (p. 329). In other words, for Dewey, a truly progressive pedagogy marshaled experience in service of communal inquiry and eschewed instrumental ends that sort children into stratified vocational categories.

Though the Progressive Education Association folded in the late 1950s, Dewey's politically-infused approach to experiential learning continued to enjoy favor among certain educational stakeholders. Notable among these were the scholars, teachers, and activists whose work extended the counter-cultural critiques of the 1960s and 1970s to the sphere of school reform: Paul Goodman, John Holt, Neil Postman, Herb Kohl, and Ivan Illich, among others. Goodman (1966, for example, argued against compulsory education, in part, because he believed its institutional structure could not sustain the open-endedness that a true student-centered, "learning by doing" pedagogical method required. Postman and his collaborator Charles Weingartner (1969), likewise, suggested that such modes of experiential learning might necessitate a utopian re-imagining of school spaces entirely—offering the hypothetical of a school with no formal building or classes, but rather, a community support network that would allow students to explore their curiosities and learn about civic-engagement in real-world environments. Such viewpoints reinforced Dewey's contrast between experiential pedagogies tied to democratic participation and those tied to more utilitarian outcomes.

While it would be anachronistic to suggest that these 19th and 20th century iterations of material-based "learning by doing" were direct examples of *making*, or even proto-*making*, this is the genealogy that Maker Media publications frequently invoke to contextualize and justify the proliferation of *making* in education reform and practice. AnnMarie Thomas (2014), makes this connection when she writes,

> The Maker Movement, and the self-identified makers who are at the heart of it, are celebrating many of the qualities and actions that educators have long been trying to promote: lifelong learning, self-directed learning, communication, collaboration, creativity, and design. (2014, p. x)

Others in the movement make more explicit links to this history of experiential pedagogy. Most Maker Media book publications, for example, include a chapter on the origins of *making* that maps its development onto the lineage of progressive pedagogy—usually starting with John Dewey, extending through Goodman, Postman, and the "radical romantic" reformers of 1960s and 1970s, and culminating in the work of computer scientist Seymour Papert, who is often fashioned as "the godfather of *making*" (e.g. Gabrielson, 2013; Martinez & Stager, 2013).

Researchers, likewise, have explored the intersection of *making* and the history of experiential learning. In his book, *Making is Connecting* (2011), media scholar David Gauntlett devotes an entire chapter to elucidating the ties between the modern-day Maker Movement and the work of anarchist philosopher and educational theorist, Ivan Illich. Gauntlett interprets the core of Illich's best-known educational works, *Deschooling Society* (1970) and *Tools for Conviviality* (1973) as centering on the distinction between the processes of "schooling"—that is, "conformity to rules and memorization of a set body of knowledge without necessarily learning or understanding"—and a true education, which is driven by the impulses of experiential, "learning by doing" (Gauntlett, 2011, p. 168). Gauntlett argues that *making* is an extension of this latter category—a "humanizing" activity that allows individuals to express themselves in "authentic" ways that are not normally sanctioned by industrialized environments, like schools. Similar beliefs—and connections to Illich's work—are echoed throughout the literature of Maker Media, Inc. (e.g. Lang, 2013; Thomas, 2014). Even the slogan for the company's education initiative—"Every Child a Maker"—suggests that there is an inherently human value in *making* that necessitates that it be developed in every student. By invoking this history of progressive, experiential education, then, the Maker Movement aligns itself with this larger tradition that aims to produce students who see learning as something valuable for its own sake, tied to democratic engagement, and distinct from any larger utilitarian outcome.

STEM-based Learning

In addition to experiential learning, many within the Maker Movement highlight the capacity of *making* to aid in the instruction of particular disciplinary content—especially forms associated with science, technology, engineering, and mathematics (STEM). In some respects, this pairing follows logically: *making* activities are well-aligned with STEM education as both increasingly foreground the use of emerging digital and physical prototyping technologies, such as 3D-printers, microcontrollers, and laser cutters. Because of this explicit and visible link, many of the practitioners and researchers who celebrate the promise of the Maker Movement for education reform are those whose work already sits at the intersection of STEM and education—researchers in the learning sciences, for example, or those who train pre-service and practicing PK-12 STEM teachers.

But the connection between STEM and *making* is not limited to university research and classrooms, it also circulates in broader policy discussions—especially as philanthropic and federal initiatives continue to seek education reforms that might stimulate national growth in STEM enterprise (cf. Honey &

Kanter, 2013). Of course, this is not a new phenomenon: there is a long history of post-war and Cold War policies that configure science and technology curricula and instruction for the purpose of bolstering labor in STEM sectors—perhaps most famously in the National Defense Education Act of 1958 (Cohen-Cole, 2014; Phillips, 2015; Rudolph, 2002; Urban, 2010). But initiatives with similar goals have continued to flourish in subsequent decades—from the emphasis on rigorous content standards and the growth of standardized testing in the 1990s (Atkin & Black, 2003) to the more recent publication of the National Research Council's (2012) *Framework for K-12 Science Education.*

Importantly, these efforts to reform and expand STEM education are not only concerned with improving the quality of disciplinary content and instruction, they are often framed in terms that align such reforms with the future of the nation itself—in particular, its competitiveness in the global market. For example, Thomas Kalil (2013), the Deputy Director for Technology and Innovation in the Obama administration, has written about the potential for *making* to not only improve students' learning, but to promote the "economic well-being" of the United States by training up future STEM workers. He argues that the *makers* produced through such reforms will not only help to address the "grand [science and technology] challenges of the 21st century"—such as the decline of fossil fuels and the proliferation of cancer rates—but will also help increase the potential for technological innovations that create new industries and jobs and, by extension, promote national economic growth (pp. 13–15).

Perhaps because of how persistent this alignment of education reform, STEM development, and national competition has been, many of those within the Maker Movement have staked its promise on its ability to generate student interest in STEM careers. Not only is this delineated among the chief goals of Maker Media Inc.'s education initiative (Thomas, 2014), but the company has very publicly aligned itself with government programs designed to build and reinforce pipelines between formal and informal education and STEM industries. These include the Defense Advanced Research Projects Agency's (DARPA) Manufacturing Experimentation and Outreach (MENTOR) and AmeriCorp's Maker VISTA programs, which bring *making* to high school students in urban and rural districts for the purpose of training a more robust and diversified national STEM workforce (Kalil, 2013).

Even those *maker* activities that are not explicitly related to science or technology—designing, playing, and tinkering, for example—are often framed in the movement's literature as a means to increased engagement with STEM disciplines. Bennett and Monahan (2013), for example, argue,

> Design is not necessarily an efficient way to teach specific STEM content. It is, however, a powerful way to kindle a desire to learn that content ... At the heart of every good design problem is the opportunity to bump up against rich STEM content. (p. 36)

Thomas (2014), likewise, argues that "the Maker Movement presents real opportunities for increasing technical literacy and reintroducing people to the arts of *making* and tinkering" at a time of "increased emphasis on STEM in PK-12 curriculum" (p. x). In this way, even as advocates of *making* highlight its potential for undirected, experimental play, this ethos is often justified by highlighting its capacity to lead students toward more directed engagement with STEM activities. In *Design, Make, Play: Growing the Next Generation of STEM Innovators* (2013), editors Honey and Kanter argue that "presenting science as a creative, hands-on, and passionate endeavor" can "enable young people to fall in love with science and technology" (p. 2). Kalil's (2013) contribute to the same book even suggests that "companies concerned about the lack of students with strong STEM and manufacturing" should "support makerspaces in schools and afterschool programs" so that "more students will become excited about excelling in STEM subjects and pursuing STEM-related fields" (p. 15). As such, even when it is not explicitly focused on STEM content, *making* as an educational activity can be seen to support the larger project of STEM education, and its associated goal of nation-building through the maintenance and development of STEM industries.

Entrepreneurship

While not as prominent as the emphasis on experiential learning or STEM development, there is a third undercurrent in the literature of the Maker Movement that foregrounds individual entrepreneurship and social mobility as an educational end. Chris Anderson, the former editor of *Wired Magazine*, highlights this promise in his book, *Makers: The New Industrial Revolution* (2012). Anderson's argument, as his sub-title suggests, is that *making* signals a new form of business enterprise, and that "the industrialization of the Maker Movement" will lead away from traditional mass-manufacturing models and toward an emergent model he terms "personal manufacturing" (p. 41). This theory runs through the Maker Media, Inc. literature as well. David Lang (2013), co-founder of OpenROV—a *maker* community focused on underwater robotics—builds on Anderson's idea, saying, "In the new maker economy, a makerspace membership is the new entry-level job" (p. 155). Elsewhere, he elaborates:

> Large corporations are watching this [*making*] trend, too, and making big bets that this new form of distributive, small-batch manufacturing takes hold ... In a

> time when job and career uncertainty are at an all-time high, it's refreshing to see a budding industry ... with so much potential. (p. 14)

In accounts like these, *making* is positioned as an inoculation against a changing industrial economy—one that positions individuals to "move up" and take advantage of this volatile landscape of production for personal financial gain. This idea is further reinforced in Maker Media Inc. book publications, where the final chapters are often devoted to strategies for taking products to market and building a consumer base (e.g. Kemp, 2013; Lang, 2013). Similarly, the link between *making* and manufacturing can also be seen in the expansion of commercially-focused makerspaces that have spread through United States cities in recent years. Many of these explicitly identify their role not in terms of STEM training or experiential learning, but rather, as support systems to develop start-up businesses (e.g. NextFab in Philadelphia or the Staten Island Makerspace), or as idea incubators for local industry (e.g. the GJMakerspace in Grand Junction, CO).

Importantly, the alignment of *making* and personal manufacturing also extends to the ways the Maker Movement literature discusses education reform. Kalil (2013), for example, links market innovation with *making* in schools. He writes, "Makers are becoming entrepreneurs, leading the development of industrial robots, 3D printers, and smart devices that integrate hardware, software, sensors, and Internet connectivity," and as such, he argues that by engrafting *making* into the curriculum and practice of classrooms, "more students will be empowered not just to get a job, but to create the industries and jobs of the future" (pp. 14–15). In this way, the aims of *making* as expressed in the movement's literature are not only concerned with developing a skilled workforce that can join already-existing modes of economic growth, but also with producing neoliberal subjects capable of creating their own economic opportunities in the national and global marketplace.

Internal Conflicts

In parsing these three threads in the literature of the Maker Movement—experiential learning, STEM development, and personal entrepreneurship—we can see how some of their competing impulses might have implications as *making* is integrated into education research, policy, and pedagogy. While the stated aim of the movement is the cultivation of students who embody "the maker mindset" (Dougherty, 2013), the vagueness of *making* in its intransitive form allows it to be taken up by diverse stakeholders in service of multiple contradictory purposes. We conclude by analyzing these inconsistencies in order to better understand their consequences when applied in educational contexts.

One of the clearest demonstrations of how movement's internal incongruities have been layered together and presented as a coherent educational vision is in Dougherty's (2012) own articulation of *making* education. In an article section titled "Expanding to education, business, government"—itself a testament to the ways these disparate institutions are tangled together in the literature—Dougherty begins by citing the influence of Deweyan progressive education on the movement, saying, "[Dewey] extolled the virtues of learning by doing, and contemporary science of the brain confirms the importance of tactical engagement and of using our hands in the learning process" (p. 12). He then transitions to discuss how the Maker Movement has gone about expanding this sort of experiential learning, detailing Maker Media Inc.'s partnership with DARPA to prepare "student populations that are not well-served" for future participation in STEM careers (p. 13). Dougherty then concludes with the story of a 14-year old who exhibited a product at a Maker Faire that was similar to a home-automation system that a large semiconductor company had spent millions of dollars developing. According to Dougherty, there were two morals to this story. First, that STEM-companies can benefit enormously by "embracing the Maker Movement"—in particular, by attending Maker Faires and providing STEM jobs to those students who inhabit the *maker* ethos. And second, that a 14-year old with these abilities "doesn't need a big manufacturing facility or lab anymore" in order to become a personal entrepreneur. In the span of just a few paragraphs, we can see all three of these differing purposes for education—experiential learning, STEM training, and personal entrepreneurship—layered together and presented as mutually supporting facets of *making* education.

Yet despite the surface-level coherence that Dougherty grants to these tangled purposes, there is danger in conflating their aims without recognizing the ways they are tied to conflicting tensions—not just in the Maker Movement, but in the larger history of education reform. According to sociologist David Labaree, there has always existed a fundamental ambivalence about whether American education "should be considered primarily a public good (one that is inclusive, providing shared societal benefits) or a private good (one that is exclusive, providing selective individual benefits)" (1997, p. 5). Labaree argues that the history of United States education has can be narrated as an effort to negotiate the contradictory pressures these impulses exert on schools, teachers, and students. Such a framing makes legible some ways that educational *making*, as articulated in the literature of the movement, is also braided together with these discordant aims. For example, while it may appear that education for entrepreneurship and social mobility is compatible with education for developing a robust STEM workforce, the former is rooted in the idea of education as a private good, poised to improve the fortunes of individuals by equipping them to compete for jobs, credentials, and income. The latter, by contrast, is grounded in public concerns—albeit, utilitarian ones—for meeting demands of economic efficiency by filling the oft-feared "STEM worker shortage" in the United States (though, it is worth noting that some have questioned the legitimacy of such concerns—e.g. Charette, 2013; Fallace & Fantozzi, 2013; Freeman & Goroff, 2009). Further, both of these perspectives are at odds with the progressive educators so often referenced in the historiography of the movement. Dewey's (1916) insistence that the purpose of experiential pedagogy is not easily reconciled with concerns over industry and job training, for example, suggests an uneasy alignment of *making* with progressivism. Likewise, Ivan Illich—who Gauntlett (2011) cites as the inspiration for *making* culture—openly denounced the tendency for curricula and school reform to be reduced to economic ends—be it job preparation or personal financial security.

In this way, the literature of the Maker Movement brings together diverse and contradictory educational aims and packages them with an attendant vocabulary and set of practices that allow it to appear as a coherent and harmonious whole—one that can be leveraged to do different kinds of work for different stakeholders. For policymakers, *making* increases and diversifies the STEM workforce, even as it encourages personal economic mobility. For educators, it provides a way to align district and state mandates for rigorous STEM programming with the progressive pedagogical practices to which most teachers aspire (or, at least, claim to; cf. Labaree, 2004). And for researchers, it extends the philosophies of educational progressivism to new technological landscapes and, by extension, to the resources and financial supports that are often paired with STEM scholarship. The result is something eminently palatable to all stakeholders. Dougherty himself suggests this when he writes, "It is significant that [*making*] gets little resistance from superintendents and principals" (2012, p. 13)—this is, perhaps, because the concept is malleable enough to be, in a sense, all things to all people.

And yet, taken together, this diversity of aims—each catering to different stakeholders—also exerts contradictory demands on the teachers and students who are asked to carry out these ideals in the day-to-day work of the classroom. Teachers interested in integrating *making* in a school setting, for example, may do so differently if their understanding of the concept more closely aligns with STEM learning or personal entrepreneurship rather than experiential education. And of course, because these competing purposes are, as Labaree (2012) suggests, embedded in the structure of systemic education itself, they are not easily untangled—and the momentum of the historical processes that braid them together prevents us from neatly dis-embedding one purpose from the others. This suggests the importance of reflexivity in research and teaching with regard to educational *making*. Such an approach promotes sustained critical inquiry into the different kinds of work that *making* is being asked to do when invited into school settings, and the ways its implementation reinforces and undermines the purposes for teaching and learning that may be most important to particular educators and students, schools and communities.

We believe an important step in this direction involves highlighting these internal contradictions—especially at a time when *making* programs are finding steady uptake in formal and informal educational contexts. With such a broad array of purposes—experiential learning, STEM development, and personal entrepreneurship—different stakeholders are, in effect, deploying the same terms and practices for dramatically different ends: to serve politics, markets, and individuals; to promote equity and to give certain people a competitive edge; to meet collective and personal needs. And crucially, as education historians have argued, there is a strong precedent for such tensions in the public and private aims of schooling to ultimately favor the latter (e.g. Katz, 1975; Labaree, 2012). Already, scholars across disciplines have warned that *making* could yield a similar outcome—especially considering the vast majority of those who currently participate in *maker*-based activities are white and from middle-class backgrounds (Bean & Rosner, 2014; Rose, 2014). Without a careful and intentional parsing of these conflicts then, *making* risks reproducing or reinforcing some of the same inequities that have persisted through (or been exacerbated by) previous educational movements, revolutions, and reforms. For this reason, we argue that it is necessary to not only remain vigilant in identifying the ideological conflicts between the public and private educational aims of *making*, but to actively work against the neoliberal impulses that may privilege certain aims—and by extension, certain individuals and communities—over others. Only then can the movement truly claim to live up to its ideal of making "every child a maker."

References

Anderson, C. (2012). *Makers: The next industrial revolution*. New York, NY: Business Books.

Armitage, K. (2009). *The nature-study movement*. Lawrence, KS: University of Kansas Press.

Atkin, J. M., & Black, P. (2003). *Inside science education reform: A history of curricular and policy change*. New York, NY: Teachers College Press.

Bean, J., & Rosner, D. (2014). Making: Movement or brand? *Interactions, 21*(1), 26–27.

Bennett, D., & Monahan, P. (2013). NYSci design lab: No more bored kids!. In M. Honey & D. Kanter (Eds.), *Design, make, play: Growing the next generation of STEM innovators* (pp. 17–33). New York, NY: Routledge.

Blikstein, P. (2013). Digital fabrication and 'making' in education: The democratization of invention. In J. Walter-Herrmann & C. Büching (Eds.), *FabLabs: Of machines, makers, and inventors*. Bielefeld: Transcript Publishers.

Bourdieu, P. (1990). *Reproduction in education, society, and culture*. Thousand Oaks, CA: Sage.

Bronson, P., & Merryman, A. (2010). The creativity crisis. *Newsweek*. Retrieved from http://www.newsweek.com/creativity-crisis-74665

Charette, R. N. (2013). The STEM crisis is a myth. *IEEE Spectrum, 50*(9), 44–59.

Cohen-Cole, J. (2014). *The Open Mind*. Chicago, IL: University of Chicago Press.

Cremin, L. (1961). *The transformation of the school*. New York, NY: Vintage.

Dewey, J. (1916). *Democracy and education*. New York, NY: Macmillan.

Dewey, J. (1938). *Experience and education*. New York, NY: Free Press.

Dougherty, D. (2012). The maker movement. *Innovations: Technology, Governance, Globalization, 7*(3), 11–14.

Dougherty, D. (2013). The maker mindset. In M. Honey & D. Kanter (Eds.), *Design, make, play: Growing the next generation of STEM innovators* (pp. 17–33). New York, NY: Routledge.

Egan, K. (2004). *Getting it wrong from the beginning*. New Haven, CT: Yale University Press.

Fallace, T., & Fantozzi, V. (2013). Was there really a social efficiency doctrine? The uses and abuses of an idea in educational history. *Educational Researcher, 42*(3), 142–150.

Foucault, M. (1977). *Discipline and punish*. New York, NY: Vintage.

Freeman, R. B., & Goroff, D. L. (2009). *Science and engineering careers in the United States*. Chicago, IL: University of Chicago Press.

Gabrielson, C. (2013). *Tinkering: Kids learn by making stuff*. Sebastopol, CA: Maker Media.

Gauntlett, D. (2011). *Making is connecting: The social meaning of creativity from DIY and knitting to YouTube and Web 2.0*. Malden, MA: Polity Press.

Goodman, P. (1966). *Compulsory mis-education and the community of scholars*. New York, NY: Random House.

Halverson, E. R., & Sheridan, K. M. (2014). The maker movement in education. *Harvard Educational Review, 84*(4), 495–504.

Honey, M., & Kanter, D. E. (2013). *Design, make, play: Growing the next generation of STEM innovators*. New York, NY: Routledge.

Illich, I. (1970). *Deschooling society*. New York, NY: Harper and Row.

Illich, I. (1973). *Tools for conviviality*. New York, NY: Harper and Row.

Kalil, T. (2013). Have fun—Learning something, do something, make something. In M. Honey & D. Kanter (Eds.), *Design, make, play: Growing the next generation of STEM innovators* (pp. 17–33). New York, NY: Routledge.

Kalman, C. S., & Aulls, M. (2003). Can an analysis of the contract between pre-Galilean and Newtonian theoretical frameworks help students develop a scientific mindset? *Science & Education, 12*, 761–772.

Katz, M. (1975). *Class, bureaucracy, and schools: The illusion of educational change in America*. New York, NY: Praeger.

Kemp, A. (2013). *The makerspace workbench: Tools, technologies, and techniques for making*. Sebastopol, CA: Maker Media.

Kohlstedt, S. G. (2010). *Teaching children science*. Chicago, IL: University of Chicago Press.

Kress, G. (2010). *Multimodality: A semiotic approach to contemporary communication*. New York, NY: Routledge.

Labaree, D. (1997). *How to succeed in school without really learning*. New Haven, CT: Yale University Press.

Labaree, D. (2004). *The trouble with ed schools*. New Haven, CT: Yale University Press.

Labaree, D. (2012). *Someone has to fail*. Cambridge, MA: Harvard University Press.

Lang, D. (2013). *Zero to Maker: Learn (just enough) to make (just about) anything*. Sebastopol, CA: Maker Media.

Lankshear, C., & Knobel, M. (2009). *New literacies: Everyday practices and social learning*. New York, NY: Open University Press.

Martinez, S. L., & Stager, G. (2013). *Invent to learn: Making, tinkering, and engineering in the classroom*. Torrance, CA: Constructing Modern Knowledge Press.

National Research Council. (2012). *A framework for K-12 science education: Practices, crosscutting concepts, and core ideas*. Washington, DC: National Academies Press.

National Writing Project. (2013). Writing as making/making as writing. *Connected Learning*. Retrieved from http://connectedlearning.tv/writing-makingmaking-writing

Papert, S. (1980). *Mindstorms: Children, computers, and powerful ideas*. New York, NY: Basic Books, Inc.

Papert, S. (1996). An exploration in the space of mathematics educations. *International Journal of Computers for Mathematical Learning, 1*(1), 95–123.

Phillips, C. (2015). *The new math*. Chicago, IL: University of Chicago Press.

Postman, N. (1970). The politics of reading. *Harvard Educational Review, 40*, 244–252.

Postman, N., & Weingartner, C. (1969). *Teaching as a subversive activity*. New York, NY: Delta.

Rose, M. (2014, May 15). The maker movement: Tinkering with the idea that college is for everyone—Truthdig. *TruthDig*. Retrieved from http://www.truthdig.com/report/item/the_maker_movement_why_college_isnt_for_everyone_20140515

Rudolph, J. (2002). *Scientists in the classroom*. New York, NY: Palgrave Macmillan.

Schneider, J. (2014). *From the ivory tower to the schoolhouse*. Cambridge, MA: Harvard Education Press.

Stornaiuolo, A., & Nichols, T. P. (2018). Mobilizing audiences in high school makerspaces. *Teachers College Record, 120*(8), 1–38.

Thomas, A. (2014). *Making makers: Kids, tools, and the future of innovation*. Sebastopol, CA: Maker Media.

Tyack, D., & Tobin, W. (1994). The "grammar" of schooling: What has it been so hard to change? *American Educational Research Journal, 34*(3), 453–479.

Urban, W. J. (2010). *More than science and Sputnik: The National Defense Education Act of 1958*. Tuscaloosa, AL: University of Alabama Press.

Vossoughi, S., Hooper, P. K., & Escudé, M. (2016). Making through the lens of culture and power: Toward transformative visions for educational equity. *Harvard Educational Review, 86*(2), 206–232.

White House. (2014, June 18). FACT SHEET: President Obama to host first-ever White House Maker Faire. Retrieved March 23, 2015 from http://www.whitehouse.gov/the-press-office/2014/06/18/fact-sheet-president-obama-host-first-ever-white-house-maker-faire

Zilversmit, A. (1993). *Changing schools: Progressive education theory and practice, 1930–1960*. Chicago, IL: University of Chicago Press.

2. *Kevin Mitnick,* The New York Times, *and the Media's Conception of the Hacker*

MOLLY R. SAUTER
McGill University

Hacker Mitnick: Threat or Menace?

Kevin Mitnick, often identified as the "Most Wanted Hacker in the US," was arrested in 1995 after a three-year FBI manhunt. His capture, trial, and imprisonment was exhaustively covered by the news media, particularly in the *New York Times*, where reporter John Markoff spearheaded the initial coverage and later published two books on Mitnick. In total, from 1994 to 2012, *The New York Times* published 47 articles that mention Kevin Mitnick.

The *New York Times* coverage of Mitnick reflects the changing popular conception of hackers, as well as legitimatizing and solidifying those evolving images and stereotypes. In this chapter, I examine how over the course of the this coverage the term "hacker" shifts from an identifier of a particular technological subculture to a stand-in term for criminality. Kevin Mitnick is most consistently referred to as a "hacker" throughout the *New York Times* coverage, but from 1994 to 2012, the use of "criminal" terms to modify "hacker" terms decreases, until eventually "hacker" terms are used a pure synonyms for "criminal" terms, as well as referencing general hacking culture in contexts where criminality is assumed. This transition is reinforced by the introduction of the term "ex-hacker" following Mitnick's release from prison as a theoretically rehabilitated member of society.

The "hacker equals criminal" move is reinforced by a variety of persistent metaphors, characterizations, narratives, and other repeated references that appear throughout the *New York Times* coverage. Following Lakoff & Johnson (1980), I show how the repeated use of metaphors of "house and home" to describe Mitnick's illegal acts characterize the actions of hackers as a threat to individuals, not just states or corporations. The news articles also repeatedly describe Mitnick and the hacker community at large in the context of anti-social behavior, insanity, addiction, and the cultural narrative of juvenile delinquency, resulting in a persistent association of "hacking" with bad or broken actors. This aspect in particular follows conclusions previously made by Helen Nissenbaum (2004) on the use of certain associations to create social enemies and out-groups.

I further argue that through the extensive press coverage of his arrest, trial, imprisonment, and post-release life, Kevin Mitnick became a mainstream representation of the hacker within a redemptive social melodrama, particularly due to the reformative values presented by the "ex-hacker" figure. This life cycle begins with the youthful juvenile delinquent, who grows into the accomplished hacker/criminal (also manifest as the Super User figure postulated by Paul Ohm) (2008). What follows is a performative hunt/detective narrative (here enabled by the counter-figure of Tsutomu Shimomura, the independent computer security expert who aided the FBI in their search and was a major source for *New York Times* reporter John Markoff), culminating in a cathartic encounter with the criminal justice system, which further tags the Mitnick-as-"hacker" figure as a bad actor to be purged. However upon release from the criminal justice system, Mitnick sheds the "hacker" label, becoming an "ex-hacker" or "former hacker," and taking on the professionalized role of "computer" or security consultant. This progression can also be seen in the news industry's changing use of Mitnick from 1994 through 2012, first as a subject, then as a referent or symbol for hackers at large, and finally as an expert or source himself.

Through the tools of metaphor, frame, and content analysis, and building on the analytical work of George Lakoff (1980), James Aho (1984), Helen Nissenbaum (2004), and Deborah Halbert (1997), this chapter shows how the conception of the "hacker" as reflected in *The New York Times* coverage of Kevin Mitnick from 1994 through 2012 merges semantically with the concept of "criminal" via certain textual choices, including repeated use of certain metaphors, characterizations, narrative arc, and other references. The chapter further argues that the "hacker-criminal to ex-hacker-member-of-society" trajectory as reflected in the full *New York Times* corpus of Kevin Mitnick coverage portrays a socially idealized life cycle for the "hacker" figure, as the malignant "hacker" is transformed via the criminal justice system into a professionalized "security consultant."

History of Mitnick's Case and Times Coverage

Kevin Mitnick is one of the most widely known hackers in the US, thanks to his high profile pursuit and arrest in 1995, and subsequent controversies surrounding his case and terms of imprisonment. Mitnick led the FBI on a two-and-a-half year merry chase through several jurisdictions, engaging in acts of phone phreaking, social engineering, and computer hacking along the way, mostly to aide his escapes and to taunt the FBI. The Department of Justice ultimately claimed that Mitnick had compromised dozens of computer networks, cloned hundreds of cellular phones, stolen proprietary code from phone companies such as Pacific Bell, and broken into private voice and email accounts. He was charged with 14 counts of wire fraud, eight counts of the possession of unauthorized access devices, unlawful interceptions of wire or electronic communications, unauthorized access to a federal computer, and causing damage to a protected computer. He pled guilty in 1999 to four counts of wire fraud, two counts of computer fraud, and one count of unlawful intercept, and received a sentence of 46 months in prison (partially time served), plus 22 months for violation of his parole for an earlier computer fraud conviction.

The terms of Mitnick's imprisonment were particularly controversial and led to his case receiving a great deal of attention from the burgeoning hacker community. Mitnick spent four and half years in pre-trial detention, and a total of eight months in solitary confinement without access to a phone or other technology due to concerns that he could "start a nuclear war by whistling into a pay phone" (Mitnick, 2008) There were also widespread rumors that as a teenager in 1983, Mitnick had hacked into computers at NORAD for a lark. However, these rumors were untrue. They can most likely be traced back to the fact that the popular film *War Games*, featuring a teenage Matthew Broderick hacking into the computer at NORAD for a lark, was released in 1983. In actuality, Mitnick's adolescent arrest record stemmed from his illicitly accessing the Pacific Bell billing system to alter telephone bills. Further, after his release in 2000, prosecutors initially tried to restrict his use of communications technology to nothing other than a landline telephone until 2003, a restriction Mitnick defeated in court. In 2000, Kevin Mitnick began work as a paid security consultant with the eponymous firm Mitnick Security Consulting LLC.

From 1994 to 2012, the *New York Times* published a total of 47 articles that mention Kevin Mitnick by name. Of these, seven were written by Peter Lewis, and six were written by John Markoff. Markoff's feature-length articles, several of which closely followed computer scientist Tsutomu Shimomura, who consulted with the FBI on Mitnick's pursuit, set the tone for the *Times*'s treatment of Mitnick's case. This chapter performs a close reading of those 47 articles, identifying key metaphors and narratives that persist throughout the nearly two decades of coverage. Due to the infamy of Mitnick's case, the depth and scope of the *Times*'s coverage, and the ability of the *New York Times* to set journalist agendas and norms nationally, this chapter argues that the metaphors, narratives, and assumptions present in the *Times*'s coverage of Kevin Mitnick influenced conceptions of Mitnick and hackers in the broader culture.

Metaphor and Metonym in Mitnick Coverage

In his book, *Metaphors We Live By*, co-authored with Mark Johnson, George Lakoff articulates the definitional power of metaphors in every day speech and thought. In their analysis, metaphors operate systematically, linking the thing which is being explained (the "target domain") to a different, theoretically already understood object or concept (the "source domain") by which linkage the first is better grasped. The "conceptual metaphor" delineates the operating space for the metaphorical expressions, defining the scope of metaphoric references that are available to describe a situation, while "metaphorical expressions" are the actual phrases and words deployed in speech and writing. The conceptual metaphor is present, though at a higher level of abstraction from the everyday, and is often unquestioned. Rather, the conceptual metaphor is makes up a defining, background conception of the way things are or should be. This can in turn foreclose other interpretive frames for the concepts and actions at play (Lakoff & Johnson, 1980).

Though conceptual metaphors and metaphoric expressions are a constant in our interactions with the world, each other, and ourselves, they are most powerful when deployed to clarify new, complex, and potentially threatening topics and events. In the modern era, new technology is particularly apt to be explained through metaphor: think of the "information superhighway," or your computer's "desktop" with its different "folders" and "trashcan" to the "emptied," and its susceptibility to "viruses" and "worms." These different metaphoric expressions refer back to broader conceptual metaphors of "the internet as public infrastructure," and "computer as office/home/body" which in turn exercise a significant influence over how people interact with and think about these technologies.

In the *New York Time*'s coverage of Kevin Mitnick and of hackers in general, conceptual metaphors of "house and home" are most prevalent, though the repeated use of terms like "break in" (Lewis, 1995a, 1996a; Markoff,

1995a, 1995b, 1995c), references to home burglaries (Markoff, 1995a), vandalization (Lewis, 1995a; Markoff, 1995d), and familiar physical security protocols, like locked doors (Flynn, 1995; Hamilton, 2006) or fences (Markoff, 1995c). Over the course of 47 articles, metaphors of house and home appeared 22 times, the most of any conceptual metaphor employed. In one article, John Markoff described the emotions that resulted from theft of computer files as "the feelings of outrage and violation normally provoked by a burglar's rifling their home" (Markoff, 1995d). This personalizes the threat posed by Mitnick and hackers in general, portraying them a specific and immediate danger to the personal wellbeing of the news-reading audience.

Another aspect of Lakoff's metaphor theory most relevant here is metonymy. Metonymy is the use of an aspect or attribute to act as a conceptual stand in for the thing referred to. Common examples include referring to a business man as a "suit" or the phrase "good heads" to refer to "intelligent people." "Metaphor and metonymy are different *kinds* of processes. Metaphor is principally a way of conceiving of one thing in terms of another, and its primary function is understanding. Metonymy, on the other hand, has primarily a referential function, that is, it allows us to use one entity to *stand for* another." (Lakoff & Johnson, 1980) Further, metonymic functions allow one to select which *aspect* of the whole should be held to be preeminent. Regarding the metonymic use of "good heads" for "intelligent people": "The point is not just to use a part (head) to stand for a whole (person) but rather to pick out a particular characteristic of the person, namely, intelligence, which is associated with the head." (Lakoff & Johnson, 1980)

From 1994 to 2012, the *New York Times* published 47 articles that mentioned Kevin Mitnick by name, 27 of which were direct coverage of Mitnick's case or life. The remaining twenty covered a wide variety of subjects, but all invoked Mitnick's name as a metonymic symbol for hackers in general, acting as a personal embodiment of the threat posed by both vulnerable yet invasive technology and the new criminal class eager to exploit it. Mitnick's name was repeatedly invoked both in quotations from security experts and in the narrative texts of articles themselves. John Gilmore, a Cygnus software executive, was quoted in a 1995 article by Peter Lewis as saying, "Unless the [ban on crypto exports] policy is changed, we will continue to see plenty of large and small scale intrusions like Kevin Mitnick's" (Lewis, 1995b). He was name-dropped in articles about the Computer Freedom and Privacy Conference (Lewis, 1994) and DEFCON 2 (Markoff, 1994b) before his arrest. After his arrest, his name popped up in articles on the generally poor state of online security (Lewis, 1995a); columnist Frank Rich's personal experience of the World Wide Web (Rich, 1995); the phenomenon of internet addiction

(O'Neill, 1995); the export of cryptographic software (Lewis, 1995b); a discussion of computers in popular film (Chen, 1995); the controversial release of the vulnerability-scanning tool SATAN (Flynn, 1995); different types of computer viruses (Lewis, 1995c); credit card numbers being traded online (Lewis, 1995d); an article about reverse engineering (Haftner, 1999); and a second article about hackers in popular film (McKinley, 1999). His name cropped up in a book review by Barbara Ehrenreich: "But sooner or later some hacker, like Kevin D. Mitnick, the recently apprehended computer security breaker, will come along and set [data] loose just for fun" (Ehrenreich, 1995). After his release from prison, he was referenced in articles on Frank Abagnale, the conman subject of the film *Catch Me If You Can* (Schwartz, 2003a); sentencing practices for computer crimes (Richtel, 2003); illicit tests of airport security (Schwartz, 2003b); smart homes (Hamilton, 2006); the hacker/snitch Adrian Lamo (Cohen, 2010); and LulzSec's announced retirement (Richmond & Bilton, 2011). Through references to "people like Kevin" and repeated invocations of his name is otherwise unrelated articles, Mitnick's life story, personal character, and even his physical body serve as handy stand ins for an entire antagonistic subculture.

Hackers Reified as Enemies

The news cycle surrounding the pursuit, arrest, trial, and incarceration of Kevin Mitnick served to help solidify the hacker as one of the prime social enemies of technological age. Deborah Halbert describes how, starting with *Operation Sundevil*, an FBI raid targeting a distributed group of phone phreakers in 1990, five years before Mitnick's arrest, media coverage of hackers, in the news and in popular television and film, contributed heavily to the establishment of hackers as a popular social enemy or folk devil (Halbert, 1997). Following James Aho, Halbert identifies the "stages of reification" by which this identification is achieved: first hackers are *labeled* as a threat; then that threat is *legitimated* or *validated*; *mythmaking* presents "biographical or historical accounts of defamed persons showing why it is inevitable necessary and predictable they act as they do" (Aho, 1994); *sedimentation* represents the stage at which the threat described becomes a threat in and of itself, regardless of the actual dangers present; finally *ritual*, or the playing out of "appropriate" reactions by "good guys," usually law enforcement (Halbert, 1997).

Mitnick is identified or named as a criminal threat early in the *Times*'s coverage. The first article published but the *Times* to mention Mitnick, a general interest article on data encryption, refers to him as "fugitive computer hacker, Kevin Mitnick" (Lewis, 1994). Later in 1994, John Markoff refers to

Mitnick as a "grifter" (Markoff, 1994d). The reality of the threat posed by Mitnick was legitimated simultaneously, as computer security experts interviewed in the articles testified to the seriousness of the threat posed by hackers generally. The mythmaking step of Aho's reification process was present in the persistent pathologizing of Mitnick's behavior. Markoff referred to him in an early article as a "computer programmer run amok" and said he "has an addiction to computers" which was compared to drug or gambling addiction. (Markoff, 1994) Similarly, Mitnick was described as "obsessed" with power and computers (Markoff, 1995b). Computer hackers in general were described as "paranoid." (Lewis, 1994; Markoff, 1994b) Douglas Thomas also notes that descriptions of Mitnick's physical body, which painted him as "overweight" (Lewis, 1996a; Markoff, 1994a), "clumsy" (Markoff, 1995a), ungroomed (New York Times, 1995), and a "scowling, bespectacled young man" (Fallows, 1996) indicate a belief that the hacker's presumed over-identification with technology has alienated him from his own experience of physical embodiment, aggressively othering him from "normal" people (Thomas, 2002). The meta-process of "sedimentation" occurs as articles about the existential threats presented by hackers and computer crime become a normalized beat, with Mitnick established as a familiar reference point (see Ehrenreich, 1995; Gelernter, 1996; Haftner, 1999; O'Neill, 1995 for examples of the ease with which references to Mitnick were inserted into stories on a variety of computer topics).

The "ritual" step in Aho's reification process is established in the parts of the *Times*'s coverage, particularly those early pieces written by John Markoff, that cast Mitnick's story as a battle over the ethics of computer use between the threatening hacker fringe and white-hat-esque professionals, here metonymically embodied by Tsutomu Shimomura, a computer security expert, FBI consultant, and Markoff's primary source. Shimomura served as Mitnick's foil and mirror image throughout Markoff's coverage of the story. Where Shimomura was described as a "brilliant cybersleuth" and part of a "tightly knit community of programmers and engineers who defend the country's computer networks" (Markoff, 1995c), in the same article Mitnick is described as a "computer outlaw" and "invader" (Thomas, 2002). Through the metonymic use of Kevin Mitnick, as the *New York Times* coverage hits these major steps, it proffers Mitnick and the hacker class he represented as a new folk devil or paradigmatic informatic deviant of the computer age, the one who marks the boundaries of appropriate and ethical behavior by crossing them (Halbert, 1997).

Repeatedly invoking Mitnick, and through him, hackers in general, in the context of crime, deviance, trespass, and burglary "establishes a new prototype," as noted by Helen Nissenbaum. "The more times people hear about hackers in these terms, the more they are led to see these hackers not as the exceptions but as the rule. A category shift occurs not as a result of revised formal definitions, nor at the edges where boundaries are carved, but at the center where the typical hacker is drawn" (Nissenbaum, 2004). Through repetition, Kevin Mitnick became the "typical hacker" of the *New York Times*'s news coverage, the center around which all other computer criminals orbited.

This process can be directly observed in the use, and subsequent discarding of, criminal modifying terms used with the word "hacker" when identifying Mitnick and others. Early *Times* coverage characterized Mitnick in either purely criminal terms or as a "hacker" modified with other terms of criminality: a "fugitive computer hacker" (Lewis, 1994); "grifter" (Markoff, 1994a); "31 year old computer outlaw" (Markoff, 1995a). However, this compound characterization gives way to the noun "hacker" being used as a direct synonym for criminality, or in situations where criminality is assumed: "… Kevin Mitnick, the hacker who devoured the internet" (Rich, 1995); "the computer hacker arrested last month" (Flynn, 1995); "hackers making off with credit card numbers and computer passwords" (Lewis, 1995b); "the country's most notorious hacker" (Ravo, 1996); and simply "convicted computer hacker" (AP, 1996). Here it can be seen how the concept of the "hacker" does the work of the previously appended criminal modifying terms. External criminality no longer needs to be specified. After sufficient association, the criminality of the hacker can now be assumed.

After being established through such coverage as a metonymic symbol, Mitnick's presence and action as an individual had a definitional impact on the social group he represented. In that capacity, his decision following his release to re-enter the field of computer security as a security consultant and "ex-hacker" entered into the popular understanding of technology and technology adepts as more than simply a personal career choice. Rather Mitnick's progression into respectability became a melodramatic arc, an idealized lifecycle of the hacker.

Persistent Metaphors, Characterizations, and Narratives

Over the course of the *Times*'s converge, Mitnick's life story is continually invoked in the context of the familiar narrative of the anti-social juvenile delinquent. Much is made of his childhood being raised by a single mother (Fallows, 1996; Markoff, 1994a), his early dabbling in ham radio (AP, 2002), his troubles in high school (Fallows, 1996; Markoff, 1994a, 1995b), and his first arrest at the age of 17 (Markoff, 1994a, 1995b). Early "pranks" involving conning free rides out of the San Bernardino transit system are read as forecasting an inevitable downfall: "Even if one is unfamiliar with Mitnick's life story, it's kind of obvious where he's heading here, and it's far beyond the bus routes around San Bernadino County" (Biersdorfer, 2011). Markoff highlights Mitnick's "spending time with a loosely knit group of 'phone phreaks,' young people whose hobby was illegally mastering the inner workings of the telephone switching system" (Markoff, 1994b) as if the young phreakers were a street gang. Anti-social characterization of Mitnick and hackers in general are played up: Mitnick is described as a "shy loner who found delight and a sense of power" though technology (Markoff, 1994a), as keeping "nocturnal habits" (Markoff, 1995b), and as being "locked away from other people" (O'Neill, 1995). Hackers in general are described as "paranoid" and a "shadowy community of computer users" with "nose rings and dyed hair" (Markoff, 1995a), and as having the "self-ostracizing habit of turning people on and off" (Rich, 1995).

The juvenile delinquent narrative undergirds a secondary characterization present in these stories, one that Paul Ohm identifies in other contexts as the Superuser (Ohm, 2007). Ohm defines the Superuser as a mythical character present primarily in reactive storytelling about the rampant dangers of the networked age. The Superuser is a technological wizard with skills so advanced they might as well be magic, surpassing those of normal users and even of law enforcement (Ohm, 2007). Mitnick's technological exploits are referred to as "legendary" (Markoff, 1994a), and he himself is dubbed a "cult figure" (AP, 2002) and "folk hero" (Schwartz, 2003a). In stories about his arrest, authorities are quoted lamenting the fact that Mitnick's capture "does not *necessarily* solve all the recent internet crimes" (Markoff, 1995b), and Markoff blithely repeats the federal prosecutor's position that "Mr. Mitnick with a telephone is a menace to society" (Markoff, 1995c). The Superuser characterization is augmented by the ways in which Mitnick was further characterized as a technological trickster of sorts, of the species described by Gabriella Coleman in her work on the hacker as modern trickster (Coleman, 2012). Coleman describes the trickster as "[a]lthough clever, some are irreverent and grotesque. They engage in acts of cunning, deceitfulness, lying, cheating, killing and destruction, hell raising, and as their name suggests, trickery. Sometimes they do this to quell their insatiable appetite, to prove a point, at time just to cause hell, and in other instances to do good in the world" (Coleman, 2012). Paralleling this description, Mitnick is described as "a master at fooling and eluding the authorities" (Markoff, 1994a), having a "misguided spirit of adventure" (Markoff, 1995d), "just playing" (Gelernter, 1996), and is at multiple points described as "taunting" the FBI and Shimomura in particular. The *Times*'s coverage even invokes the ways the trickster traditionally plays with physical form, shape shifting, and identity, referencing Mitnick "worm[ing] his way into" systems (Markoff, 1995b), getting "caught in his own web" (Markoff, 1995b), and using a computer to alter his voice to leave those "taunting messages" for his pursuers (Markoff, 1995a).

By casting Mitnick as a trickster with Superuser abilities, the *Times*'s news coverage endowed him, and via him hacker populations in general, with wizardly powers over technology and unpredictable, anti-social, troublesome motivations. The *Times*'s narratives placed Mitnick outside the realm of the everyday, depicting him as an unpredictable threat with "murky" (Markoff, 1995b) motivations. The tenacity of the *Times*'s coverage and Mitnick's metonymic role meant that the hacker as extra-human threat remained a primary concept in popular coverage of technology. Nor does Mitnick's ultimate personal development into an "ex-hacker" and security consultant rehabilitate the concept of the hacker. Rather, the hacker becomes a stage that must be passed through and transcended, and thus in and of itself, it remains a threat to public security.

The Rise of the Ex-Hacker Figure

A coherent narrative can be seen in the *New York Times*'s coverage of Kevin Mitnick, both in the personal history presented in individual articles and in the arc of the full coverage. Mitnick progresses from a juvenile delinquent, messing around with ham radios and running with gangs of phone phreakers, to being an accomplished hacker and criminal, engaged in extended cat-and-mouse games with the FBI and Shimomura. This leads directly to an encounter with the criminal justice system, through which Mitnick's criminal past is redeemed and he emerges as a "former" or "ex-hacker" (Cohen, 2010; Hamilton, 2006), applying his formerly malicious skills to a respectable career as a paid security consultant (Biersdorfer, 2008; Perlroth, 2012; Schwartz, 2003a). As Mitnick establishes himself as the "ex-hacker"/security consultant, his role in the news production process also shifts: he moves from being the symbolic, metonymic subject of the news production process to acting as an expert or source himself. As he moves away from a criminal role, he is permitted more agency in those articles he features in, being personally interviewed (Hamilton, 2006; Thompson, 2003), having his public talks covered (Perlroth, 2012), and having his book reviewed (Biersdorfer, 2011).

As noted in the previous section, the emergence of the "ex-hacker" as a concept does not rehabilitate the concept of the "hacker." Rather, it freezes the hacker, further isolating him from mainstream society as an anti-social life stage that must be transcended, but cannot, in and of itself, be redeemed: Mitnick moves past the hacker, but the hacker figure is left behind. By virtue of that transcendence, the criminal associations that had become inherent in the idea of the hacker become further engrained. In the narrative present the *New York Times* coverage, the hacker himself cannot be rehabilitated as a law abiding, positive figure. Rather, to achieve mainstream respectability, Mitnick can only stop being a hacker. The hacker figure continues to appear in the *Times*'s coverage, this time in concert with the new figures of "ex-hacker" and "security consultant." Mitnick's transition from "hacker" to "security consultant" also resolves a tension present in the *Times*'s coverage regarding Mitnick's motivations. Because Mitnick never performed his hacks and pranks for money, authorities and the *Times* itself were at a loss to explain his motivations. This comes up repeatedly as the police, prosecutors, and various reporters collectively throw up their hands, calling his motives "murky" (Markoff, 1995b), "unclear" (New York Times, 1995), and "unusual" (Lewis, 1995b) because they do not fit into a capitalist logic of crime. By shifting into a professionalized role, Mitnick's motives become intelligible and acceptable to the broader capitalist society. Where the hacker's motives had been opaque and threatening, the "security consultant" is transparent and easily assimilated into the technology market structure. The "security consultant" is not only the hacker all grown up, but is the hacker's moral opposite number.

This denouement is a reinforcement of the central drama of the Mitnick story as written by John Markoff: his pursuit and capture by Tsutomu Shimomura. In both his articles and the book he subsequently authored on the subject, Markoff depicts the conflict between Shimomura and Mitnick as a "battle of values" (Thomas, 2002) over the future of computing. Shimomura is depicted as a "driven" and "brilliant" computer scientist (Markoff, 1995a), part of a "tightly knit community" of network administrators (Markoff, 1995a) with ties to the offline community as well: much is made of his "beach cottage near San Diego" and his stint with the volunteer ski patrol at Lake Tahoe (Markoff, 1995a). Shimomura is the moral white hat, representing community engagement and the ability to balance a physically active, offline life with a successful career in technology. In Markoff's telling, Shimomura has a place in the world, online and off. Conversely, Mitnick's only community is the "loosely knit group" of phone phreakers he falls in with in high school (Markoff, 1994a). His physical whereabouts during the investigation are a mystery, in one article only revealed through impersonal cellphone pings (Markoff, 1995a). He "prowl[s]" the "ethereal world known as cyberspace" (Markoff, 1995a). In contrast to Shimomura, in these articles Mitnick has no ties to the offline community, no respectability, no physicality to separate him from the technologies he exploited. Mitnick is the disembodied threat lurking in the network (or the "Ghost in the Wires" as he ultimately titled *his* book), whereas Shimomura is a pillar of the community.

Impacts on Technology Discourse and Reporting

Through the repeated use of "house and home" metaphors, narratives of juvenile delinquency, and references to insanity, pathological and anti-social behavior, and the deployment of the "ex-hacker" figure, the *New York Times*'s coverage of Kevin Mitnick painted him as a metonymic representation of hacker culture at large, reifying Mitnick and other hackers as deviant "folk devils" of the information age. The hacker character, as partially defined through Mitnick, serves as a central sink for generalized anxiety about the increasingly central role of complex infrastructural and personal technology in our lives.

The hacker folk devil plays central roles in the "discourse of danger" and "culture of fear" regarding the internet. David Campbell identified the "discourse of danger" as a discourse used to create insecurity from real and imagined threats within and without the United States, used to shape foreign policy and American identity (Campbell, 1992). Central to the functioning of a discourse of danger is the fixing of an othered enemy. In the mid 1990s, Mitnick's metonymic hacker stepped neatly into this role, serving as a target for fears, anxiety, and ultimately laws and policy decisions aimed at controlling the "threat" the folk devil poses to national security (Halbert, 1997), in addition to extreme and unreasonably harsh penalties for Mitnick personally. The language used to report on Mitnick, reinforced by the sheer persistence of reporting on his activities, helped to personalize a culture of fear surrounding computers and the networked technologies seen to be insinuating themselves into people's homes. By repeatedly using metaphors of home invasion and home- and body-based crime, the *New York Times*'s content contributed to overwrought expectations on the part of the public of the actual risk at hand, leading to widespread fear of crime and danger regardless of the actual level of risk. This fear is exacerbated by the perceived gap between the reported hacker threat in news media and popular culture and the rate of perceived rate of response from law enforcement. The sensational manner in which stories are covered can exert a disproportionate impact upon beliefs and attitudes when compared with whatever "objective" consequences the actions reported on might have, further raising level of fear and justifying over-the-top reactions on the part of law enforcement and the judiciary. Mitnick's absurd punishment could also be read as a response to his metonymic role in the news media: quite simply, he was being made an example of.

Though Kevin Mitnick was able to personally progress past his metonymic role as the Every Hacker, the hacker folk devil is still present in modern tech journalism and media. When the hacker folk devil is used as a main character in news media coverage, commentary, and public discourse around the current state of technology, the result is often that levels of fear, anxiety, and paranoia rise, while the quality of debate is lowered. Technologically complex stories are simplified to find existing narratives of anxiety, as with stories regarding potential SCADA hacks or the just-over-the-horizon "digital Pearl Harbor." Complicated computer and network-based actions are translated into easily-grasped physical world metaphors with seemingly little regard for whether or not these metaphors are appropriate. The *Time*'s use of Kevin Mitnick as a metonymic, anti-social folk devil potentially displaced useful discussion of developing technological culture and its increasingly omnipresent use in every day life. This contributed to the current state of discourse regarding technology and hackers, one that undergirds an environment that supports, confirms, and encourages existing fears about the internet and technology's role in our modern lives.

New York Times *Articles Cited*

AP Wire Service. (1995, April 21). Dismissal asked in hacker case. *The New York Times*. Sec. D, Page 8.

AP Wire Service. (1995, July 2). Hacker is said to agree to plea bargain. *The New York Times*. Sec. 1, Page 22.

AP Wire Service. (1996, October 1). New fraud charges denied by hacker. *The New York Times*. Sec. A, Page 16.

AP Wire Service. (2002, December 27). FCC lets convicted hacker go back on air. *The New York Times*. Sec. A, Page 16.

Biersdorfer, J. D. (2008, September 11). Q&A tech talk. *The New York Times*. Sec. C, Page 8.

Biersdorfer, J. D. (2011, August 14). The happy hacker. *The New York Times*. Sec. BR, Page 9.

Chen, V. (1995, May 28). Your home computer would scoff at these plots. *The New York Times*. Sec. 2, Page 11.

Cohen, N. (2010, June 28). Ex hacker who accused suspect of army leak is still talking. *The New York Times*. Sec. B, Page 3.

Ehrenreich, B. (1995, May 7). Surfing the third wave. *The New York Times*. Sec. 7, Page 9.

Fallows, J. (1996, February 4). An outlaw in cyberspace. *The New York Times*. Sec. 7, Page 14.

Flynn, L. (1995, March 12). The executive computer: Software that pits alarmists against devil's advocates. *The New York Times*. Sec. 3, Page 11.

Gelernter, D. (1996, February 27). Cyberwar's literary fallout rivals the cyberwar. *The New York Times*. Sec. C, Page 15.

Haftner, K. (1999, June 17). Reinvent the wheel? This software engineer deconstructs it. *The New York Times*. Sec. G, Page 13.

Hamilton, W. L. (2006, November 23). You're not alone. *The New York Times*. Sec. F, Page 1.

Lewis, P. H. (1994, March 26). Collisions in cyberspace on data encryption plans. *The New York Times*. Sec. 1, Page 37.

Lewis, P. H. (1995a, February 22). Security is lost in cyberspace. *The New York Times*. Sec. D, Page 1.

Lewis, P. H. (1995b, April 10). Between a hacker and a hard place. *The New York Times*. Sec. D, Page 1.

Lewis, P. H. (1995c, September 4). Computers beware! New type of virus is loose on the net. *The New York Times*. Sec. 1, Page 1.

Lewis, P. H. (1995d, November 13). On the net: Today's Willie Suttons mug corporate users, not the little guy from hackensack. *The New York Times*. Sec. D, Page 5.

Lewis, P. H. (1996a, April 24). Leading computer hacker strikes a plea agreement. *The New York Times*. Sec. A, Page 17.

Lewis, P. H. (1996b, April 28). A super hacker enters a plea bargain, in person. *The New York Times*. Sec. 4, Page 2.

Markoff, J. (1994a, July 4). Cyberspace's most wanted: Hacker eludes FBI pursuit. *The New York Times*. Sec. 1, Page 1.

Markoff, J. (1994b, July 25). Computer underground comes out of the cold. *The New York Times*. Sec. A, Page 10.

Markoff, J. (1995a, February 16). How a computer sleuth traced a digital trail. *The New York Times*. Sec. D, Page 17.

Markoff, J. (1995b, February 16). A most wanted cyber thief is caught in his own web. *The New York Times*. Sec. 1, Page 1.

Markoff, J. (1995c, February 17). Hacker case underscores internet's vulnerability. *The New York Times*. Sec. D, Page 1.

Markoff, J. (1995d, February 19). Hackers and Grifters duel on the net. *The New York Times*. Sec. 4, Page 1.

McKinley, J. (1999, December 12). Suddenly hackers are sexy, hip and evil. *The New York Times*. Sec. 4, Page 5.

O'Neill, M. (1995, March 8). The lure and addiction of life on line. *The New York Times*. Sec. C, Page 1.

Perlroth, N. (2012, March 5). Warning and buzzwords at security event. *The New York Times*. Sec. B, Page 8.

Pollack, A. (1999a, March 19). Deal said to be in works that would free hacker. *The New York Times*. Sec. C, Page 6.

Pollack, A. (1999b, March 27). Famed computer intruder gets prison term. *The New York Times*. Sec. C, Page 15.

Ravo, N. (1996, September 27). New indictment is filed by US in hacker case. *The New York Times.* Sec. A, Page 24.

Rich, F. (1995, March 5). Journal: Bit by bit. *The New York Times.* Sec. 5, Page 15.

Richmond, R., & Bilton, N. (2011, June 27). Saying it's disbanding, hacker group urges new cyberattacks. *The New York Times.* Sec. B, Page 1.

Richtel, M. (2003, January 21). Barring web use after web crime. *The New York Times.* Sec. A, Page 1.

Schwartz, J. (2003a, January 5). Hacking away, long before there were hackers. *The New York Times.* Sec. 4, Page 4.

Schwartz, J. (2003b, October 26). Hacking into airline security, box cutters and all. *The New York Times.* Sec. 4, Page 2.

Thompson, C. (2003, January 12). The way we live now: Questions for Kevin Mitnick. *The New York Times.* Sec. 6, Page 13.

References

Aho, J. (1994). *This thing of darkness.* Seattle, WA: University of Washington Press.

Campbell, D. (1992). *Writing security.* Minneapolis, MN: University of Minnesota Press.

Coleman, G. (2012). Phreaks, hackers and trolls. In M. Mandiberg (Ed.), *The social media reader* (pp. 99–119). New York, NY: New York University Press.

Halbert, D. (1997). Discourses of danger and the computer hacker. *The Information Society, 13*(4), 361–374.

Lakoff, G., & Johnson, M. (1980). *Metaphors we live by.* Chicago, IL: University of Chicago Press.

Mitnick, K. (2008). The last HOPE: Keynote. Retrieved from https://www.youtube.com/watch?v=5YWu8IBpCkc

Nissenbaum, H. (2004). Hackers and the contested ontology of cyberspace. *New Media Society, 6*,195, 195–217.

Ohm, P. (2007). The myth of the superuser: Fear, risk, and harm online. *UC Davis Law Review, 41*(4), 1327.

Thomas, D. (2002). *Hacker culture.* Minneapolis, MN: University of Minnesota Press.

3. *Making Civic Media in the Post-Fukushima Japanese Media Ecology*

YASUHITO ABE
Komazawa University

On March 11, 2011, a massive 9.0 magnitude earthquake and an ensuing tsunami devastated northeastern coastal regions in Japan. The combined natural disasters knocked out critical cooling systems at the Fukushima Daiichi nuclear power plant, which released a tremendous amount of radioactive material into the air. The Fukushima Daiichi nuclear disaster turned out to be the largest nuclear disaster since the Chernobyl nuclear disaster of 1986. In the wake of the disaster, the Japanese government repeatedly emphasized that no immediate health effects would be caused by the radiation released from the power plant; unfortunately, citizens also suffered from a serious lack of practically useful data related to radiation in their everyday lives (Fukushima genpatsujiko dokuritsu kenshō iinkai, 2012). This discrepancy provided opportunities for collective action.

As in the case of the Chernobyl disaster (e.g., Abe, 2015), the Fukushima disaster ultimately created its own public, who can be roughly described as a *networked measuring public*; this public became a fundamental aspect of the post-Fukushima social landscape in Japan. Within the first week after the disaster, a wide variety of citizens not only measured the radiation levels in the air using their own dosimeters but also *circulated* the resulting data among themselves through digital media, reshaping themselves as a specific public.[1] Among the diverse kinds of networked measuring public, this study focuses on investigating a specific project: Safecast. Safecast emerged immediately after the Fukushima Daiichi nuclear disaster, designating itself as "a global project working to empower people with data, primarily by mapping radiation levels and building a sensor network, enabling people to both contribute and freely use the data collected" (Safecast, n.d.-a). From its start,

Safecast allowed volunteer dosimeter users to submit their collected data to its online database and thus engaged in participatory digital media practices that enabled citizens to get involved. Perhaps more importantly, Safecast makes the resulting measurement data available for anyone to use for free, under a Creative Commons 0 (CC0) designation (Safecast, n.d.-a), encouraging citizens or organizations to create their own media that are tailored to their individual contexts (Abe, 2015). As Zuckerman (2013) correctly points out, Safecast thus exemplifies civic media, which can be broadly defined as "any medium which fosters and enhances civic engagement" (Jenkins, 2007). By the end of 2014, Safecast proudly reported that its database contained more than 25 million data points from different parts of the world (Azby, 2014), marking it as a key global-scale network of measuring publics.

Much scholarship has focused on conceptualizing Safecast as a "hacker group," partly due to its original connections with Tokyo Hacker Space (THS) (e.g., Hemmi & Graham, 2012; Kelly, 2014; Murillo, 2013). In light of these claims that Safecast is a case of a group of hackers, an in-depth analysis of Safecast as an alternative media outlet could do much to illustrate the participatory nature of its digital media practices; its primary goals involved not only providing powerful visibility to imperceptible radiation through the use of dosimeters but also making its collected data accessible to the public through digital media. This study therefore examines not only how Safecast, a group of networked measuring publics, generates "raw" data related to radiation in the air though its dosimeters but also how the network re-presents its collected data as a fundamental resource for making its own media.

While scholars of communication, media studies, anthropology, and science, technology and society (STS) have discussed forms of digital media practices that proliferated following the nuclear disaster (e.g., Abe, 2014a, 2014b, 2015; Jung, 2012; Slater, Nishimura, & Kindstrand, 2012; Utz, Schultz, & Glocka, 2012), it is a necessary prerequisite to investigate these practices in relation to a broader media landscape because it is critical to take into account those who missed digital media practices in the wake of the disaster and those who were affected by the disaster in particular (Abe, 2015; Tanaka, Shineha, & Maruyama, 2012). This study thus focuses on examining both the opportunities and challenges that Safecast faced as an alternative media outlet in the post-Fukushima Japanese media ecology.

To investigate Safecast's civic media practices in relation to the Japanese media ecology, this study draws on Media System Dependency (MSD) theory and ultimately suggests that Safecast engaged in alternative participatory digital media practices that effectively supplemented the Japanese media system. The central focus of this study is to describe why and how Safecast became relevant as a civic media outlet in the Japanese media landscape after the Fukushima Daiichi nuclear disaster. This study first elaborates on MSD theory and discusses the Japanese mainstream media's structural dependency. Next, this study describes post-Fukushima Japanese society as a kind of risk society and uses this lens to investigate Safecast's participatory digital media practices. Finally, this paper discusses its key findings and their implications for future research.

Media System Dependency (MSD) Theory

This study examines Safecast as alternative media from the perspective of MSD theory. Originally proposed by Ball-Rokeach (1974) and developed by Ball-Rokeach and DeFleur (1976), MSD theory conceptualizes media as an information system in control of limited information resources and lays out the macro- and the micro-level of analysis of audience members' dependency relations with media from an ecological perspective. This approach allows for an analysis of the dynamic and evolutionary nature of the interdependent relations between media and other social systems (e.g., Ball-Rokeach, 1985; Ball-Rokeach & Jung, 2009). In the macro-level analysis, MSD theory is concerned with media's interdependent relations with other social systems, namely structural media dependencies (Ball-Rokeach, 1985). Structural media dependencies are analyzed by examining "the goals of each party to a relation that are, to one degree or another, contingent upon the resources of the other party(ies)" (Ball-Rokeach, 1985, p. 490). As suggested by Ball-Rokeach (1985), it is particularly important to investigate the media's interdependent relations with the politico-economic system. In contrast, the micro-level analysis of MSD theory is concerned with individual audience members' relations with the media, investigating individual audience members' capacities to construct their own meanings in their own contexts. Combining these different levels of analysis, MSD theory critically accounts for the ecological constraints that may limit audience members' interpretative power (Ball-Rokeach, 1998). Ball-Rokeach (1985) notes that:

> Structural dependencies upon the media system that are logically prior to individual media-system dependencies, place the media in the position of being an essential link between individuals and their social environs. *Indeed, the media system, through its social construction of reality activities and products, fundamentally shapes the contours of the audience's perceptions of the degree of ambiguity and threat in those social environs that are beyond its capacity to observe directly or indirectly through interpersonal reports.* (p. 501, emphasis added)

As such, MSD theory highlights the need to conceptualize the interpretative power of audience members in relation to the media and communication environment in which they are embattled. With this theoretical framework in hand, much research has examined individuals' dependency relations with the media during and after various disasters and emergencies (e.g., Hirschburg, Dillman, & Ball-Rokeach, 1986; Jung, 2012; Tai & Sun, 2007). For instance, Loges (1994) conducted a survey on the intensity of media systems dependency in relation to perceptions of threats in the environment, and he demonstrated that the level of perceived threat is positively related to the intensity of dependency relations with mass media, such as newspapers, radio, and television.

Furthermore, many scholars have adopted MSD theory as a theoretical framework to analyze the relations between individuals and the media in "new" media environments, despite the fact that the theory was proposed in the 1970s (Brough & Li, 2013; Jung, 2012; Ognyanova & Ball-Rokeach, 2015). Given the relative media abundance in "new" media environments, Ball-Rokeach and Jung (2009) note that "the most serious criticism of MSD theory came not from other scholars, but from the realities of a changing media environment" (p. 509); in fact, MSD theory is relatively silent about the capacities of individuals as media users to organize and respond to specific risks that are not discussed in the mainstream media. Drawing on MSD theory, this study thus examines Safecast's civic media practices in relation to the Japanese media system from an ecological perspective, seeking to contribute to MSD-guided scholarship. In the following section, this study analyzes the structural constraints that shaped the Japanese media system.

Structural Dependency Relations in Japan

Space does not permit a detailed examination of the interrelation between the Japanese media system and other social systems; this section instead focuses on outlining a salient characteristic that shaped the Japanese mainstream media's structural dependencies on politico-economic systems before and after the Fukushima disaster. Much research has shown that the Japanese media system shares a close relationship with politico-economic systems in Japan (e.g., Freeman, 2000; Hall, 1998; Kasza, 1988; Krauss, 2000; Pharr & Krauss, 1996). For instance, Sasaki (1999) investigates the interrelations between Japanese mainstream newspapers and the Japanese state from the latter half of the 19th century to the end of World War II and suggests that most Japanese mainstream newspapers were more or less supportive of the Japanese state. By the same token, Watanabe and Nohara (2000) critically analyze the Japanese mainstream media's structural interdependencies with the state and major business organizations, and they show strong historical links among the three parties. More importantly for this research, the Japanese mainstream media, such as national and local newspapers and television stations, were more or less supportive of the nuclear industry in Japan before the Fukushima disaster (e.g., Abe, 2013; Arima, 2006, 2008; Jōmaru, 2012, Shibata, 2013; Yamakoshi, 2013; Yamamoto, 2012).

While much scholarship has indicated that the Japanese media system is interconnected with politico-economic systems, including the nuclear industry, in many ways, there is one definitive system that most scholars mention when discussing the Japanese mainstream media's interrelations with politico-economic systems. Many critical scholars and journalists focus on the press club system or "kisha kurabu" system as a key component that strengthened the links between the Japanese mainstream media and Japanese politico-economic systems (e.g., Feldman, 1993; Freeman, 2000; Hall, 1998; McCargo & Lee, 2010; Reporters without Borders, n.d.).

In defining the press club system of Japan, the Japan Newspaper Publishers and Editors Association (2006), which is an umbrella organization for the Japanese mainstream media, characterizes the press club as "'a voluntary institution for news-gathering and news-reporting activities' made up of journalists who regularly collect news from public institutions and other sources." In contemporary Japanese society, press clubs are everywhere; these clubs are located in most major organizations, including every government ministry, political parties, business organizations, sports organizations, and even in universities (Freeman, 2000).

However, many scholars agree that the press club system negatively impacts Japanese journalism in general. For instance, Freeman (2000), a scholar of political science, critically analyzes the interdependent relations between the Japanese mainstream media and the state, with a particular focus on political aspects of the press club, and she argues that Japanese politicians, bureaucrats, and the mainstream media *collaborate* with each other to promote what she calls "information cartels," which in turn protect their mutual interests. Her critical analysis of information cartels indicates that the press club system contributes to promoting "closed" reciprocal relationships between major mainstream media and politico-economic systems. More specifically, the press club system limits membership to an exclusive group of large Japanese news organizations, allowing those organizations exclusive access to their news sources (Feldman, 1993; Freeman, 2000; Hall, 1998). In contrast, the state and major business organizations have tactically drawn on the system to control the news by excluding freelance and investigative journalists from their press conferences (Feldman, 1993; Freeman, 2000; Hall, 1998). The press club system can thus be seen as one of the key factors responsible for shaping the Japanese mainstream media's structural interdependency relations with Japanese politico-economic systems; both the Japanese media system and Japanese politico-economic systems have capitalized on the press club to promote their interests with each other by excluding freelance and investigative journalists.

The advent of the Internet and social media certainly provided rich resources for freelance and investigative journalists in Japan (e.g., Shiraishi, 2012), but this increase in resources does not mean that the Internet has revolutionized the Japanese media system; it is important to note that the emergence of the Internet and social media should be understood in relation to the existing Japanese media system's interdependent relations with other social systems. Ball-Rokeach (1998) explains the implications of "new" media for media system dependency research as follows:

> The MSD theorist is predisposed to an evolutionary, not revolutionary perspective vis-à-vis new communication and information technologies … the Internet thus intrudes on traditional relations by being integrated into an expanded media system that may expand the reach of understanding, orientation, and play goals that individuals, groups and organizations may attain through media dependency relations. (p. 32)

In fact, for instance, the Japanese state reportedly excluded freelance journalists from attending its press conferences by using the press club system in the wake of the disaster. In 2014, the World Press Freedom Index by Reporters without Borders (n.d.) further critically described the characteristics of the press club:

> Censorship of Fukushima
>
> Arrests, home searches, interrogation by the domestic intelligence agency and threats of judicial proceedings—who would have thought that covering the aftermath of the 2011 Fukushima nuclear disaster would have involved so many risks for Japan's freelance journalists? The discrimination against freelance and foreign reporters resulting from Japan's unique system of Kisha clubs, whose members are the only journalists to be granted government accreditation, has increased since Fukushima.
>
> Often barred from press conferences given by the government and TEPCO (the Fukushima nuclear plant's owner), denied access to the information available to the mainstream media (which censor themselves), freelancers have their hands tied in their fight to cover Japan's nuclear industrial complex, known as the "nuclear village."

Whereas the Internet and social media have provided investigative journalists and freelancers with the resources to express their individual viewpoints for various audiences, the advent of the Internet and social media did not drastically transform the Japanese mainstream media's structural interdependencies with politico-economic systems. The next section describes a post-Fukushima situation in which citizens constructed their own meaning of the Fukushima disaster, which provides the backdrop for an analysis of Safecast as a civic media outlet.

Post-Fukushima Japanese Society as a Risk Society

In the wake of the disaster, many citizens were left with one critical question: "Are we really safe?" In other words, citizens suffered from an ambiguous situation in which they desperately needed information about their environment. This section describes an ambiguous post-Fukushima situation that paved the way for citizens to engage with digital media practices.

While the Japanese state announced that there would be no immediate health effects following the disaster, there is no scientific consensus on the health effects of low-dose radiation. More specifically, the health effects of low-dose radiation are *scientifically* unobservable, partly due to a lack of data (Torii, Shozugawa, & Watanabe, 2012). In fact, the International Commission on Radiological Protection (ICRP), the most authoritative international organization that provides recommendations on radiation protection, *hypothesizes* that there are no thresholds below which low-dose radiation is harmless for human bodies; in essence, the health effects of low-dose radiation are *scientifically incalculable* (Abe, 2015). Perhaps because of the incalculability of its health effects, Japanese scientists provided different views on the health effects of low-dose radiation (e.g., Abe, 2015); in contrast, the Japanese public apparently struggled to access comprehensive and reliable information about their safety.

Closely related to the incalculability of the health effects of low-dose radiation is the concept of risk society that was formulated by Ulrick Beck. Beck (1992, 1995, 2008) illustrated key characteristics of contemporary society in terms of new types of risks, and he offered a new theoretical framework of the risk society. Describing the notion of a risk society as "a constellation in which the *idea* of controllability of decision-based side effects and dangers which is guiding for modernity has become questionable" (Beck, 2008, p. 15), Beck insisted that the development of science and technology, which was once one of the key factors that contributed to the modernization of society, is now responsible for producing *unprecedented* risks whose temporal and spatial side effects are technically and scientifically incalculable. More relevant for this study, the health effects of low-dose radiation are likely to be open to various interpretations in post-Fukushima Japanese society, which can be seen as a kind of risk society (Beck, 2011).

In this scenario, it is reasonable to assume that immediately after the disaster, the Japanese public suffered from what Ball-Rokeach (1973) describes as "pervasive ambiguity," which occurs "when individuals or collectives are unable to define a social situation" (p. 378), not only because they lack practically useful information about low-dose radiation in their immediate surroundings but also because the health effects of low-dose radiation are scientifically incalculable. Ball-Rokeach (1973) further elaborates on the situation of pervasive ambiguity, stating that:

> When persons experience pervasive ambiguity, they are unable to determine the relationships between themselves and other elements in that social environment and are unable to identify the contextual meaning of the situation which ties the elements or parts of a situation into a meaningful unit-forming whole. (p. 379)

Much research indicates that ambiguous or threatening situations affect the intensity of individuals' media system dependencies (e.g., Ball-Rokeach, 1985; Loges, 1994); to define an ambiguous or threatening situation, citizens use media tactically and develop their dependency on the media accordingly. Furthermore, Shibutani (1966) contended that ambiguous situations could be one of the factors responsible for improvised interpretations:

> If such information is not forthcoming from formal news channels, demand for news remains. Men still have to meet the situation; unless they can put together some kind of definition, they cannot act. In such contexts they are likely to pool their intellectual resources and to improvise an interpretation. (p. 174)

Thus, Shibutani suggests that people actively participate in rendering post-disaster situations less ambiguous by improvising a tentative interpretation. From his perspective, the mainstream media is just one of several intellectual resources that citizens use to redefine post-disaster situations in a less ambiguous manner.

In the case of the Fukushima disaster, it is not far-fetched to assume that the Japanese mainstream media failed to provide practically useful information about low-dose radiation, in part because of their structural interdependencies with politico-economic systems; much scholarship indicates that the Japanese mainstream media uncritically reported the official announcements of the Japanese national government and Tokyo Electric Power Company (TEPCO) (e.g., Hizumi & Kino, 2012; Segawa, 2011). Perhaps more importantly, the Japanese mainstream media could not provide location-specific information about low-dose radiation for their various audiences (e.g., Sato, 2012). In contrast to other emergency situations (e.g., volcano eruptions), after a nuclear disaster, the mainstream media may not necessarily provide intellectual resources through which individual audience members can render their immediate surroundings less ambiguous. Along with the structural constraints that shaped the Japanese mass media's interdependency relations with other systems (e.g., the press club system), the characteristic of low-dose radiation as a key source of pervasive ambiguity apparently made it difficult for the Japanese mainstream media to provide practically useful information for various audiences.

This section describes an ambiguous situation in post-Fukushima Japanese society as a case of a risk society. Even if the Japanese mainstream media conveyed information about low-dose radiation in the air, this does not necessarily mean that they provided practically useful intellectual resources for various citizens to reasonably "improvise" their own interpretations of their immediate situation. A lack of practically useful information not only created opportunities for various citizens to seek alternative media but also generated a need for creative spaces in which citizens could "improvise" their own media. In the next section, this study further examines individuals' media dependency relations by analyzing Safecast as an alternative media outlet.

Safecast as Civic Media in Post-Fukushima Japanese Society

In the wake of the disaster, a wide variety of networked measuring publics viewed data as a prerequisite for their own interpretations of their immediate environments and accordingly engaged in the practice of data production using dosimeters. Those individuals then actively shared "raw" data related to radiation in the air through digital media, apparently contributing to breaking "the interpretative monopoly of governments" (Baack, 2015) and mainstream media. As a result, vast amounts of data ultimately became available online, but it is important to note that measurement readings collected with different dosimeters could differ (Shozugawa, 2014). In fact, various types of dosimeters, including shoddy ones, were available immediately after the disaster (Dokuritsu gyōsei hōjin kokumin seikatsu sentā, 2011). Thus, it was not necessarily easy for the networked measuring public to use the tremendous amount of data effectively in the wake of the disaster. As a result, one of the most challenging problems facing the networked measuring public was to create a practically useful radiation map that would allow various citizens to use the data effectively in their everyday lives (Abe, 2015).

In designing its radiation maps as alternative media, Safecast tactically used a standardized dosimeter, a Geiger-Müller counter named "bGeigie," paired with Global Positioning System (GPS) technology. In doing so, Safecast not only made the data points collected with bGeigie more or less comparable to one another but also made it possible to obtain accurate information concerning the locations of individual data points. Safecast thus focused on enhancing the *compatible accuracy* of its data because such data allow its audiences to compare their data with other data collected in different parts of the world (within its radiation map) (Abe, 2015).

In addition, Safecast standardized its method for visualizing the level of low-dose radiation in the air around the globe. This act is perhaps one of the most significant contributions that Safecast made as a global-scale civic media outlet because institutionalized agencies, such as governmental agencies, do not provide information that is outside their scope. For example, the radiation maps provided by Japan's Nuclear Regulation Authority do not offer any information about the level of low-dose radiation outside of Japan (Genshiryoku kisei iinkai, n.d.). In comparison, Safecast's civic radiation map consists of *globally comparable* data that are standardized through the use of its visualization method, allowing its Japanese audiences to see the levels of low-dose radiation outside of Japan in comparison to the levels in their immediate surroundings in Japan (Abe, 2015). In September 2012, the Fukushima Prefecture adopted a portion of Safecast's data as a useful resource (Tanaka, 2012), in large part due to its institutionalized world radiation map. This move allowed its audience to see the level of low-dose radiation outside of Japan, which in turn indicated the power of Safecast's radiation map as a global-scale civic media outlet.

As such, as alternative media, Safecast's radiation map can be seen as a type of civic technology, which Baack (2015) described as "small-scale, specialized applications that aim to 'connect people'" (p. 7). Baack (2015) elaborates the concept of civic technologies as follows:

> Civic technologies can be described as alternative ways of fulfilling functions traditionally described as "journalistic" … or of accessing and using public services … In other words, these applications are developed *independently outside* professional journalism or public institutions, but at the same time are trying to fulfill *similar functions*. (p. 7)

Rather than depending on mainstream media and governmental institutions, Safecast, as a civic media outlet of globally networked measuring publics, played a role in providing practically useful data that would help various citizens define and discuss their immediate environments. While the health effects of low-dose radiation are scientifically unobservable, Safecast created an alternative space for citizens to "improvise" a *data-based* interpretation of their surroundings.

While Safecast designed its global radiation map to allow the Japanese public to see the levels of low-dose radiation in their immediate surroundings in comparison with other parts of the world, much research indicates that the Japanese public did not necessarily decrease their dependency relations with the mainstream media (e.g., Hirai, 2013; Jung, 2012; Takano, Yoshimi, & Miura, 2012; Yoshioka, 2011). In short, Safecast by no means replaced the mainstream media in Japan. As civic media, Safecast's global radiation map provided intellectual resources with which citizens could make data-based judgments about their safety, but it would have been difficult for non-Internet users to take full advantage of Safecast's map. Furthermore, it could also be difficult for certain citizens to base their informed judgment on Safecast's civic media alone; citizens may also need experts' advice about their safety. Indeed, it is noteworthy that Safecast focuses exclusively on providing data as a resource for its audiences to define their environments. Safecast's website reads, "Safecast is apolitical, and takes no stance for or against nuclear power. Safecast is pro-data and committed to giving people accurate information with which they can draw their own conclusions" (Safecast, n.d.-b). Identifying itself as a "pro-data" network, Safecast thus provides no comment on its data for its various audiences.

That said, as civic media, Safecast has shared with individual audiences what mainstream media cannot provide: data related to low-dose radiation in the air in their immediate surroundings. Rather than replacing the Japanese mainstream media, Safecast thus supplemented the mainstream media by providing alternative intellectual resources upon which audience members could draw when making judgments about their health and safety. Accordingly, Safecast not only provided its free global radiation map, which consists of practically useful data related to radiation, for various audiences but also allowed its audiences to visualize the consequences of the Fukushima disaster *globally*, in terms of the level of radiation in the air (Abe, 2015). In fact, Safecast went one step beyond supplementing the mainstream media; it enhanced citizens' capacities to imagine alternative media in Japanese media ecology, as Giddens (1990) points out:

> ... the circularity of social knowledge ... affects in the first instance the social rather than the natural world. In conditions of modernity, the social world can never form a stable environment in terms of the input of new knowledge about its character and functioning. New knowledge (concepts, theories, [and] findings) does not simply render the social world more transparent, but alters its nature, spinning it off in novel directions. (p. 153)

Conclusion

In July 2014, *the Fukushima Minyū*, a local newspaper in the Fukushima Prefecture, began to publish Safecast's data in a new section titled "Radiation Dosage in Major Cities around the World" (Fukushima Minyū, 2014); a Japanese mainstream newspaper thus alerted its readers to Safecast's data as a useful resource for their everyday lives. Given that the Fukushima disaster is ongoing, it may be too early to define the role of Safecast as alternative media in relation to the Japanese media ecology, but three key findings emerged from this study.

First, as a global network of measuring publics, Safecast actively sought and created data related to low-dose radiation in the air in the wake of the Fukushima disaster. To understand an ambiguous post-Fukushima situation, Safecast "improvised" a standardized dosimeter paired with GPS and created a radiation world map as alternative media for various audiences to use freely. This finding supports Ball-Rokeach's (1973) study that demonstrates that individuals are capable of dealing with pervasive ambiguities.

Second, Safecast did not necessarily "hack" the Japanese mainstream media; rather, Safecast engaged in alternative participatory media practices and supplemented the Japanese mainstream media accordingly. Safecast provided practically useful data for various audiences and ultimately achieved what the Japanese mainstream media could not. However, this achievement does not mean that Safecast replaced the Japanese mainstream media, in part because it is difficult for non-Internet users to take advantage of Safecast as alternative media. Thus, this study thus illustrates the opportunities and challenges that face civic media outlets following a nuclear disaster.

Finally, future MSD-guided research should account for the cause of the ambiguity or threat. Unlike earthquakes or volcano eruptions, the unknown health effects of exposure to low-dose radiation make it difficult for the Japanese mainstream media to provide their various audiences with practically useful information, not only because the mainstream media cannot convey location-specific information but also because of the lack of scientific agreement on the issue. It is therefore reasonable to assume that risk-specific MSD research could be one future direction for MSD-guided research.

While this study is based solely on anecdotal evidence concerning media systems dependency following the Fukushima Daiichi nuclear disaster, it is the first study to investigate media dependency relations following the Fukushima disaster, by analyzing the networked measuring public in general and Safecast in particular. The key findings of this study not only demonstrate the opportunities and challenges that face both the mainstream media and the networked measuring public but also enhance our capacities to imagine alternatives to the current Japanese media system.

Note

1. In Japan, the Internet has a long history as a space for narratives about emergency situations that predates the nuclear disaster (Murai, 2015). For instance, the Great Hanshin earthquake of 1995 witnessed the emergence of various narratives about post-earthquake situations online (Ogasawara, 1999). However, in 1995, there were limited communication spaces in which diverse types of citizens could get involved in sharing their own stories about the disaster by using digital media. In comparison, the recent proliferation of digital media sources paved the way for citizens to use those sources in different ways in the wake of the Fukushima disaster (e.g., Abe, 2014b; Murphy, 2014; Takano, Yoshimi, & Miura, 2012; Tanaka, Shineha, & Watanabe, 2012). According to *The White Paper: Information and Telecommunication in 2011* by the Ministry of Internal Affairs and Communication, approximately 80% of the entire population in Japan was able to access to the Internet by the end of 2010 (Sōmushō, 2011).

References

Abe, Y. (2013). Risk assessment of nuclear power by Japanese newspapers following the Chernobyl nuclear disaster. *International Journal of Communication, 7*, 1968–1989. Retrieved from http://ijoc.org/index.php/ijoc/article/view/1848/982

Abe, Y. (2014a). Safecast or the production of collective intelligence on radiation risks after 3.11. *The Asia-Pacific Journal: Japan Focus, 12, 7*(5). Retrieved from http://www.japanfocus.org/-Yasuhito-_Abe_/4077/article.html

Abe, Y. (2014b). "Pray for Japan": Reinventing "Japanese national character" after the 2011 Tohoku earthquake, tsunami, and nuclear crisis. In L. Cui & M. H. Prosser (Eds.), *Social media in Asia* (pp. 351–376). Lake Oswego, OR: World Dignity University Press.

Abe, Y. (2015). *Measuring for what: Networked citizen science movement after the Fukushima nuclear accident* (Unpublished doctoral dissertation). University of Southern California, Los Angeles, CA.

Arima, T. (2006). *Nihon terebi to CIA: Hakkutsu sareta Shōriki fairu* [Nippon television and the CIA: The discovered Shoriki file]. Tokyo: Shinchōsha.

Arima, T. (2008). *Genpatsu Shōriki, CIA* [Nuclear power, Shōriki, and the CIA]. Tokyo, Japan: Shinchōsha.

Azby. (2014, December 23). 25 million! [Web log post]. Retrieved from http://blog.safecast.org/2014/12/25-million/

Baack, S. (2015). Datafication and empowerment: How the open data movement re-articulates notion of democracy, participation, and journalism. *Big Data & Society, 2*(2), 1–11.

Ball-Rokeach, S. J. (1973). From pervasive ambiguity to a definition of the situation. *Sociometry, 36*(3), 378–389.

Ball-Rokeach, S. J. (1974). *The information perspective.* Paper presented at the Annual Conference of the American Sociological Association, Montreal, Canada.

Ball-Rokeach, S. J. (1985). The origins of individual media-system dependency: A sociological framework. *Communication Research, 12*(4), 485–510.

Ball-Rokeach, S. J. (1998). A theory of media power and a theory of media use: Different stories, questions, and ways of thinking. *Mass Communication and Society, 1*(1/2), 5–40.

Ball-Rokeach, S. J., & DeFleur, M. L. (1976). A dependency model of mass-media effects. *Communication Research, 3*(1), 3–21.

Ball-Rokeach, S. J., & Jung, J.-Y. (2009). The evolution of media system dependency theory. In R. N. Nabi & M. B. Oliver (Eds.), *Sage handbook of media processes and effects* (pp. 531–544). Los Angeles, CA: Sage.

Beck, U. (1992). *Risk society. Towards a new modernity.* Thousand Oaks, CA: Sage.

Beck, U. (1995). *Ecological politics in an age of risk.* Cambridge: Polity Press.

Beck, U. (2008). *World at risk.* Cambridge: Polity Press.

Beck, U. (2011). Kono kikai ni: Fukushima, aruiwa sekai risuku shakai ni okeru Nihon no mirai [On this occasion: Fukushima or Japan's future in world risk society]. In U. Beck, M. Suzuki, & M. Itō (Eds.), *Risukuka suru Nihon shakai: Ururihhi Bekku to no taiwa* [Japanese society at increasing risk: A dialogue with Ulrich Beck] (pp. 1–12). Tokyo: Iwanami Shoten.

Brough, M., & Li, Z. (2013). Media systems dependency and human rights online video: The "Saffron Revolution" and WITNESS's Hub. *International Journal of Communication, 7*, 281–304. Retrieved from http://ijoc.org/index.php/ijoc/article/view/1423/854

Dokuritsu gyōsei hōjin kokumin seikatsu sentā. (2011, December 22). Hikakuteki anka na hōshasen sokuteiki no seinō [Quality of relatively cheap radiation measurement devices]. Retrieved from http://www.kokusen.go.jp/pdf/n-20110908_1.pdf

Feldman, O. (1993). *Politics and the news media in Japan.* Ann Arbor, MI: The University of Michigan Press.

Freeman, L. A. (2000). *Closing the shop: Information cartels and Japan's mass media.* Princeton, NJ: Princeton University Press.

Fukushima genpatsu jiko dokuritsu kenshō iinkai. (2012). *Fukushima genpatsu jiko dokuritsu kenshō iinnkai: Chōsa/kenshō hōkokusho* [Independent investigation commission on Fukushima Daiichi nuclear accident: Investigation report]. Tokyo, Japan: Discover 21.

Fukushima Minyū. (2014, July 6). Sekai shuyō toshi no hōshasenryō [Radiation dosage in major cities around the World]. *Fukushima Minyū*, p. 24.

Genshiryoku kisei iinkai (Cartograher). (n.d.). Hōshasenryō sokutei mappu. [Radiation dose measurement map]. Retrieved November 1, 2015 from http://radioactivity.nsr.go.jp/map/ja/

Giddens, A. (1990). *The consequences of modernity.* Stanford, CA: Stanford University Press.

Hall, I. P. (1998). *Cartels of the mind: Japan's intellectual closed shop.* New York, NY: W. W. Norton.

Hemmi, A., & Graham, I. (2012). Hacker science versus closed science: Building environmental monitoring infrastructure. *Information, Communication & Society, 17*(7), 1–13.

Hirai, T. (2013). Genpatsujiko to intānetto: Hōshasen busshitsu no kakusan ni kansuru jōhō o jirei to shite [Nuclear accident and the Internet: A study of information on the diffusion of radioactive materials]. In Kōeki zaidan hōjin Shimbun tsūshin chōsakai (Eds.), *Daichinsai genpatsu to media no yakuwari* [Big earthquake, nuclear power and the role of media] (pp. 73–82). Tokyo: Kōeki zaidan hōjin Shimbun tsūshin chōsakai.

Hirschburg, P. L., Dillman, D. A., & Ball-Rokeach, S. J. (1986). Media system dependency theory: Responses to the eruption of Mount St. Helens. In S. J. Ball-Rokeach & M. G. Cantor (Eds.), *Media, audience, and social structure* (pp. 117–126). Newbury, CA: SAGE Publications.

Hizumi, K., & Kino, R. (2012). *Kenshō Fukushima genpatsu jiko kisha kaiken: Tōden seifu wa nani o kakushita noka* [Investigation of the press conference on the Fukushima nuclear accident: What did TEPCO and the government cover up?]. Tokyo: Iwanami shoten.

Jenkins, H. (2007, October 3). What is civic media? [Web log post]. Retrieved from http://henryjenkins.org/2007/10/what_is_civic_media_1.html

Jōmaru, Y. (2012). *Genpatsu to media* [Nuclear power and media]. Tokyo: Asahi Shimbun Shuppan.

Jung, J. Y. (2012). Social media use and goals after the Great East Japan Earthquake. *First Monday, 17*(8). Retrieved from http://firstmonday.org/ojs/index.php/fm/article/view/4071/3285

Kasza, G. J. (1988). *The state and the mass media in Japan 1918–1945.* Berkeley, CA: University of California Press.

Kelly, A. R. (2014). *Hacking science: Emerging parascientific genres and public participation in scientific research* (Doctoral dissertation). Retrieved from http://repository.lib.ncsu.edu/ir/handle/1840.16/9367

Krauss, E. S. (2000). *Broadcasting politics in Japan: NHK and television news.* Ithaca, NY: Cornell University Press.

Loges, W. E. (1994). Canaries in the coal mine perceptions of threat and media system dependency relations. *Communication Research, 21*(1), 5–23.

McCargo, D., & Lee, H.-S. (2010). Japan's political tsunami: What's media got to do with it? *International Journal of Press/Politics, 15*(2), 236–245.

Murai, J. (2015). *Intānetto no kiso* [The foundation of the Internet]. Tokyo: Kadokawa gakugei shuppan.

Murillo, L. F. R. (2013, May). *New expert eye over Fukushima: Open source response to the nuclear crisis in Japan.* Paper presented at the STS Forum on the 2011 Fukushima/

East Japan Disasters, Berkeley, CA. Abstract Retrieved from http://fukushimaforum.wordpress.com/workshops/sts-forum-on-the-2011-fukushima-east-japan-disaster/manuscripts/session-3-radiation-information-and-control/new-expert-eyes-over-fukushima-open-source-responses-to-the-nuclear-crisis-in-japan/

Murphy, S. M. (2014). Grassroots democrats and the Japanese state after Fukushima. *Japanese Political Science Review, 2*, 19–37. Retrieved from http://www.jpsa-web.org/eibun_zassi/data/pdf/JPSA_MURPHY_final_July_9_2014.pdf

Ogasawara, H. (1999). *Living with national disasters: Narratives of the Great Kanto and the Great Hanshin earthquakes* (Doctoral dissertation). Retrieved from http://www.proquest.com/en-US/

Ognyanova, K., & Ball-Rokeach, S. J. (2015). Political efficacy on the Internet: A media system dependency approach. In L. Robinson, S. R. Cotten, & J. Schulz (Eds.), *Communication and information technologies annual: Politics, participation, and production* (pp. 3–27). Bingley: Emerald Group Publishing Limited.

Pharr, S., & Krauss, E. S. (Eds.). (1996). *Media and politics in Japan*. Honolulu, HI: University of Hawai'i Press.

Reporters without Borders. (n.d.). World Press Freedom Index 2014. Retrieved November 1, 2015 from https://rsf.org/index2014/en-asia.php

Safecast. (n.d.-a). About Safecast. Retrieved November 1, 2015 from http://blog.safecast.org/about/

Safecast. (n.d.-b). FAQ. Retrieved November 1, 2015 from http://blog.safecast.org/faq/

Sasaki, T. (1999). *Nihon no kindai 14: Media to kenryoku* [The modernity of Japan vol. 14: Media and power]. Tokyo: Chuokōron Shinsha.

Sato, T. (2012). Genpatsu jiko o watashitachi wa dō tsutaeta ka: Kazoku ga chiiki ga hikisakareteiku nakade media wa sono yakuwari o hatasetaka [How did we report the nuclear accident? Did media fulfill its responsibility when families and regional communities were split up [by the disaster]]. In Y. Niwa & M. Fujita (Eds.), *Media ga furueta: Terebi rajio to higashi nihon daishinsai* [The media quaked: Television and radio after the Great East Japan Earthquake] (pp. 241–276). Tokyo: Tokyo Digaku Shuppankai.

Segawa, S. (2011). Genpatsu hōdō wa 'daihonei happyō' dattaka: Asa, Mai, Yomi, Nikkei no kijikara saguru [Were media report on nuclear power like the announcement of wartime Japan's Imperial General Headquarters?: An analysis of articles from the Asahi, the Mainichi, the Yomiuri, and the Nikkei newspapers]. *Journalism, 255*, 28–39.

Shibata, T. (2013). *Genshiryoku hōdo* [Coverage of nuclear power]. Tokyo: Tokyo denki daigaku Shuppankyoku.

Shibutani, T. (1966). *Improvised news: A sociological study of rumor*. Indianapolis, IN: The Bobbs-Merrill Company, Inc.

Shiraishi, H. (2012). Our Planet TV wa genpatsujiko o dō hōjitaka [How did Our Planet TV report the nuclear accident?]. *Days Japan, 9*(4), 210–212.

Shozugawa, K. (2014). *Minna no hōshasen sokutei nyūmon* [Introduction to radiation monitoring for everyone]. Tokyo: Iwanami shoten.

Slater, D. H., Nishimura, K., & Kindstrand, L. (2012). Social media in disaster Japan. In J. Kingston (Ed.), *Natural disaster and nuclear crisis in Japan* (pp. 94–108). London: Routledge.

Sōmushō. (2011). *Jōhō tsūshin hakusho* [White paper on information and telecommunication]. Tokyo: Nikkei Insatsu.

Tai, Z., & Sun, T. (2007). Media dependencies in a changing media environment: The case of the 2003 SARS epidemic in China. *New Media & Society, 9*(6), 987–1009.

Takano, A., Yoshimi, S., & Miura, S. (2012) *311 Jōhōgaku: Media wa nani o dō tsutaetaka* [Information studies of March 11: What and how did media report?]. Tokyo: Iwanami shoten.

Tanaka, K. (2012). Fukushima Prefecture adapts Safecast worldwide radiation map on official prefectural website [Web log post]. Retrieved from http://blog.safecast.org/2012/09/fukushima-prefecture-adapts-safecast-worldwide-radiation-map-on-official-prefectural-website/

Tanaka, M., Shineha, R., & Maruyama, K. (2012). *Saigai jakusha to jōhō jakusha: 3.11 go naniga misugosareta noka* [The vulnerable to disaster and to a lack of information: What's missing after March 11th]. Tokyo: Chikuma shobō.

The Japan Newspaper Publishers and Editors Association. (2006, March 9). Kisha club guidelines. Retrieved from http://www.pressnet.or.jp/english/about/guideline/

Torii, H., Shozugawa, K., & Watanabe, Y. (2012) *Hōshasen o kagakuteki ni rikaisuru.* [Understanding radiation scientifically]. Tokyo: Maruzen Shuppan.

Utz, S., Schultz, F., & Glocka, S. (2012). Crisis communication online: How medium, crisis type and emotions affected public reactions in the Fukushima Daiichi nuclear disaster. *Public Relations Review, 39*(1), 40–46.

Watanabe, T., & Nohara, H. (2000). Shiryō kara yomitoku media to shakai kenryoku (dai ni bu): Hōsōhō no seiritsu kaitei unyō o chūshin to shite [A study of the enactment and amendments of the Japanese Broadcasting Law, Part 2: Political economy of Japanese TV broadcasting and economic development]. *Hyōron shakai kagaku, 63*, 49–235. Retrieved from https://doors.doshisha.ac.jp/duar/repository/ir/2166/h06302.pdf

Yamakoshi, S. (2013). Genshiryoku seisaku hōdō to chihōshi [Coverage of nuclear policy and local newspapers]. In Kōeki zaidan hōjin shimbun tsūshin chōsakai (Ed.), *Daishinsai genpatsu to media no yakuwari* [The Great Earthquake, nuclear power and the role of media] (pp. 30–42). Tokyo: Kōeki zaidan hōjin shimbun tsūshin chōsakai.

Yamamoto, A. (2012). *Kaku enerugī gensetsu no sengoshi 1945–1960* [A postwar history of discourses on nuclear energy from 1945 to 1960]. Kyoto: Jinbun shoin.

Yoshioka, H. (2011). Masu media to "sanjū no kabe" [Mass media and "three walls"]. *Shimbunkenkyū, 770*, 49–53.

Zuckerman, E. (2013, August 29). Citizen science versus NIMBY? [Web log post]. Retrieved from http://www.ethanzuckerman.com/blog/2013/08/29/citizen-science-versus-nimby/

4. *Project Chanology and the Formation of Anonymous as an Activist Movement*

Rhea Vichot
University of Wisconsin-Whitewater

> Gentlemen, this is what I have been waiting for. Habbo, Fox, The G4 Newfag Flood crisis. Those were all training scenarios. This is what we have been waiting for. This is a battle for justice. Every time /b/has gone to war, it has been for our own causes. Now, gentlemen, we are going to fight for something that is right. I say damn those of us who advise against this fight. I say damn those of us who say this is foolish. /b/ROTHERS, THE TIME HAS COME FOR US TO RISE AS NOT ONLY HEROES OF THE INTERNETS, BUT AS ITS GUARDIANS.
>
> —Anonymous, "It is time."

> This is gay. I just want to fix my code and shoop things onto Dakota Fanning's face. Idiots. ...
>
> —Response by Anonymous

Anonymous, as a group, has been difficult to describe as a whole. As Gabriella Coleman notes, "Anonymous resists straightforward definition as it is a name currently called into being to coordinate a range of disconnected actions, from trolling to political protests" (2011). Since 2009, Anonymous has emerged as a notable hacktivist group involving themselves online in causes ranging from defending Wikileaks from banking services which froze donations to the site, promoting and bringing media attention to the teen rapists in Steubenville, Ohio, and finding information on police officers involved in police brutality as part of the online Black Lives Matter movement. That said, Anonymous has also been known for activities ranging like DDoS attacks, doxxing individuals, compromising websites and social media accounts, and spreading information over multiple internet platforms ranging from YouTube, Twitter, and IRC. Oftentimes, individuals who have denounced or ridiculed the group have found themselves on the receiving end of griefing activity by Anonymous, such as was the case with Aaron Barr, then CEO of cybersecurity firm HBGary Federal, after he claimed he could obtain data from hacker groups like Anonymous.

While academic attention has been placed on emergence of Anonymous as a hacktivist group and its work related to hacking governmental and financial agencies (Coleman, 2014; Phillips, 2011; Schwartz, 2008), comparatively little attention has been paid to the formation of Anonymous as an online community. The context for the shift towards hacktivism began in late 2007 with the beginnings of protests against the Church of Scientology. While Anonymous had developed a sense of itself as an organization, utilizing the Anonymous moniker (created via a misreading of the affordances of the imageboard platform the group utilized for its activities), its early activities were comprised primarily of griefing and trolling players on online games, creating memes and other forms of media, and doxxing individuals over social media for the purpose of harassment or, in a few cases such as that of Chris Forcand (J. Jenkins, 2007), to report them to authorities. It was the interruption of its activities by Scientology, and the nature of Scientology's organization that allowed Anonymous to effectively protest the Church online and offline. It was also the first moment that Anonymous utilized skills it had practiced on other activities to effectively mount such a protest. After the main energy of these Scientology protests (dubbed Project Chanology) ebbed, other causes began to emerge as rallying points for Anonymous.

This chapter aims to look at a point in time where Anonymous—the griefing, meme creating community—and the group that became involved in activist causes first differentiated themselves. While an important conversation thread among Anonymous members in the period was the understanding that such a distinction was being made, it should be emphasized is that this argument is not to suggest a teleological development from Anonymous as a community sited along various related online imageboards and communities to Anonymous as a hacktivist collective (or even necessarily a splintering off of individuals from one to the other), but as two identifiable groups which emerge from the same online spaces that have, since the protests against Scientology, divergent goals, attitudes, and activities, though still an overlap in membership. Some scholars have utilized notation such as "little a" Anonymous and "big A" Anonymous in order to distinguish from the two, but during this period, the distinguishing term utilized by Anonymous as Project Chanology, a reference to the way that earlier raids/DDoS/harassment/griefing attacks were named. It was only after the transition from Project Chanology to other political and activist activities have issues of what to call the group engaging in these endeavors over the group which continued in similar griefing/trolling activities emerged.

Historical Context for the Emergence of Anonymous

What has become Anonymous can be traced back to Japanese message board 2-Channel ("ni-channel", 2ch for short). Established in 1999 by a Japanese expatriate Hiroyuki Nishimura who was a student at the University of Arkansas, the board has grown in the largest message board in the world with an estimated 8 million posts daily.

The major difference between 2ch and other message board and forum post software was twofold. First, forum posts fall into one of hundreds of boards dedicated to topics as wide ranging as cooking, venture capital, and social news. Among these boards are several topics. These topics are organized around an original posting, usually centered around topics such as "iPod Touch Part 71" or "Best Wheat for Pizza: 3rd Slice" were organized vaguely by topic, without any threading of replies and with a topic limit of 1000 messages, organized by tripcode.[1] When a topic reaches the limit, a new post must be made to continue discussion. These topics are then 上げ(aged or "bumped") or 下げ(saged or "lowered") depending on whether participants like the topic or not.

Secondly, and perhaps most important, all posts not made by the forum owner of moderators are done anonymously. Nishimura claims the reason for this was the he wanted to "create[d] a free space, and what people did with it was up to them ... No major corporations were offering anything like that, so I had to." The result of the anonymization was that people on the boards felt free to comment on everything from celebrity gossip, to politics, to teachers and classmates. Many writers on 2-channel and Japanese internet culture feel that it is the anonymizing element of 2channel that affords it its real power. An article in *Japan Media Review* (2009) describes this dynamic as "an anarchic and free alternative to Japan's mainstream press and uncompromised by the main media's networks of press clubs, political and corporate allegiances, and consensus-minded stances. Channel 2 can't rival the mainstream media for authority or accuracy, but it is obvious why the Japanese media sense a threat."

Nishimura, when talking about the site, also uses a rhetoric of counterpublics as the motivation behind forced anonymity. Nishimura notes that:

> There is a lot of information disclosure or secret news gathered on Channel 2. Few people would post that kind of information by taking a risk. Moreover, people can only truly discuss something when they don't know each other. Under

> the anonymous system, even though your opinion/information is criticized, you don't know with whom to be upset. Under a perfectly anonymous system, you can say, "it's boring," if it is actually boring. All information is treated equally; only an accurate argument will work. (2009)

As a spin-off to 2-Channel, Futaba Channel[2] was created August 30, 2001 as both "an image-based counterpart to the text-only 2ch" and as a refuge from 2-Channel when the site was threatened with closing down. Similar to 2ch, Futaba-Channel (or Futaba-chan) operates using boards that range in topic from local and international news to toys, ramen, and insect collecting though the format of the site was such that images could be posted along with text. Futaba-Channel, as well as other imageboards to follow, borrowed from 2ch the system of anonymous posting as the default method of posting, though allowing users to make up a username if they wanted to. Also an important difference is the creation of the 二次元裏 (nijikenura or "underside") board, which is a collection of random, off topic discussion that is fast paced and carnivalesque in content and tone.

In 2003, the imageboard *4-chan* was created by then 15-year old Christopher Poole. Poole copied the codebase of Futaba-Channel, translated the interface into English, and hosted the site himself. Poole, using the handle moot from his days as a regular poster in the forums of Internet humor site Something Awful, organized the imageboard with topics separated between "worksafe" and "not worksafe" sections.[3] Of the various boards on 4chan, who like its predecessors range from the whimsical (/po/, papercraft and origami) to obscene (/d/"Alternative" Hentai porn), the one that has drawn the most coverage is the Random board, otherwise known as /b/. As with the Futaba-Chan version, the board is filled with random discussion topics, ranging from questions about consumer electronics, requests for pornography, to amateur philosophy.[4] Members of the board, who are referred to as /b/tards, often use the mostly rules free board[5] as a testing ground for offending each other, posting images meant to shock or provoke anger or humor. This activity is usually limited to /b/, but seeps out occasionally to other boards for the purpose of trolling[6] non-/b/readers.

From 4-chan, many spin-off imageboards were made that specialized in certain aspects covered only broadly by 4-chan such as moe,[7] drug culture, and pornography. Usually these boards had an analog to /b/, though these ranged from iterations of /b/that were simply off topic conversation to boards such as 7chan's /i/or invasion board where many of the more well-known raids (such as the Habbo Hotel Raid[8]) were planned out. These raids range from organized griefing of online games and virtual worlds, to online and off-line harassment and trolling of specific people or subcultures.[9]

During August of 2006, an Anonymous member going by the name of Captain Cornflake wrote and posted "Declaration of /b/Independence" on /b/. This message was ostensibly written in response to friction between board members of /b/and moderators of the board, particularly once 4chan moderators began deleting threads related to raids in the wake on the raids on online game *Habbo Hotel* and the website *Wikifur*. The declaration coincided with a move of many self-identified Anonymous members to a new website, called 7chan and can be seen as solidifying Anonymous as a multi-sited internet community, existing in what became a whole host of *chan style imageboards, various non-imageboard websites, and other communities based on other internet protocols such as IRC. This multi-sited nature is used in much the same way as the Anonymous label is used. When Anonymous members raid or are blamed for raiding, they will often times claim to be from unaffiliated sites, specifically hated websites such as EBaum's World and Gaia Online in order to divert attention away from Anonymous sites.

Scientology, Online Schism, and the Formation of Project Chanology

In December of 2008, a video meant for internal distribution among Scientology churches was leaked onto YouTube. The video consists of an interview of perhaps the most well-known celebrity Scientologist, Tom Cruise. In the interview, Cruise makes wild claims such as, "[Scientologists] are the authorities on getting people off drugs. We are the authorities on the mind. We are the authorities on improving conditions. Criminon.[10] We can rehabilitate criminals. We can bring peace and unite cultures." After the video was leaked the Church of Scientology asked it to be pulled from YouTube and claimed the video was taken out of context from a supposed 3 hour version.

The removal of the original video from YouTube prompted various others to simply re-upload the video onto YouTube several times, giving various titles in order to make permanent removal nigh impossible. Certain blogs, most famously celebrity news blog Gawker posted it and its editor, Nick Denton, rebuffed Scientology's demands to remove it by stating the video "[is] newsworthy, and we will not be removing it." On January 16, 2008, the Church of Scientology issued a copyright violation claim against YouTube for hosting material from the Cruise video.

This was the moment in which Anonymous took notice of the situation. The video was spread among the various imageboard sites during the month of December and many readers of the board were among those who helped to re-upload the video on YouTube and other online video sites. The consensus to begin what would be called Project Chanology has been attributed to readers of 4chan, 711chan, the Anonymous-related wiki partyvan.info, and countless IRC channels. At the same time, other members of Anonymous were reading information on established anti-Scientology organizations online, such as Enturbulation.org and Operation Clambake. During this period, noted anti-Scientologists published videos and posts to Anonymous. While some were intrigued at the new influx of attention towards Scientology, others condemned their methods.[11]

These methods included several standbys that Anonymous members have used for previous episodes of harassment. Distributed Denial of Service (DDoS) attacks on Scientology servers were made. Prank phone calls were made using VoIP services such as Skype. Unending "black faxes",[12] pizzas, and other services were directed at Scientology. These activities brought down Scientology's servers for more than a week.[13] These attacks were haphazard at best and, while the main Scientology servers were attacked, some innocent people were mistakenly targeted.[14] The Church attempted to find and press charges on the perpetrators of the DDoS attacks, but to date only one person, who pled guilty, has been convicted.

Beyond the initial direct attacks on Scientology, There were several attempts at raising general consciousness over Church and its practices. There was a campaign to highlight Scientology information sites on aggregation sites such as Digg. A "google bomb"[15] was used to make Scientology the first hit in a search for "dangerous cult" and anti-Scientology site Xenu.net the first link to a search for Scientology. Online, there are many sources Anonymous used to educate individuals about Scientology. Users who were already familiar with Scientology pointed others to longstanding online sources of information such as Operation Clambake, an online community and Norway-based non-profit created in 1996 for the purpose of collecting Scientology materials and publishing criticism of the Church of Scientology. In addition, there are many sites which are much more specialized such as ex-Scientology Kids, Why are They Dead, and Fair Gamed. In addition, users pointed people to YTMND, a site devoted to normally humorous audio/image/video juxtapositions, to view a work created in 2006 entitled "The Unfunny Truth About Scientology". In these ways, information was quickly spread to users to gain momentum needed for the initial raids, but more importantly, for the first global protest that was held afterwards. Information and media were spread quickly through Anonymous locations, such as the instance where the Mark Bunker videos to Anonymous were discussed on /b/.

On January 21st, a group of people speaking for Anonymous and calling themselves "Chan Enterprises" released a press statement declaring that

Anonymous was prepared to engage in a long-term war with Scientology. The video caused much controversy, and members of Anonymous posted a message to several of their websites proclaiming war against Scientology. Soon after, Anonymous struck at the church; they blocked access to its website, made prank calls, organized protests, distributed anti-Church pamphlets and information, and extracted secret files from the Church of Scientology and its parent company, the Religious Technology Center. Anonymous' members cited several reasons for their actions against the Church of Scientology: many have stressed the alleged human rights violations under the auspices of the Church. Others accused the Church of fraud due to its costly ceremonies, while some merely sought the entertainment they refer to as "lulz," a corruption of the Internet slang "LOL," or "laugh out loud." Most members, however, were concerned with the threat to free speech that the Church posed. This was most evident in the recent attacks on websites such as Digg and YouTube, where the Church filtered anti-Scientology comments. Concurrently with this press release, a video, entitled "A Message to Scientology", was placed on YouTube. In it, a computer generated voice, over stock footage of clouds moving, described the Cruise video debacle as but a first step in a war between Anonymous and Scientology in the name of free speech and liberation of members who were being financially exploited by the church. By February 8, the video had been seen 2 million times.

However, the computer-based attacks on Scientology were rather brief. By January 28th, 2008, moderators of 711chan asked that that DDoS attacks, black faxes, and other online-based harassment cease. Instead, a call for non-violent legal protest was made. A new video, "A Call to Action" was released, espousing similar sentiments and it was during this period that protests against Scientology began to organize for a series of global protests that would take place in the middle of February 2008. Partyvan.info, an Anonymous-affiliated website, used its bandwidth to act as an information clearing house, informing Anons who may not have been keeping up to date of the events since December as well as provide links to sub forums on Enturbulation.org, an anti-Scientology page previously unaffiliated with Anonymous, with which local planning and discussion could occur.

On February 12, 2008, the first set of "global raids" took place. These protests took place over 168 cities in North America, Western Europe, Japan, Australia, and New Zealand. These protests have continued, though smaller in size and number of protests. While this timeline has attempted to show the history and actual events that occurred in the early part of the Chanology movement, I have not yet shown the context with which both Anonymous and Project Chanology draw their ability to organize and create an environment for protest.

The initial organization of Project Chanology occurred in two well-known foci of Anonymous activity: 4chan's /b/and 7chan's /i/. On these forums instructions to raiders were given such as the motivation for the raid, what to do (in general terms), and precautions to take such as masking IP addresses, claiming that raiders came from EBaum's World, and other actions that were typical of earlier raids. An example of some of the discourse during the raiding period is below. These are a few posts among 587 replies to one thread, entitled "project chanology" on January 16, 2008:

> this anon has a question: HOW THE FUCK DOES MAKING THEIR OFFICIAL WEBSITE NOT WORK ELIMINATE SCIENTOLOGY???
>
> It's their hilarious insane reaction to being DDOSed or hacked that contributes to their downfall.
>
> Wow, the site is down. Great. For what? 10 minutes?
> WE NEED TO FUCKING RUIN IT /b/.
> FILL IT WITH INTERNETHATEMACHINE STUFF.
> Pr0n OR SOMETHING?!!
> LET'S FUCKING MAKE THEM SHAT BRIX.
> LET'S SHOW FOX WHO ANONYMOUS ACTUALLY IS /b/.
>
> THIS IS AN EBAUMSWORLD RAID.
> Ebaums raid! Do your coordination off-site, don't link back, and wait until the 18th to put this on places like Digg.

After the two waves of raiding conducted by Anonymous, which were deemed very successful as evidenced by both the disruption of various Scientology websites and centers, as well as with the flurry of media attention gained. This prompted Anons to produced work such as image macros which proclaimed expressions such as "1/21/08 NEVAR FORGET" and "WHY MY WEBSITE NO WERK?" accompanied with images of Tom Cruise and of Scientology symbols.

The shift in focus from traditional raiding to actual protest organization was the result of several actions. One was the recording of videos by people involved in anti-Scientology movements directly targeted towards Anonymous. Of particular note is Mark Bunker, whose XenuTV videos directed towards Scientology were disseminated on forums such as /b/and noted with generally positive reactions. Some Anons agreed with the premise that the illegal raiding strategies are not sustainable and that the long term goal should be to get Scientology's tax-exempt status reformed. Other sentiments were more negative, ranging with feeling that Bunker did not understand Anonymous's motivations[16] to outright mockery and dismissal.[17]

Another motivating factor may have been the amount of mainstream press that covered Anonymous and the initial actions against Scientology. One post on /b/echoes both the confusion and sense accomplishment echoed by the mainstream media press regarding both the raids and the organization of the first protest planned for February of 2008:

> ok so at first i thought this scientology thing was a fun thing, but nothing too serious. even thought of showing up on the 10th in chicago … but then was like whatever.
>
> im reading my weekly subscription of the london economist, not fox news, not the sun, not a tabloid, THE ECONOMIST. articles from here are about israeli foreign policy and economic policies of the UN. Not puppies.
>
> So i skip to the international section. The article reads "an online onslaught against scientology"
>
> so it goes on … talks about how the tom cruise video got out. i take a breath, ok nothing to be excited about. and then i shat bricks and decided to post on /b/ for the first time in 6 months. they even talk about the 10th.
>
> HOLY SHIT. WHAT HAVE YOU DONE /b/?
>
> WHY IS THIS MAKING INTERNATIONAL NEWS?
>
> SHOULD I BE WORRIED ABOUT YOU DEAR /b/?

This mix of sentiments also added to the perception that the February 2008 protest was more an "IRL raid" of Scientology than a traditional demonstration. There is also the perception, among Anonymous supporters of the protests that this was the way to do significant damage to Scientology. As a poster puts it:

> The more people we get on our side whether /b/tards //i/insurgents or not, the more we can get away with in terms of "protesting" and fucking their shit up. The IRL raids have to be the main part, DDoSing, fax DDoSing and ordering them tons of pizzas etc is just a small part and to piss them off. Getting them to react/flip may just be what we need.

While enturbulation.org was used a site of information and some organization, the first protest's planning was performed in the pages of Anonymous wiki partyvan.info.[18] The haphazard nature of organization led to an idiosyncratic variety of protests styles based on both local laws regarding protests as well as who was organizing the event. In one excerpt for the March 2008 protest in Boston, the organizer mentions attendees, "DO bring cameras, camcorders and flyers and signs—the more the better. DON'T bring obnoxious retards. This isn't about shouting memes." However, in other locations, such as London, the use of memes was prevalent.

However, idiosyncratic the organization was at a local level, a coherent set of rules was created as a set of universal protesting guidelines, which were spreading utilizing memes, as shown in Figure 4.1. These rules focused on having protesters maintain their cool,[19] following the local laws and directions

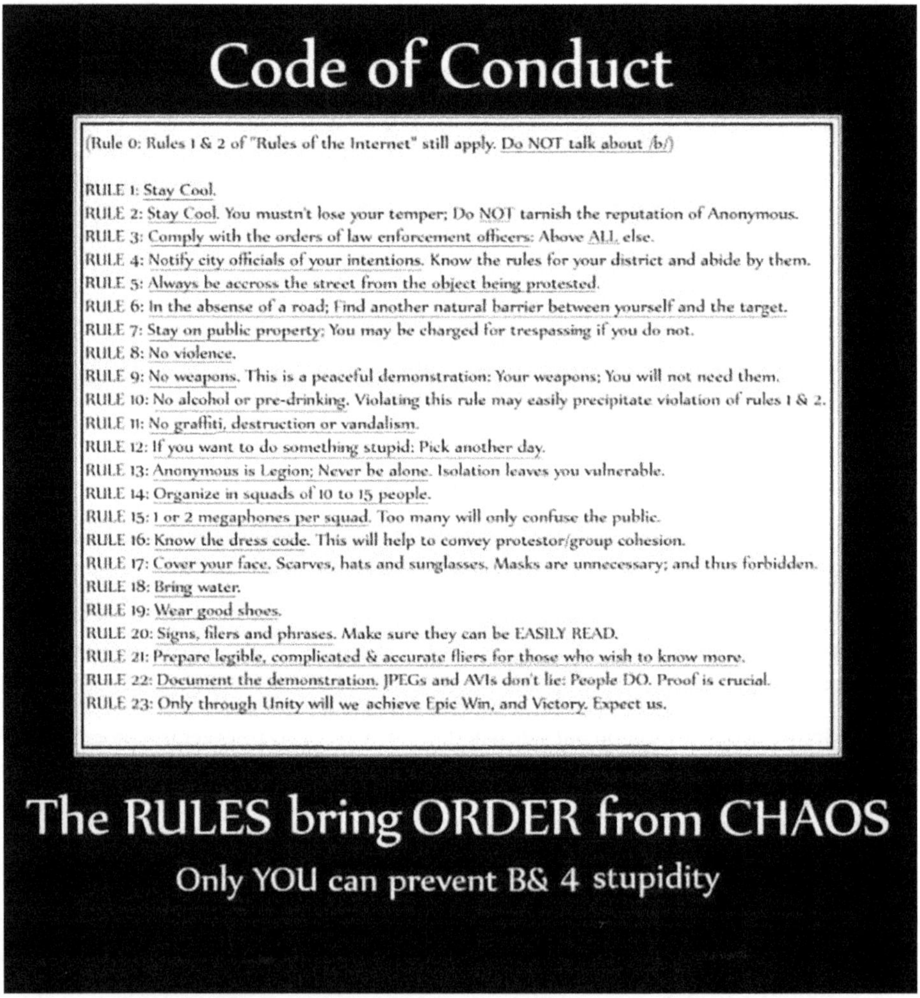

Figure 4.1. The generalized Code of Conduct that was created for the protests distributed online using the demotivational poster meme.

of police, keeping comfortable and safe, and documenting the protests. In fact documentation using cameras became a paramount rule, with embers of Anonymous stating, "JPEGs and AVIs don't lie: People DO. Proof is crucial." Throughout the years that the real-world protests occurred, much of the larger scale organizational work has been routinized with general announcements as well as "think tank" planning occurring at sites such as *Why We Protest*, which emerged after the old enturbulation.org went down due to internal problems with moderators separate from Project Chanology. The site operates several geographically organized sub-forums in order to have local organization happen though other platforms of the time, such as Ning, were utilized on a location by location basis.

Anonymous as a Tactical Media Group

In its current incarnation as hacktivist group, Anonymous has become rooted in broad scale efforts against injustice against the public, whether that is rooted in issues of government secrecy, police violence, or the blanket erasure of criminal activity by government and media. As Project Chanology, Anonymous's activist goals were analogous, focused on discrediting Scientology in the public's imagination, attempting to dry up the organization's ability to gain money, and destroying the church's organizational infrastructure by convincing members to leave the church. The goal of these effects overall was to shape public opinion against Scientology as well as inform the public about the system of belief and the various costs and abuses perpetrated by the church. Much of the media produced by Anonymous can be said to rely heavily on détournement for its impact. Détournement is a style of political art which was created by the Letterist International, a Paris-based leftist group which was created by French Marxist theorist Guy Debord and existed in the mid-1950s and was later adapted by Situationist International, another radical leftist group in Paris. The technique, which stems from Dadaist and surrealist movements, involves reworking and recontextualizing elements of culture (whether high or popular) for critique of the state and its practices. Situationist International described détournement as "The integration of present or past artistic productions into a superior construction of a milieu," (1958). Détourned works were used quite often in the leftist protests of 1968 in Paris as well as in Debord's work, such as the film version of *The Society of the Spectacle* (1973).

In one example of Anonymous's use of détournement, a flyer advertising the February 2008 protest utilized Maoist-era propaganda poster art to rally protesters. Another example involved the media for the June 2008 protest, which targeted the controversial Scientology group SeaOrg. Using pirate and sailing imagery, ranging from simple juxtapositions or reworkings of the Jolly Roger sign to screenshots of the Pirates of the Caribbean films and images of pirate ships made from Lego blocks. Figure 4.2 utilizes détournement, in this case borrowing from the aesthetic of film posters and science-fiction, to target and mock Scientology, its terminology, and how Scientology's initial response to Anonymous.

One example of Anonymous as a Tactical Media Organization occurred at a London protest in April of 2008, two months after the initial offline protests which Anonymous had incorporated into its overall tactics. Outside of the Church of Scientology, the protesters, which numbered in the few dozen, played Rick Astley's "Never Gonna Give You Up" and singing the lyrics in front of the church. This use of détournement turned the original context and use for the song (generic love song which, with its synthesized sound and drum loops, echoes the late 80s/early 90s pop aesthetics) into a message that the Anonymous protest would not be a short-lived phenomenon. A second example of tactical media usage by Project Chanology occurred in the October protest. In London, members of Anonymous announced a "Day of the Dead". Members of the protests appropriated the imagery of movie posters for the George Romero film of the same name as t-shirts and signs. These protesters then carried a coffin down the street, which represented the various people anti-Scientology protesters claim have been killed by the church and its practices. The invocation of the Romero film was enhanced further, with signs proclaiming that the church brainwashes people into zombies, mindlessly attacking Suppressive Persons. A third example is the use of YouFoundTheCard.net. Protesters, using printed business cards, hand these to passersby. On the card is the phrase "You found the card", a small snippet of information about Scientology, and a link to a clearing house site which offers links to most the various online sites for information about abuses perpetrated by the church. This technique is useful because it relies on its iconography as a marker of importance and business and relatively small size of the to make it a more effective, not to mention longer lasting, object to give out information than a flyer and its relatively dismissive iconography and bulky size.

Aside from these specific moments, the Project Chanology group in general uses a wide variety of popular imagery to further their cause. This use of imagery evokes the patterns and usage of Culture Jammers, particularly in media produced for the protest such as the business card mentioned above, but also in flyers and posters advertising the monthly protests. Of note is the use of the Guy Fawkes mask. Invoking the anti-totalitarianist V from the V for Vendetta comics and film, Chanologists utilize that role as an agent for compelling people to question and free themselves from the church. Similarly, Project Chanology uses elements from the Scientology scripture itself. For example, some Chanologists have referred to themselves as the "Marcab Confederacy". In Scientology,[20] the Marcab Confederacy is the organization led by Xenu who, as galactic overlord 75 million years ago, killed various alien races by throwing them into volcanoes on earth along with hydrogen bombs from spaceships which looked like DC-8 aircraft. The use of this bit of theology began when a post on alt.religion.scientology posited that, based on sources inside the church; David Miscavige had accepted an explanation for the Anonymous protests: that they were acting as an advance force for a incoming fleet of Marcabians. The humor in this possibility led many members of Anonymous and Project Chanology to assume this role. By using the rhetoric of the Marcab Confederacy and embracing that theology as their own, Chanology is not only assumed the role of enemy of Scientology, but also promoted the theology of the church, which can be construed as tenuous at least and patently absurd at worst. Figure 4.2, through its sci-fi movie poster aesthetic, highlights the absurdity of this element of Scientology doctrine, the former by casting Anonymous themselves as the Marcab Confederacy and the latter to mock the convoluted nature of this element of Scientology doctrine.

The techniques of both griefing and tactical media here were particularly effective to protest against Scientology, since the religion's tenets involve actively attacking so-called Suppressive Persons or SPs. In practice, at least in the Atlanta-area protests, the Church attempts to ignore with Chanology participants referring to "Curtain Tech", or the act of drawing blinds and turning off lights during protests. This use of Scientology scripture against them is an attempt to make ignoring Project Chanology participants an untenable strategy. Chanologists used it as a derisive dismissal of the Church's attempts to ignore to dismiss them, much in the same manner of trolling someone online, where the goal is to goad or further push someone into acting irrationally. Tactics such as the Marcab Confederacy further push Scientology into paying attention and to have members of the church question both the tenets of the religion as well as the economic, legal, and human rights practices of the church. The ultimate goal of these actions by Project Chanology is the revoking of tax-exempt status in countries where it enjoys the privilege, such as the United States, and, ultimately, bankrupting the church and its collective organizations. The use of these techniques, whether it be various amounts of détournement, tactical media, or culture jamming, have been crucial to both shaping Project Chanology's message as well as exposing Scientology's real belief system.

Figure 4.2. A détourned movie poster meant to play up and mock the rumor that Scientology believes Anonymous were members of the so-called Marcab Confederacy.

Attitudes Between Anonymous and Project Chanology

Even from the beginning, though perhaps more pronounced as Project Chanology continued, the attitudes towards the project among non-Chanology participants in Anonymous changed. The general attitude ranged from simple claims that protesting Scientology (as opposed to raiding or trolling) was boring. Others felt it was either a fruitless effort or even worse, antithetical to what Anonymous is. The main tensions that were brought about by the formation of Project Chanology centered on a few factors: whether Project Chanology's goals are tenable, whether their goals and methods are in keeping with those of Anonymous itself, and whether, in becoming more visible in the name of a social justice cause, they become something that members of Anonymous would usually mock or troll rather than celebrate. One instance of this sentiment lies in the *Encyclopedia Dramatica* description of Project Chanology. *Encyclopedia Dramatica* was a wiki-like site that attempted to document topics and events related to Anonymous. As with other wiki platforms, much of the content was debated and debatable, displaying in a manner Coleman (2014) describes as "the moral kinetics of trolling". The site itself was not neutral or objective by any means, but that made the ambivalence of Anonymous members towards Project Chanology even more stark. While the site kept good records of the events that have occurred as well as the cultural productions related to the group, the group was described as "Anonymous had nothing better to do than get their collective panties in a twist because there is a cult masquerading as a church." The page also noted that, "Project Chanology is now considered gay by most chans. Posting about it will get you b& [banned], your IP posted," which indicates the mixed feeling Anonymous has towards the current status of Project Chanology.[21] Using language associated with memes familiar to Anonymous as well as referencing previous non-activist activities and memes made by Anonymous, the overall ambivalence about Project Chanology was expressed. At the conclusion of the ED article:

> PC CoSplay is argued to be an epic IRL win, and an utterly amazing failure … Nevertheless, (and) nothing of value was lost; for somewhere hidden on ebaumsworld the next phase was being enturbulated. Fortunately, fans and critics alike will always have Magoo, The Geterator, Gas Mask Girl, Wise Beard Man, and Raidfag Wench to fall back on just in case the future raids miscarry.

Or in other words, even if Project Chanology or, indeed, any future activism was deemed a failure by the community, there would always be the group's past as trolls, griefers, and hackers to fall back on and rely on as the true foundation of Anonymous's community.

It should also be noted that counter-critiques of anti-Chanology Anons emerged almost simultaneously. These counter-critiques focused primarily on refuting the narrative of Anonymous as having a special place of mystery and power on the internet as defined by the icon of the faceless man in a suit. Through text posts and visual media, the critics of Anti-Chanology Anons argued that Anonymous was not the shadowy powerful figure that had been created as a form of self-image, but instead consisted of prototypical image of white male nerds whose collective interest lay more in video games, internet pornography, and online piracy than having any semblance of power over the internet. Thus attempts to become invested in activism, such as protesting Scientology would not undercut Anonymous's status, because that status was merely an ironic facade created as a joke in the first place.

Conclusion

Ultimately, the protest latent in Project Chanology ebbed as members burned out and attention was placed on other issues, though protests have continued sporadically in certain locations such as Atlanta, Georgia. However, the project was successful in raising media attention around Scientology and its practices. Interestingly, Project Chanology remains the only Anonymous activist action that had a real-world component to it as subsequent actions by the group have been online only. These actions have been, in certain respects just as if not more successful (or at the least more newsworthy) than the actions of Project Chanology.

Overall the impact of Project Chanology has been in its impact on Anonymous itself. It showed that Anonymous as a group showed it could rally around a cause and use techniques it had developed in the course of its other activities to affect change or bring attention to an issue. This was also the moment where the community showed a direct ambivalence about performing activist or political activities under the moniker of Anonymous. The historical context for the emergence of Anonymous offers a chance to understand how online communities can utilize their resources to affect political actions. At present, Anonymous has been one of the forefront movements online in echoing messages from campaigns such as Black Lives Matter as well as using skills breaking into websites and cybersecurity to indict police for their remarks and, sometimes, affiliations with racist organizations. This of course must be juxtaposed against the racist, anti-Semitic, misogynist, homophobic, and otherwise offensive dialogue that Is part and parcel of the larger Anonymous community's dialogue and practices. The events and context of Project Chanology give a picture at how that juxtaposition between positive political action and negative social activity is talked about, contested, and enacted within the larger online community.

Notes

1. A tripcode is a number that represents what number message on the server it is, based on history. Particularly large milestone tripcode are noted, even celebrated on 2ch-like message boards.
2. Futaba Channel's name is a play on the reading of the number two and the word for bud or sprout. In certain contexts, 二(ni) can be read as futa. Futaba means sprout or bud.'
3. While Futaba Channel separated topics, it was between 2-D images and CGI or Photographic images rather than by content.
4. Matthias Schwartz, in an article for the New York Times Magazine, describes a discussion on /b/that asked "What makes a bad person? Or a good person? How do you know if you're a bad person?"
5. The rules for /b/are inherited from the major rules of the site. In General, no Child pornography, no posts calling for invading other sites, and readers of /b/are assumed to be of the age of majority. These rules have been iterative over time as moderators and moderation scripts have increased over time to enforce these rules.
6. Trolling is the act of intentionally deceiving or riling a person or persons in a message board or similar environment, such as Usenet.
7. 萌え (moe, pronounced "Mo- eh") is a trend in character design that has developed over the last ten years in Japanese animation and comics. The word stems from 萌える(moeru, to blossom). It refers to female characters who posses child-like innocence and naivete, though sometimes this extends to even a child-like appearance.
8. Utilizing the collision mechanics of *Habbo Hotel* (which have since been changed) they blocked players from entering an in-game pool, re-enacting the "Pool's Closed" meme. Also, since the characters they used were all black men with suits and large afros, they coordinated themselves into human swastikas in an effort to offend the tween demographic.
9. Of these subcultures, furries are a common target and anti-furry memes such as a still image from Disney's Robin Hood where the titular character is caught in a fire, shocked in overlaid with text proclaiming "Yiff in Hell Furfag" are common.
10. Criminon is an organization operated by Scientology which purports to rehabiolitate criminals using methods derived from Scientology. It is one of several similar organizations such as Narconon and Citizens Commission of Human Rights
11. As an example, *alt.religion.scientology* posts about Anonymous ranged from rationalization: "I think that more people will be harmed and even possibly killed in the time it takes to bring down CoS legally than it may take using more extreme methods like Anonymous is trying to do," to nihilistic condemnation: "So why promote illegal actions? It won't work-----------ever---------long term, and they then become what they say they're fighting."
12. Taping two sheets of black paper to form a continuous loop through a fax machine in order to waste fax toner.

13. The time period of Scientology server downtime is estimated to have been between January 18th until January 25th.
14. A group associated with the project, called "g00ns," mistakenly targeted a 59-year-old man from Stockton, California. They posted his home telephone number, address and his wife's Social Security number online for other people to target. They believed that he was behind counter-attacks against Project Chanology-related websites by the Regime, a counter-hack group who crashed one of the Project Chanology planning websites, partyvan.info. The group allegedly attempted to gain personal information on people involved in Project Chanology to turn that information over to the Church of Scientology, though other people felt that the group was merely trolling the Project. After discovering they had wrongly targeted the elderly couple, one of the members of the g00ns group called and apologized.
15. A Google bomb is a directed attempt to have a site listed at or near the top of a Google search, especially for specific terms. Usually this involves some amount of scripting or manipulating the algorithms Google uses in assigning page rank to its searches.
16. One comment: "Well, the thing is, his arguments are mostly based on morality, and I don't really give a shit about morals. Sure, sabotaging their shit may seem 'wrong', but who cares? I'm only in it for the lulz."
17. Perhaps the harshest comment of the thread: "Why don't you all go suck his cock then? Srsly, this guy came out of the fucking blue. He's never gotten media coverage, he's never gotten scientology websites shut down. He's a spec on the windshield. If you're going to be a faggot about it, then don't DDOS in the first place. And if you do DDOS, don't turn around and BAWWWWW after some random guy with a beard calls you on your shit. Make up your fucking minds."
18. The original text was deleted following the first protest and the move to enturbulation.org and more secure places. However, *Encyclopedia Dramatica* maintains an archive of the protest information.
19. A common tactic that Scientologists perform is provoking protesters so that, in a emotional moment, they do something to provoke law enforcement officials.
20. At least at the OTIII level, which one achieves after hundred of thousands of dollars are spent in "auditing" and training courses.
21. One such admittance of the mixed opinion is also on this page where the entry notes "PC is also notable for being considered serious business; when in fact it is whatever you make it. You can help by realizing PC's broad subversive and ultimately satirical nature."

References

!3GqYIJ3Obs, O. K. (2008). An irreverent guide to the many boards of Futaba channel. Retrieved from http://www.bluethree.us/futaba/

(2008, January 27). Anonymous has no ideology. Message posted to alt.religion.scientology. Retrieved from http://groups.google.com/group/alt.religion.scientology/browse_thread/thread/13a2f5b4d5b85aa0?q=anonymous+group%3Aalt.religion.scientology&pli=1

(2008, February 15). Madness, absolute madness. Thread on alt.religion.scientology. Retrieved from http://groups.google.com/group/alt.religion.scientology/browse_thread/thread/c0976de8b6d1c384/ca7608a087a8a6f0?pli=1

(2008, February 4). The economist on project chanology. Post # No.52886416. Retrieved from http://4chanarchive.org/brchive/dspl_thread.php5?thread_id=52886416&x=the+economist+on+project+chanology

(2008, January 27–28). Message to anonymous from scientology. Retrieved from http://groups.google.com/group/alt.religion.scientology/browse_thread/thread/a2b66118696fb639?pli=1

(2008, January 27). More on anon vs scientology. 4chan.org. Retrieved from http://4chan-archive.org/brchive/dspl_thread.php5?thread_id=52158619&x=more+on+anon+vs+scientology

(2008, January 27). My message to anonymous. alt.religion.scientology. Retrieved from http://groups.google.com/group/alt.religion.scientology/browse_thread/thread/3d822406e4735748?q=anonymous+group%3Aalt.religion.scientology&pli=1

(2008, January 16). Transcript of Tom Cruise on Scientology Video. *The Times Online*.

/b/day. (n.d.). Retrieved December 2, 2015 from The LURKMORE wiki: http://www.lurkmore.com/view//b/day

Coleman, E. G. (2014). *Hacker, hoaxer, whistleblower, spy: The many faces of Anonymous*. London: Verso.

Coleman, G. (2011a). Anonymous: From the Lulz to collective action. *The New Everyday*. Retrieved from http://mediacommons.futureofthebook.org/tne/pieces/anonymous-lulz-collective-action

Coleman, G. (2011b). Hacker politics and publics. *Public Culture, 23*(3), 511–516.

Coleman, G. (2012). Anonymous. In C. Wiedemann & S. Zehle (Eds.), *The glossary of network ecologies*. Amsterdam: Institute of Network Cultures.

Denton, N. (2008, January 15). The cruise indoctrination video scientology tried to suppress. *Gawker*. Retrieved from http://gawker.com/5002269/the-cruise-indoctrination-video-scientology-tried-to-suppress

Jenkins, J. (2007, December 07). Man trolled the web for girls: Cops police seek possible victims after undercover sting. *The Toronto Sun*. Retrieved from http://cnews.canoe.ca/CNEWS/Crime/2007/12/07/4712680-sun.html

Phillips, W. (2011). LOLing at tragedy: Facebook trolls, memorial pages and resistance to grief online. *First Monday, 16*(12). DOI:10.5210/fm.v16i12.3168

Schwartz, M. (2008, August 3). The trolls among us. *The New York Times Magazine*.

Section II. Politics Introduction

ANDREW R. SCHROCK

Hacking has long been connected with politics through its hybridization with activism. The portmanteau "hacktivism," popularized by Tim Jordan, referred to a genre of resistant politics expressed online. Many recent texts trace how distributed politics becomes embedded in technologies or find a home online. For example, in *The Coming Swarm*, Molly Sauter traced the emergence of distributed denial of service (DDoS) attacks that made online resources unusable. They persuasively argued that DDoS attacks reflected a legal and theoretical framework of civil disobedience drawn from history. Another way of looking at hacker politics, then, is as socio-technical assemblages that emerge from particular historical trajectories and reflect local situations and institutions, with varying degrees of political resistance and advocacy.

This section expressly steers clear of the internet as the sole object of study. "Not all studies about contemporary politics and culture," wrote Michael Schudson in *The Rise of the Right to Know*, "are about the internet." He was partly defending his historical approach against those who would misuse it. After all, he was frustrated that his concept of the monitorial citizen had been misinterpreted as a normative ideal. The argument of his book was pushing back against concepts such as the public's "right to know" being attached solely to digital technologies. The public at large came to view transparency as their right. The emergence of this political understanding required a widely understood ethos, institutional frameworks, and everyday politics. Using this tripartite lens, we can unpack how political perspectives, goals, and practices become associated with particular cultures.

Quite a bit has been made of the "ethos" of hacking. Most writing on hackers starts off with a gesture to the "hacker ethos" that Steven Levy promoted in his 1984 book *Hackers: Heroes of the Computer Revolution*. Despite this widespread usage, it is important to understand that Levy's definition as much articulated values to the hacker community as much as it captured an inner essence. Levy was, and some thirty years later still is, primarily a long-form journalist and book author. In *Hackers* he produced a compelling story. Like Clifford Stoll's *Cuckoo's Egg*, it became a way for hackers to understand their collective identities and political practices. It was also quickly out of date, given the massive shifts in hacking during the 1980s and 1990s that spawned different practices, motivations, and histories. Relying on the hacker ethic has become a harmful crutch since it no longer unifies disparate factions. Accordingly, this section sidesteps an ethos to explore Schudson's other two components that enabled the rise of political concepts: everyday politics and institutional frameworks.

Everyday politics necessarily involves more commonplace and unexceptional forms of hacking than the spectacle of Anonymous or state-backed security teams active on an international stage. Nathanael Bassett's chapter on "Conscientious Hacking and the Weak Collective" examines the forms that hackathons take in shaping a relationship with political action. He promotes the concept of a "conscientious hackathon," or "a place where one could hack for a day, to reinvent what was possible … in a creative and innovative way." His concept of a "weak collective" evident in conscientious hackathons can be imagined as a typology of free form, mixed methods, and organizational structures. The afterward of this book further explores what it means to conceptualize hacking and making as "ordinary."

The following three chapters in this section describe what happens when hackers collaborate to strengthen political institutions—law and government. An institutionally collaborative turn may seem surprising to readers who think of hackers as expressly involved with agonistic or adversarial politics. Arne Hintz presents the relationship between hacking and policy-making, particularly through policy hackathons and development initiatives. He notes that the hybridization of hacking with policy-making might be mutually beneficial; Policy-makers tend to neglect prefigurative strategies in media activism and technical cultures, which often seek deeper systematic change by developing technical formats, code, and programs to put pressure on corporations and bureaucracies. He concludes that the combination of legal and technical expertise takes a more open approach to government that values experimentation and digital platforms.

Morgan Currie evaluates predictions about "civic hacking" through an astute ethnography of Los Angeles' burgeoning tech scene. To Currie, the novel contribution of civic hacking is that it encourages participation in administrative functions of government, which "many theorists argue are beyond politics." In particular, government sees seeing civic hackers as desirable for creating technical prototypes for city services. The idea of civic hacking is quite new, and many projects are oriented more along agonistic politics and monitorial citizenship. The implications of civic hacking for democracy are therefore unclear, but do not easily align with either predictions by boosters or staunch critics.

The Chaos Computer Club (CCC) has long collaborated with the German government on a number of pressing social issues. Sebastian Kubitschko suggests that the group's ability to assemble very diverse groups has much to do with how hackers regard space and place. Through physical proximity, direct exchange, and face-to-face meetings, the CCC was able to translate technical issues to the political establishment, and advise the public at large through the media. He describes the civic society group as engaged with "constructive, yet critical, mode of engagement." Developing relationships with government requires political *presence*, not just mere connectedness. The CCC can be seen as a novel form of direct political participation that can be contrasted with weaker, more symbolic forms of political affiliation.

Despite the quite different approaches of these chapters, helpful connections can be drawn. Each author encounters terrains where interpretations of "the political" are actively under construction. Morgan Currie describes just such difficulties of researching emerging topics, reminding us that the intersection between hacking and politics is still highly unsettled. Hackathons were present in most papers, integrated into distinct and localized scenes. Bassett writes of conscientious hackathons as handily aligning with the Occupy movement, while Currie found they were a form of administrative planning. Finally, "civic tech" (evolving from "civic hacking") has emerged to unite these efforts that promise cooperation across ideological boundaries, and institutional change, by reforming policies and processes. These claims should be treated warily, much as authors in this section have.

5. *Conscientious Hacking and the Weak Collective*

Nathanael Bassett
University of Illinois at Chicago

Much of what drew me to hackathons emerged from the political climate of 2011. Optimism dangerously bordering on digital utopianism suggested anything was possible when autonomous and individual activists worked together via technology. At the same time, political ambitions seemed to depend on the use of technology. The Arab Spring and Tahrir Square demonstrations inspired the Occupy Wall Street protest movement, which produced an abundance of self-documentation (DeLuca, Lawson, & Sun, 2012). The proclivity of Occupiers to produce and share their own media content could make it one of the most documented events in history (Meehan, 2011). A self-awareness of the independence of participants characterized Occupy as a "leaderful" rather than a leaderless movement. This lead to strong tensions within the community between "open/closed" positions, or hierarchical vs non-hierarchical organization (Costanza-Chock, 2012). The non-hierarchical social arrangement of Occupy resonated with the politics of mesh networks, and how the power to define the movement was distributed among the participants, rather than being passed down from the top.

Those tensions over hierarchy mimicked another type of collaborative community work. Occupy had protestors coming together and working towards communitarian ideals, in pursuit of some collective agenda or goals. Hackathons also have people come together to accomplish personal and collective goals. Hackathons are work "sprints," from a portmanteau of "hack" and "marathon" bring together hacking (not just programming, but also addressing problems through unique and original means) and the intensive stretch of time these events span, ranging from a few hours, consecutive days, or a series of sprints.

Like a protest, there is a belief that participants have some equality, that everyone's efforts are worthwhile and the group should have some consensus. Hackathons and protests are marathon events, where endurance, stamina and adaptability are important. Unlike a protest, ideology seems to play a lesser role in the hackathon, even when people come together to "hack" for public issues and social good. Certain types of consensus may never be reached by participants. Engagement is a negotiated and nuanced practice, where participants use tactical maneuvers to determine their level of involvement, what they hope to accomplish, and how they interact with others. At a social good hackathon, "conscientious" work characterizes the hope of a successful hackathon. Ideally, people can use or learn technical skills and abilities that help them address social problems. The diminished role of ideology at the hackathon means that the civics participants enact is closer to a "entrepreneurial citizenship" (Irani, 2015) than a radicalized or democratic mode of political action. This makes the hackathon appealing as a means to commercial ends.

In studying hackathons, I was originally interested in how collective identity and agendas are realized in a horizontal network of open engagement. Without a leader to articulate who their followers are, what their goals are, how is this decided? The initial similarities between Occupy and hackathons suggested collectivity and mutualism. Instead, observation led to a more interesting question—what does it mean to be in this group, or a conscientious hacker? Irani (2015) argues that hackathons always produce entrepreneurial subjects, actants with an instrumentalist approach to politics and technology. I argue here that conscientious subjects and their acts of hacking in the collaborative setting of a hackathon involves moral motivations and values. Thinking through the typology of hackathons and doing participant observation across various events reveal a community that was not trying to make something to fix everything (Morozov, 2013), but instead were empowering, inclusive and fluid for participants, "material publics" (Marres, 2012) that produced social arrangements, not just hacker subjects.

Exploring several hackathon communities in New York City with ethnography revealed the way they function as contextual tools for different goals (Bassett, 2013). Depending on the community they serve, organizing will take on different values and methods, and participants will have specific levels of involvement, goals and results. Because of the interest in hacktivism and conscientiousness, events with a theme of social awareness were the primary focus of the study, but these shared basic features with other types of hackathons.

Theory & Background

Since hackathons are focused around the digital materiality of the work produced, they can be conceived as a type of "material public." Noortje Marres's (2012) object-oriented conception of public engagement allows us to imagine the ways in which as an object (hackathon work) mediates sociality among participants (pp. 11–15). According to Marres, subjects are constituted as political subjects by their encounters with material entities, and these objects can "activate and mobilize publics" (p. 33). Constructing the object's identity and nature are also constitutive of the public's formation (p. 42), and technologies of participation have a "performative flexibility" (p. 154) with varying levels of involvement among that public. This also evokes Christopher Kelty's (2008) concept of "recursive publics." Free software and open source communities (whose ideology resonates with the hackathon ethos) are "vitally concerned with the material and practical maintenance and modification of the technical, legal, practical, and conceptual means of its own existence as a public" (p. 3). They perform this maintenance by being collective, independent and creating alternatives to existing forms of power. Interactions between attendees at a hackathon are not mediated by the internet. In a physical space, the maintenance of their public becomes hyper-social, leading to spontaneous weak collectives and personal empowerment among participants. The "weak ties" (Granovetter, 1983) participants build through such events can persist longer than the work participants complete.

This sort of weak collective evokes Manuel Castells' (2009) "network society." Groups with horizontal and node based relationships have a different distribution of power. Optimistic over new media's social effects, Castells' writes that this morphology "created a new material basis for the performance of activities throughout the social structure" (p. 502). A new economy emerges, centered around information and technological literacy, not capital and management. Identity in the network society is either with or against what a dominant institution has established. But alternatives exist, in communal efforts to invent a socially transformative subjectivity, what Castells calls "project identity." Castells (2010) believes identities today "are not built any longer on the basis of civil societies … but as prolongation of communal resistance" (p. 16). In a hackathon, the most ideal sort of transformative work would come from this sort of radical project identity, in creating alternative identities and practices through technology. Originally, I approached the hackathon with this theoretical perspective, but found that the missing consensus and the individual negotiations of participants prevented any coordinated effort to establish project identities.

Instead, it was useful to consider de Certeau's (1984) conception of practices and tactics. The maneuvers and negotiations between organizers and participations helped connect a sense of practice with identity or purpose. During fieldwork, there was an obvious tension over the motives in goals of each event. There is a legitimate concern in those who view the hackathon as a way of exploiting cheap labor from enthusiasts for the cost of some pizza and the space. However, critiquing hackathons as ultimately co-opted spaces assumes involvement equals complicity in the organizer's motives. It ignores the awareness of participants, their primary and secondary goals. Rather than submitting to a strategy of systems, participants are tactically practicing their own agendas. The hackathon then becomes an interstitial space. Attendees can either exploit that freedom of maneuvering to their own ends, or find mutually beneficial outcomes.

What Hacking Does

This sort of tactical maneuvering is not unfamiliar to those unfamiliar with the genealogy of hacking. Castells' (2009) suggestion that "Culture is constructed by the actor, self-produced and self-consumed" (p. 21) points to how defined roles of media producer and consumer have become less distinguished. In a network system, subjectivities of transmitter/receiver were radically restructured and some measure of power has been transferred from the elites of yesterday. There is a level of democratization in the way cultural acts and the "apparatuses of cultural production" are decentralized and controlled by participants with the right literacy and access to technology (Poster, 1997, p. 267). As a result, Castells (2000) claims there is a "fragmentation of culture" which "leads to the individualization of cultural meaning in the communication of networks" (p. 22).

However, it's important to not mistake this form of democratization as a radical collectivizing technique for communitarian politics. Castells (2000) acknowledges the conflict for power taking place over flows and codes in networks themselves. Social change must take the place of either "cultural communes" or alternative networks. While alternative networks produce "new history making," self-contained communitarian groups are fragmented and isolated. Reprograming the network state is unlikely. In addition, Lovink (2005) argues that networks will not help us to remediate and implement a mythical form of representation democracy. Post-democratic models move away from consensus, focusing on process instead.

With this in mind, hacking can be thought of as a way of examining existing networks and developing alternative social and informational infrastructure.

A precursor to the hackathon, the "hacklab" emerged as a way of accounting for changing subjectivities. As Maxigas (2012) writes, "this work was seen as in continuation with overturning those property relations in the area of media, culture and technology" (p. 3). Hacking as a subversive act furthers the reorganization of power relations. This new process mimics redistributive qualities of distributed/horizontal networks, particularly regarding access to information databases, knowledge production and participation. The civic engagement aspect of a hacklab is not just in spirit, but comes from their "historical situatedness in anticapitalist movements and the barriers of access to contemporary communication infrastructure." They "tended to focus on the adoption of computer networks … for political uses … championing folk creativity" (p. 4). Maxigas contrasts hacklabs as deriving from the anti-institutional ethos while hackerspaces are legitimized by their institutional associations. The hackathon, as an event which takes place in both institutional formats and in autonomous spaces, with attendees of unknown affiliations, has an uncertain political heritage.

Hacking certainly engenders certain types of political and social activism, but not always in an instrumentalist fashion of "build this, change that." This study focuses on individual practices as tactics and maneuvers, in de Certeau's (1984) terms. Working for social change can be about localized and cumulative efforts, flexible instances which contribute to overarching aims (Hands, 2011, p. 37). Technology does not change the world, but people who use technology do. Hackers are a specialized form of technology users, who not only define themselves in their use practices (Oudshoorn & Pinch, 2003), but their community as well (Kelty, 2008). For this study, hackathons function as field sites for examining both the physical co-presence and mediated work of individuals collaborating towards common goals.

What Hackers Do Alone and Together

Whether or not hackers have common goals, hacking practices seem to be individualist exercises of ability and agency. This speak to the way Galloway (2006) imagines the power of a node who resists protocol (pp. 147–142). The technocratic libertarianism of a resister or a hacker also has its roots in Turner's (2006) record of the techno-utopian "New Communalist" perspective. The individualist ethic in tech is not exclusive to hackers, but is part of the historical ethos of mainstream computer scientists. Hacking then takes the value of agency in the CS community to a level of greater social deviancy. The ethic of early hackers reflects the values of Silicon Valley technologists, but Nissenbaum (2004) identifies a "protest" theme "against encroaching systems of total order where control is complete, and dissent is dangerous" (p. 212). The incorporation of political and moral values with technical ends establishes a "culture of hacking" focused around individuals and their skills. Von Busch and Palms (2006) also trace this history's connection to DIY culture, identifying how hacking goes beyond customization and modification of pre-made things to "coloring outside the lines" altogether (p. 29). Curiosity, exploration, and unpacking processes of closed systems are its features. Hacking is then empowering, democratizing, creative and independent mastery.

As more of us discover unintended uses for sophisticated technology, Alleyne (2011) argues that "we are all hackers now." Rendering the identity of a hacker into a set of practices (no longer exclusive to revolutionary technocrats) makes it a process we can all engage with. Hacks are no longer only solitary transgressive actions. The word is now applied to clever solutions to everyday problems. Even the socially conscientious hacker is somewhat removed from Wark's (2009) hacker ethic. The hacker class is just as likely to create and depend on open data as they are the intellectual property of a "vectorialist class" which would have to be liberated. According to Söderberg (2011), Wark's view fails to consider the "everyday life of hackers." Again, post-democratic tactics of focusing on process rather than consensus means that political agendas can emerge from "outside" hacker spaces. Söderberg notes the tension in hacktivist politics between the ideologically motivated, and the "techies," people "interested in technology for it's own sake" (pp. 24–25). von Busch and Palms (2006) offer a reading of hacking that supplement's Wark's hacker ethic as not merely production and processing of information, but liberation through practice, sharing, breaking control, and "using hacking manoeuvres instead of dialectic tactics" (p. 39). Those in-group tensions can also in examples like Anonymous. Doubts of some on whether or not hacktivist activities abandoned its traditional social ethic of "for the lulz" led to rifts in the community. Coleman (2012b) distinguishes between "moralfags" (who advocated for conscientious agendas of high profile hacktivism) and "trolls" who wanted nothing more than "raucous" and obnoxious "trolling for its own sake." This sort of reflexivity also echoes Kelty's (2008) recursive publics. Approaching hackathon spaces required reconciling identity crises for both the collective and individuals.

Hackathons are temporary spaces. While we could argue no collective is a monolith with faceless members, these events are successful because of the agendas and goals of individuals. Granovetter's (1983) concept of "weak ties" here complements a framework of strong individualism, seen in the shift from fragmented weak individualism from the early web (Turkle, 1995) to today's single, unified public identity (van Dijck, 2013). The hackathon is a weak collective—a place where cooperation is more important than collectivity, where process beats consensus, and the goal is to "just finish" something with the time you have.

Methods

The methods for this research focused around a year long ethnography, in addition to semi-structured interviews, spatial analysis of the field sites, and an informal focus group in the shape of a "yackathon," what we participants jokingly called a "talking hackathon about hackathons." Besides following and observing online spaces for hackathon communities, participant observation took place in four different locations across New York City over five separate events in 2013. These included "Hack n' Jill" at the Etsy offices in DUMBO, "EcoHack" at the old Phizer building in Williamsburg, OccupyData hackathons at CUNY's James Gallery, and betaNYC/CodeForAmerica's hackathon at the NYU-Poly Incubator. Guidelines for observation included looking for signs of group membership (profession, social status, socioeconomic class, ethnicity, age, gender, etc.), studying the dynamics of interaction (who initiates creating a group, who approached by others, who consults freely, the length and tenor of interactions, etc.), and judging the familiarity of attendees (who seems like a stranger and who is well known). In addition, studying personal space, human traffic, whether or not people worked alone, together, or silently next to one another, were all helpful. Many openly self-identified or labeled themselves in terms of skill and ability, offering extensive insight into their own tactics for participation.

The length of this study allowed for embedding with the community, where others became involved and assumed the research questions as something for working within their own "weak collectives." Since the research focused on hackathon organizers and participants, interactions resulted in organizers looking for action plans to apply to their own events, and participants reaffirmed their own critical perspective as free agents with their own unique goals and motivations for involvement. The goodwill that emerged from these connections led to the opportunity for our small but very productive "yackathon" near the end of the study. This was arranged with the involvement of both hackathons organizers and participants. In this space we could talk about the hackathon model, what people's perspectives were on hacking as civic engagement, and the possibilities were for "conscientious" hackathons. In the spirit of hackathons, the "yackathon" served as a workshop and allowed the event's primary organizers (in this case, researchers) to approach the time together as a participant-run focus group with self-selected subjects.

Findings

Although the hackathons I visited and the communities I followed were structured around social issues, they were not collectivizing, in the sense of a monolithic body of hackers with shared goals and agendas. That was no single communal identity or sense of uniform outlook and purpose. Instead, there were a set of negotiations among individual participants, organizers and even audiences. Often each would have differing goals and purposes. There was an organic group formation among attendees that made those events very social and interpersonal. The instrumentalist expectation that a hackathon will address one particular goal (say, building an app for solving parking issues in the city) is short-sighted. Such expectations don't account what participants actually come for. From the organizer's perspective, hackathons offer an unparalleled level of freedom compared to other forms of volunteerism, and participants need focus and structure to keep them on task.

However, this focus should be facilitating the completion/creation of *something*, which may not be just a material output (the app), but a social product of new relationships between attendees and new skills for individual participants. It's important for organizers to balance their goals and involvement with the personal agendas of people who actually show up. This means facilitating in the form of providing options—pairing up people who come to learn with those who have experience and want to share their knowledge, making sure the audience for the projects (clients and users) are represented and involved if possible, and providing as much material infrastructure as possible, such as tools for collaboration.

One of the more important distinctions that emerged in the research were two different styles of civic engagement from communities using a hackathon model. In line with a more traditional hacker ethos were "hacktivists," who saw their projects and collaboration as a way to critique government from the outside. They have made their mark as noted in the literature above. But the activist mentality of hacktivists was removed from what I call the "civic hackers." This group was motivated to pursue plans and goals that focused on less confrontational and critical modes of engagement, such as advocacy and lobbying existing government. They would strategically approach government officials and their offices, pushing for open data, open government and cooperation. This group attracted technologists, professionals from the tech industry, and others with commercial interests in using tech to address social problems. "Public good" among hacktivists sees technology in aid of or addition to traditional activist methods of independent community organization and protest, but civic hackers were more like non-profit advocates who sometimes spoke the language of solutionism (Morozov, 2013) and relied on the city and state for help.

Both groups suffered from a sometimes-acknowledged need to problematize their solutions more deeply. Audiences for the work were not always understood well, and solutions lacked definitive answers on implementation and effectiveness. At the end of an event, something (hopefully) would exist, but how that something was going to change things was not always clear. Participants want to feel like they get something out of their work, and if a public good isn't satisfied, a sense of "worthy effort" may be lost. Making those projects live beyond the event was a repeated concern.

Lastly, open data was an invaluable resource in working with community issues. But any reflections on the credibility of public datasets were halting and tentative. There is a politics to data collection often ignored in favor of just getting something to work with. The who, how and why of collection was not always considered, yet such materials were treated as a reliable cornerstone of people's work at the event.

Typologies

Three distinct forms of the hackathon emerged from observation, conversation and interviews with the community. Each had a unique structure, common themes and an *esprit de corps* that was either competitive or collaborative. The differences between these styles complicates the diversity of events, leading to different forms of influence from event sponsors, demographics in attendance, reasons and attitudes towards participation, and the values and goals of the group.

Free Form

Perhaps the most conducive to the conscientious mode is a "free form" hackathon. The structure of these events is set up so that participants and their goals are primary. Previous projects and pitches (where participants propose an idea or a problem/solution to work on) are welcome. Organizers do their best to facilitate, and otherwise stay out of the way of attendees. Participants come first. The theme of such events then depend on the interests and goals of those participants, but there is an ideological link between the two. Lastly, these events are more likely to be collaborative between members, with working groups who continue to meet on projects between and during repeated events. A sense of success for the event relies on the feeling of mutualism and cooperation between all participants. As an example, OccupyData met in a series of hackathons and workshops between 2011 and 2013. Projects originated from participants but represented the concerns of the "Occupy" protest movement (private vs public, reflection, self-empowerment, etc). Participants would freely move between groups and self-organize as they pleased. Multiple points of entry in the event existed for participants to learn new skills or to mentor others. Barriers to entry were low, and the role of organizers seemed to support any goals that participants brought through the door. Sponsors for this example were non-existent, and the group was antagonistic towards any sense of outside influence directing their work.

Mixed Methods

As events become more structured, the role of organizers increases. "Mixed methods" hackathons provided a space where willing participants could volunteer to help work on existing projects, or introduced audiences who needed work on a project. Themes here depend on the audience introduced by the organizer. These were also collaborative events, but competition can be introduced in the form of prizes that keep work moving, rather than an incentive to just show up. EcoHack 2012 and Hack'n Jill Hacksgiving took place on the same weekend in November and both demonstrated this sort of balance between organizer's direction and participant involvement. More sponsors were present and their influence was certainly felt in the introductions at each event, but the participants retained a great deal of agency and control over the work itself.

Organizational

Increased competition and a loss of this agency came about in the last two types, which were both organizational. "Non-profit organizational" hackathons see the goals of organizers as primary to the event's success. There is a strong deference to sponsors, hosts, and other outsiders, but participants ideas are sometimes still welcomed. The themes of the events studied here were directly related to organizers goals (for betaNYC and NYCBigApps, these were still projects in the public interest, but also data advocacy and organizational issues for those groups). Competition in such events then work to attract people with skills who might not be interested in those themes. Commercial hackathons could be conceived of as a fourth category, but the focus of this research was on civic or "issue-oriented" hackathons, similar to the work of Lodato and DiSalvo (2015).

Values and Motivations at the Hackathon

The same spirit of curiosity, empowerment and creativity that emerges from the DIY and hacker ethos (Von Busch & Palms, 2006) can be found in the hackathon, driving the involvement of all. However, it is interesting though that the majority of participants did not self-identify as hackers. Instead they would call themselves a "developer" or a "technologist" or even a "data scientist." It is important to not confuse hacking and activism, even when participants are doing conscientious hacking, or when their work is oriented around pro-social themes. Attendees are complex and negotiating multiple roles and goals. To be clear, going to a hackathon does not make one a hacker (although one may be hacking at the event), hackers are not the only ones going to such events, and finally, not all hackathons result in hacktivism, even at these social good events.

Instead, organizers and participants have weak collective motivations, which only are satisfied after their strong individual motivations. These individual motivations sometimes conflict, although they don't have to. For organizers, events should achieve the goals of the sponsors or the hosts, whether this is generating ideas for products, drawing attention to the professional organization or the theme of the event, or improving solutions to some particular technology problem or social issue. Organizers want to have successful events that can be repeated in the future, so a sense of satisfaction is important. To this end, it's their responsibility to create a balance between the collective motivations and the goals of participants.

Participants have their own strong individual motivations. They may be interested in addressing issues from the theme of the event. They might also want to solve new problems and work on interesting projects for awhile. This helps to develop technical skills that might be useful outside the event, as a means of professional development. Each of those goals leads to a sense of personal fulfillment—that the hackathon was productive and beneficial to them, and a good use of their time.

Together, organizers and participants form the weak collective, which might work on those issues in the event theme (perhaps creating some technology solutions to aspects of the problem). But besides the sense of goodwill and enjoyment on all ends, "skill share" is one of the more important goals. Attendees all have something to offer each other, and the hackathon is an ideal place for freeform interaction and learning. Developers who work a forty hour week will take their free time to do what looks like more work, but for them is a chance to use their skills on projects that actually compel their attention. They can also gain from the networking opportunities gained from the event space.

Successful events will turn enthusiasts into committed community members. At repeated events, they may become regular attendees and pitch their own ideas, or add to the longevity of other projects. A successful hackathon will build on itself, creating a group of people who come together to share success repeatedly. The hackathon then complements traditional modes of industry and community involvement as a semi-professional outlet.

However, there was still a social ethic involved. Attendees interviewed at conscientious hackathons often had negative feelings toward commercial and competitive events, often citing fears of exploitation. There is a sense of personal morality to the choices a participant makes about what communities to engage with, what projects to pursue, etc. Hacking in this sense is still a moral act, and it might not be possible to work on certain social good projects in a professional setting. Being at a morally acceptable hackathon then allows them to do personally fulfilling conscientious hacking.

Common to all hackathon types described above are a set of shared values, which I have already referred to intermittently. These are free association, worthy effort, and mutual aid. They embody the character of a hackathon and its hacker ethic of creativity and "coloring outside the lines" (von Busch & Palms, 2006, p. 29). By "free association," I refer to the way the mentality of "open source" and "open data" affect the sociality of attendees. The work has a communal nature, even if it isn't collectivizing. The balance between organizers and participants is actually a dependence that helps each for social connections necessary for success. A lack of explicit tasks allows attendees to choose their level of involvement, and who they want to be involved with. A second balance exists between the attendees and the audience. At a hackathon for social issues, the liberty of free association can sometimes frustrate groups with a utilitarian approach to programming taken professional developers. The problems they are trying to address can be difficult to crystalize and the audience may not be well understood. Even worse, the figurative wheel may be reinvented if groups don't have any memory of what the audience for their work has, or what others have tried before.

This ties into the value of "worthy effort." In one interview, a study participant characterized the hackathon as a "weekend gamble." Projects may take the form of experiments or exercises, but no one wants to feel like their work was a waste of time. If they feel like their group project is not going to be successful, they will prioritize their secondary, personal goals, like learning a new skill or networking. "Mutual aid" ties back to the first value, and forms the basis of the hackathon. A loose structure and that lack of institutional memory means it can be very difficult to accomplish anything without the help of others. Participants depend on each other and a willingness to help.

Together, these values characterize what organizers and participants share as important to the success of a social good and conscientious hackathon.

Analysis

One of the primary outcomes of this research was the weak collective framework discussed above. Unsurprisingly, people bring their own agendas to a hackathon, even when it is organized around a single set of social issues. These strong individualist viewpoints make the weak collective a site of negotiated maneuvers and hacker tactics. The events have uncertain outcomes but their products include more than the apps and experiments of attendees—all benefit from new social relationships and skills to add to their future individual endeavors.

This sense of personal empowerment seemed to motivate participation more than anything else. Impassioned involvement, the earnestness of attendees eager to learn, and the visible satisfaction of others in contributing to something they enjoyed seemed to be the primary force of cohesion in working groups. At times, a sort of "telepathic communication" connected participants who were otherwise completely silent, each working alone on their separate laptops. Occasionally they would break to ask a brief and obscure question, devoid of context, yet their meaning was clearly understood and answered.

A few of the features of the weak collective have been briefly discussed above. Similarly to Kelty's (2008) recursive publics, they benefit from a type of collective intelligence. Kelly (1995) writes about a "hive mind" at a 1991 computer conference, where 5,000 computer graphic experts played a single game of Pong, controlling a paddle's motions together. More recently, "Twitch Plays Pokemon" saw 100,000 people simultaneously participating in a single game of "Pokemon Red" (Ramirez, Saucerman, & Dietmeier, 2014). We also have Rheingold's (2002) "smart mobs" to inspire us, and each undoubtedly have their own internal tensions and struggles. In strong and weak collectives, individuals are guided by the frameworks of their interaction. Nissenbaum's (2004) account of the assignment of a "protest" theme on hackers show how there are some limitations on the agency of a group to define and control its social ontology, despite how "open source" it tries to be within its ranks. Meaning is ascribed to the group, its materials and its texts, by those in and outside. Ownership within the community is a debated question. Attendees believe that whoever does the work should own it, but there is no strict way of dealing with those issues. Self-policing keeps participants from claiming work that isn't theirs, but the potential capital from projects produced at events is not protected by anything more than community values.

This spirit of mutualism may not have emerged in the ideology of the hackathon crowd, but was certainly a part of their collective identity. The concept that someone would steal or take work that wasn't their was abhorrent seemed I suggested it to participants. The goodwill and graciousness among many of those present extends to the newcomers, who may not be as familiar with the power they have to affect the group just by showing up consistently. People who try to participate with bad or incomplete ideas are not shunned, though they may elicit some incredulous looks. Instead, alternatives are suggested. People on the periphery or who lack the skills to really contribute meaningfully are not discouraged from trying, though seemingly high barriers led one attendee bluntly state, "I feel in over my head."

Radically democratic collectivity means it is difficult for individuals to accumulate power (Coleman, 2012a). The strong collective concept includes a mob mentality, operating among the dialectics proposed by Wark (2009) with a master/slave relationship. Like hard drives and computers, individuals may be "slaved," turned into "zombies" and controlled in a botnet by a botmaster. This type of hacking manipulates and leads the mob, creating consensus and cooperation by compliance. This is our traditional view of collectivity, similar to how Eagleton (1979) characterizes Sorel's view of political myth: motivating the "inert, chronically retarded creatures" to do the bidding of their betters, who are above such trickery (p. 79). Those hierarchies do not reflect a network society and horizontal collectives, and as such are inadequate to address recursive and networked publics. As Coleman (2012b) writes, "Falling back on the story of the hacker ethic elides tensions and differences that exist among hackers." Certainly, the strong collective does not represent the sort of multivocality in ethnographic work on these communities and at hackathons.

Instead, we can recognize a hackathon is descended from while not indebted to the hacker ethic. Hacks are actions and hacktivism is one possible result. Participants partake in hacking, without always buying into the "protest theme." Politics do not always preclude participation, though it may influence one's level of involvement. At the beginning of this research, questions about how events could bring disparate agendas together with a collectivist consensus unveiled the tensions and individualist motivations at each event. The insights here were only possible after abandoning assumptions a ubiquitous hacker ethic and questions of how to get temporary volunteers to do hacktivism. A grand unified theory of hackathons or hacking seems short-sighted, if we recognize the diversity among participant tactics and organizer strategies. These events have broad parameters and can ask very different things of attendees in each instance.

Facilitation goes a long way to helping all present get a sense of unity (if not collectivity). Practice will trump consensus, especially when skilled practitioners don't have to slow down for others. The mentality of "startup" culture the lucrative economy of promising big data apps is pervasive, and even finds its way into conscientious hackathons. But, as one respondent asked in our Yackathon, "if you can't fix a hackathon, whats a better method?" The civic technology community, hacktivist circles, and commercial hackathon organizers all have their eyes on different goals. But the commonality behind their method are both the tensions between participant and organizers, as well as the individual effects mentioned above. Those who would use a conscientious hackathon for the goals of a social movement should dismiss the idea that consensus and ideology will be consistent across attendees and projects.

Looking at the practices of technology use in the hackathon and elsewhere has implications for how we think about identity in network society. Collectivity encourages a rhetoric around connection on our own terms, building our own networks, creating our own systems and strategies, and coming up with a counter-public in the face of hegemony. Collaboration and the weak collective is about disconnective practices (selective engagement), resistance and limited sociality. The strong individual perspective (van Dijck, 2013) and "private spheres" we experience emerge from the balancing of private/public tensions in social networking sites (Papacharissi, 2012). This is not solely an online phenomenon, and we should not make the mistake of "digital dualism" (Jurgenson, 2012) in assuming those concepts can't be brought to bear in a holistic conception of our interaction with others. Digital monism (Vial & Stimler, 2014) is the metaphysical position that the distinction between "real" and "virtual" is obsolete. The way we receive our sense of being-in-the-world can be generated in part by technology (Vial, 2013). But is never distinct from ourselves—much of our existence depends on these technics, in the way described by Bernard Stiegler (Hansen, 2006).

Those technics, or the relationships between people and technology, are fully on display in the hackathon. Discussions about access, audience, and purpose are about technic politics, and these are central to the conscientious hackathon.

Conclusion

Ethnography is a method that requires researchers to be especially reflective and honest about their involvement with study participants, the field site, and the exit ethics of their work. The social context is also important to the atmosphere both the researcher and research is situated in. Approaching hackathons organized to address social issues in New York City, around the same time as Occupy Wall Street's demonstration, greatly influenced the sense of what was or should be possible at these events. It was my hope that the model of a hackathon could be used with the same effectiveness of other social movement tactics, helping make positive changes for the public. It was only after acknowledging hackathon attendees as independent social actors that I realized there were unique tactics at work, and that cooperation was more important than any ideological collectivity. Process trumped consensus in a way that was maybe detrimental to overarching group goals—however, the individual goals and motivations were equally empowering and significant for those in the hackathon. From this, it was possible to approach the "conscientious hackathon" as a place where one could hack for a day, to reinvent what was possible for themselves (and maybe others) in a creative and innovative way.

Appropriately, this approach had recursive consequences on the research. It became possible to "hack" what was possible in studying this field site and this population. Important distinctions emerged between organizers and participants, civic hackers and hacktivists, individual and collectives, and lastly, collectivity versus collaboration. This led to the understanding that a weak collective model was good for framing not just participants in social movements, but participation in the network society. This weak collective model is justified based on the evidence gathered through ethnography.

Coloring outside the lines of our expectations allows us to move research in new directions that can be more productive than what we originally intended. Similarly, a protest may set to challenge an injustice, but it is unfair to say the efforts are a failure if that injustice persists. Projects at a hackathon may not accomplish their original goals or result in a trite bit of solutionist technology, but the efforts of the participants are still beneficial to all involved. This research was never intended to fully define collaborative hacking, or totally understand collective identity, but I hope it will prove useful to others in thinking about hackathons in productive ways.

References

Alleyne, B. (2011). "We are all hackers now": Critical sociological reflections on the hacking phenomenon. *Under Review*, Goldsmiths Research Online. Retrieved from http://research.gold.ac.uk/6305/1/Alleyne_-_We_are_all_hackers_now_-_critical_sociological_reflection

Bassett, N. (2013). *The conscientious hacker: An ethnography of identity and community among hackathons* (Unpublished master's thesis). The New School, New York.

von Busch, O., & Palms, K. (2006). *Abstract hacktivism*. London, UK: Skyscraper Digital Pub Ltd.

Castells, M. (2000). Materials for an exploratory theory of the network society. *The British Journal of Sociology, 51*(1), 5–24.

Castells, M. (2009). *The rise of the network society*. Malden, MA: Wiley-Blackwell.

Castells, M. (2010). *The power of identity*. Malden, MA: Wiley-Blackwell.

Coleman, G. (2012a). Our weirdness is free. *Triple Canopy*, (15). Retrieved from http://canopycanopycanopy.com/15/our_weirdness_is_free

Coleman, G. (2012b). Phreaks, hackers, and trolls. In M. Mandiberg (Ed.), *The social media reader* (pp. 99–119). New York, NY: NYU Press.

Costanza-Chock, S. (2012). Mic check! Media cultures and the occupy movement. *Social Movement Studies: Journal of Social, Cultural and Political Protest, 11*(3–4). DOI: 10.1080/14742837.2012.710746.

de Certeau, M. (1984). *The practice of everyday life*. Berkley, CA: University of California Press.

DeLuca, K. M., Lawson, S., & Sun, Y. (2012). Occupy Wall Street on the public screens of social media: The many framings of the birth of a protest movement. *Communication, Culture & Critique, 5*(4), 483–509. DOI:10.1111/j.1753-9137.2012.01141.x.

Eagleton, T. (1979). Ideology, fiction, narrative. *Social Text, 2*(2), 62–80. DOI:10.2307/466398.

Galloway, A. R. (2006). Protocol. *Theory, Culture and Society, 23*, 317–320. DOI:10.1177/026327640602300241.

Granovetter, M. S. (1983). The strength of weak ties: A network theory revisited. *Sociological Theory, 1*(1), 201–233.

Hands, J. (2011). *@ Is for activism*. New York, NY: Pluto Press.

Hansen, M. B. N. (2006). Media theory. *Theory, Culture and Society, 12*(2–3), 297–306. DOI:10.1177/026327640602300256.

Irani, L. (2015). Hackathons and the making of entrepreneurial citizenship. *Science, Technology & Human Values*, 1–26. DOI:10.1177/0162243915578486.

Jurgenson, N. (2012). When atoms meet bits: Social media, the mobile web and augmented revolution. *Future Internet, 4*(4), 83–91. DOI:10.3390/fi4010083.

Kelly, K. (1995). *Out of control*. Basic Books. Retrieved from http://www.kk.org/outofcontrol/contents.php

Kelty, C. M. (2008). *Two bits*. Durham, NC: Duke University Press Books.

Lodato, T. J., & DiSalvo, C. (2015). Issue-oriented hackathons as ad-hoc design events. Presented at the Participatory Innovation Conference, The Hague, Netherlands.

Lovink, G. (2005). *The principles of notworking: Concepts in critical internet culture*. Amsterdam: HVA Publicaties.

Marres, N. (2012). *Material participation: Technology, the environment and everyday publics.* London: Palgrave Macmillan. DOI:10.1057/9781137029669.

Maxigas. (2012). Hacklabs and hackerspaces – tracing two genealogies. *Journal of Peer Production,* (2). Retrieved from http://peerproduction.net/issues/issue-2/peer-reviewed-papers/hacklabs-and-hackerspaces/

Meehan, R. (2011, December 6). Winding down: Occupy! Scenes from occupied America. *Slant Magazine.* Retrieved from http://www.slantmagazine.com/house/article/winding-down-occupy-scenes-from-occupied-america

Morozov, E. (2013) *To save everything, click here.* New York, NY: Public Affairs.

Nissenbaum, H. (2004). Hackers and the contested ontology of cyberspace. *European Journal of Communication, 6*(2), 195–217. DOI:10.1177/1461444804041445.

Oudshoorn, N., & Pinch, T. (Eds.). (2003). *How users matter: The co-construction of users and technology.* Cambridge, MA: The MIT Press.

Papacharissi, Z. (2012). A networked self: Identity performance and sociability on social network sites. In F. L. Lee, L. Leung, J. Qiu, & D. Chu (Eds.), *Frontiers in new media research* (pp. 207–221). London: Taylor & Francis.

Poster, M. (1997). Cyberdemocracy: The internet and the public sphere. In D. Porter (Ed.), *Internet culture* (pp. 201–218). New York, NY: Routledge.

Ramirez, D., Saucerman, J., & Dietmeier, J. (2014). *Twitch plays Pokémon: A case study in big G games.* Paper presented at the 2014 Digital Games Research Association (DIGRA) conference, Snowbird, UT. Retrieved from https://library.med.utah.edu/e-channel/wp-content/uploads/2016/04/digra2014_submission_127.pdf

Rheingold, H. (2002). *Smart mobs.* Cambridge, MA: Basic Books.

Söderberg, J. (2011). *Free software to open hardware: Critical theory on the frontiers of hacking.* Gothenburg: University of Gothenburg.

Turkle, S. (1995). *Life on the screen.* New York, NY: Simon and Schuster.

Turner, F. (2006). *From counterculture to cyberculture: Stewart Brand, the Whole Earth Network, and the rise of digital utopianism.* Chicago: University of Chicago Press.

van Dijck, J. (2013). "You have one identity": Performing the self on Facebook and LinkedIn. *Media, Culture & Society, 35*(2), 199–215. DOI:10.1177/0163443712468605.

Vial, S. (2013). There is no difference between the "real" and the "virtual." *Presented at the Theorizing The Web 2013.*

Vial, S., & Stimler, N. (2014). Digital monism: Our mode of being at the nexus of life, digital media and art. Retrieved from https://hal.archives-ouvertes.fr/hal-01164486/

Wark, M. (2009). *A hacker manifesto.* Cambridge: Harvard University Press.

6. *Policy Hacking: Opening Up the Code of Media and Communications Regulation*

Arne Hintz
Cardiff University

Introduction

If a typical goal of hacking is to investigate and open up closed systems, the regulatory and policy environment is a particularly obscure candidate that citizen groups and civil society organisations have long tried to crack, with varying success. From local initiatives to national campaigns and transnational networks, non-state actors have exerted pressure on policy-makers to respect communication rights, curb media concentration, allow citizen and community media to operate, reduce copyright restrictions, maintain an open internet, and implement many other demands and concerns. Scholars have analysed strategies and practices of advocacy, and disciplines such as policy and social movement studies have developed a significant body of theory that can explain motivations, tactics, necessary conditions, successes and failures. They have observed how advocates use information, norms and public pressure to convince policymakers of a certain course of action, or try to alter their interests and align them with reform agendas. Yet the role of advocacy has mostly been to provide input for a policy process that is controlled by established policy forces, typically in government and international institutions, and increasingly complemented by commercial entities.

However some civil society activists have moved beyond normative interventions and the exercise of public pressure and have instead developed regulatory proposals, written legal text, and created platforms that allow for open, collective and crowdsourced development of policies. Expanding, and potentially surpassing, classic forms of advocacy, they have thus taken a more active role in the policy process and have tried to take law-making into their own hands. Curiously, practices such as assembling, upgrading and tinkering, and the use of digital tools and platforms, are key components of this approach to policy change which I call "policy hacking".

In this chapter I will explore two different types of policy hacking—policy hackathons and policy development initiatives—and discuss their characteristics and implications for policy, advocacy and, more broadly, democracy and participation. Through several case studies I will trace and investigate the two forms of policy hacking which differ in many respects but engage with media policy change through citizen- and grassroots-based do-it-yourself policymaking.

The first part of the chapter will present theoretical and conceptual foundations, including the context of multi-actor and multi-level global governance, classic civil society and social movement approaches to policy change, and the role of technological as well as civic hacking. The second and third sections will explore the two different types of policy hacking through a set of case studies. The fourth section, finally, will analyse key characteristics and implications of policy hacking. I will investigate commonalities with technological development, review potential changes to how we understand policy change and civil society's role, explore the international dimension of policy hacking, and ask whether it allows for more direct democratic interventions or whether it may actually have problematic implications for democracy.[1]

Communication Policy, Civil Society Advocacy, and Hacking

Before we turn to the specific examples of policy hacking, we need to understand the context from which policy hacking emerges and in which it is embedded. Particularly, this concerns media and communication policy. The academic field of media policy investigates the regulatory rules and norms that shape the media landscape, and explores how they are created, based on which values and interests, and how they shift. It encompasses the analysis of existing legislation, but also the process of policy-making as political negotiation between a variety of actors and interests. It highlights interactions between social forces, the conditions and environments of interaction, and prevalent societal norms and ideologies that underlie and advance specific policy trends (see, e.g., Freedman, 2008).

National policy increasingly intersects with developments taking place at other levels and is subject to both normative and material influences by a variety of actors. Both the local and the national have "become embedded within more expansive sets of interregional relations and networks of power"

(Held & McGrew, 2003, p. 3) and policy-making is thus located at "different and sometimes overlapping levels—from the local to the supra-national and global" (Raboy & Padovani, 2010, p. 16). Policy debates such as the World Summit on the Information Society (WSIS) and the Internet Governance Forum (IGF) have experimented with new forms of multi-stakeholder processes that include non-state actors such as civil society and the business sector. The vertical, centralized and state-based modes of traditional regulation have thus been complemented by collaborative horizontal arrangements, leading to "a complex ecology of interdependent structures" with "a vast array of formal and informal mechanisms working across a multiplicity of sites" (Raboy, 2002, pp. 6–7).

Civil society advocates have been able to use this complex environment for interventions into the "consensus mobilization" (Khagram, Riker, & Sikkink, 2002, p. 11) dynamics of policy debate. They define problems, set agendas, exert public pressure, hold institutions accountable, sometimes participate in multi-stakeholder policy development, and they hold significant leverage by lending or withdrawing legitimacy to policy goals, decisions and processes. The mix of strategies applied by advocates to influence policymaking includes, among others, the ability to create successful conceptual frames to articulate the characteristics of an issue to policymakers, potential allies and the wider public; to create networks and alliances across movements and transcending the boundaries of institutional arenas; and to shift predominant institutional ideologies as well as public opinion (Keck & Sikkink, 1998). Policy scholars have observed that interventions are greatly enhanced by the existence of a "policy window", i.e. a favourable institutional, political and sometimes ideological setting that provides a temporary opening for affecting policy change (Kingdon, 1984). A crisis in the social, economic or ideological system can cause disunity among political elites and create a dynamic in which established social orders become receptive to change. "Policy monopolies"—stable configurations of policy actors—may be weakened or broken up as political constellations change and the balance of power shifts (Meyer, 2005). Traditionally, social movement scholars have distinguished between "insider" and "outsider" strategies to achieve policy change, where "insiders" interact cooperatively with power-holders through advocacy and lobbying while "outsiders" question the legitimacy of power-holders and address them through protest and disruptive action (Banaszak, 2005; Tarrow, 2005).

Strategies and action repertoires like these have been applied and observed, and favourable conditions have been exploited across different policy fields and disciplines. However this perspective on policy interventions has also been criticised as too limited to encompass the full range of activities by activists and civil society groups in affecting policy change. Two (related) sets of limitations concern a) the range of actors involved in this form of advocacy, and b) the range of tactics and approaches. As for the former, classic advocacy has typically centred around the activities of large professionalised non-governmental organisations (NGOs) which are better resourced than smaller activist groups and enjoy easier access to policy institutions. Social movement scholarship has been criticised for its strong focus on these forms of organised civil society (de Jong, Shaw, & Stammers, 2005), and networked governance arrangements that include these actors in multi-stakeholder policy processes have been described as "neo-corporatism" (Messner & Nuscheler, 2003; McLaughlin & Pickard, 2005). Those parts of civil society that are organized as informal networks, loose collaborations and temporary alliances amongst engaged individuals—which have been termed, for example, "connective action" (Bennett & Segerberg, 2012) and "organised networks" (Lovink & Rossiter, 2005)—have often been overlooked by both policy institutions and scholars.

Secondly, classic advocacy neglects the many forms of prefigurative action that exist in media activism and technological development. Technical communities have long engaged in latent and invisible "policy-making" by practically setting technical standards and developing new protocols and infrastructure that allow some actions and disallow others (e.g., Braman, 2006; DeNardis, 2014). Media activists, similarly, have focused on the creation of alternative infrastructure that bypasses regulatory obstacles, rather than lobbying against those obstacles. Rather than campaigning for privacy rights and against online surveillance, many of them develop communication platforms that respect user privacy, and rather than advocating for community media legalisation, many of them broadcast their own unlicensed ("pirate") radio. This prefigurative approach challenges the classic division into "insider" and "outsider" strategies and, instead, points to interactions with the policy environment that take place neither "inside" nor "outside" institutional or governmental processes, as they do not directly address power-holders, but "beyond" those processes by creating alternatives to hegemonic structures and procedures and by adopting a tactical repertoire of circumvention (Hintz & Milan, 2013; Milan, 2013).

"Hacking" as it is commonly understood—as changing, manipulating, revising or upgrading a technological system, such as a computer or wider communication infrastructure—plays a role in the "beyond" approach as it means to bypass restrictions and enable new uses. However the more established definitions of hacking allow for a broader perspective. According to the "requests for comments" (RFCs) that have accompanied the development of the internet, a computer hacker is "a person who delights in having an intimate understanding of the internal workings of a system, computers, and computer networks in particular" (Network Working Group, 1993). Based on curiosity, a "positive lust to know" (Sterling, 1992) and a search for innovation, a hacker thus seeks to understand a complex system and to experiment with its components, often with the aim of changing or modifying its structure. This system can be broader than a narrowly confined computer system. The trend of civic hacking, for example, has applied the notion of hacking and its practice of making, creating and tinkering as a method to change social, cultural and political aspects of one's environment. Beyond the realm of technological applications, science- and bio-hacking has encompassed amateur scientific experiments, and life sciences have opened up to do-it-yourself (DIY) approaches that exhibit similarities with hacker ethics (Delfanti, 2013). The development of alternative licences such as the Creative Commons license has applied this DIY practice to the legal realm (Coleman, 2009), and activist practices such as culture jamming and the manipulation of meanings (e.g., of advertising) have been discussed as a form of hacking of our cultural, social and political environment (Harold, 2004). Whether the main concern is the enclosure of scientific knowledge, restrictions through proprietary software, or copyright limitations to cultural practices, questions of control and participation are at the centre of hacking. Hacking thus becomes a way of discussing questions of openness and enclosure, and of exploring new ways for enhancing participation, public intervention and grassroots-based development—not just in the realm of ICT systems but in the wider reaches of society.

Beyond the more specific backgrounds of both "policy" and "hacking", the study of policy hacking is underpinned by academic concerns with the role and nature of social forces involved in the governance of (technological) systems and the interactions between technical and political governance. These are often based in science and technology studies (STS) that highlight the politics of technical architecture and the networked interactions of human and nonhuman actors, and share an interest in fluid and temporary assemblages around particular issues (Musiani, 2015). Yet they also include political-economic approaches that address the objectives and capacities of social forces in shaping the global media environment (Chakravartty & Yuezhi, 2008).

Policy Hackathons

The first type of policy hacking that I will discuss here resembles closely the practices of hacking as they are generally known from the technological realm and the related experiences of civic hacking. It is based on the idea of hackathons—meetings for collaborative work by computer programmers and technological developers, which have typically aimed at solving specific technological problems and developing software applications. Increasingly some of these events have targeted social and political concerns. Civic hackathons such as "Hack the Government" and "In Hack We Trust" have developed open data and transparency tools for government; the "Education Hack Day" served to develop solutions for education problems; and "Random Hacks of Kindness" has been devoted to disaster management and crisis response. Many other examples exist.

Policy Hackathons have been the latest addition to the growing range of civic hacking events. Their idea is to bring together interested people (often, but not always, with technological skills) to analyse and, ideally, improve policies. The open source development company Mozilla, for example, has organised several hackathons to analyse and improve the privacy policies of websites. The Brooklyn Law Incubator & Policy Legal Hackathon (BLIP) brought together advocates, lawyers and technologists in 2012 to "explore how technology can improve the law and society, and, conversely, how law can improve technology".[2] In addition to presentations and debates, BLIP encouraged participants to tackle issues of intellectual property policy and propose policy reform through collaborative online tools. On the other side of the Atlantic, a broad network of civil society groups including the European Digital Rights Initiative (EDRI), the Electronic Frontier Foundation and Transparency International have teamed up with leading Internet companies like Google and Facebook to hold an annual EU Hackathon in Brussels. The series has included, for example, "Hack4YourRights" in September 2013, bringing together 30 participants to work on "data sets from network analysis, corporate Transparency Reports and freedom of information (FOI) requests [and] create apps and visualizations that shed light on the state of government surveillance in their country".[3] The 2014 edition "Hack4Participation" served to develop tools to enhance citizen participation in European policy debates.[4]

With the concept of policy hackathons still in its infant stage, the goals of these events and their implementation practices vary. While some focus on solving specific policy problems and improving particular policies, others address the broader institutional framework of policy-making. The organisers of the EU Hackathon, for example, seek to widen citizen participation in policy processes and understand their initiative as "closing the gap" (Rucic, 2014) between citizens and the policy level. By facilitating the development of technical tools to both enable and analyse the inclusion of citizen voices, they aim at expanding and thereby institutionalising the role of citizens in policy-making. Further, they connect technological developers with the political level by bringing members of the European Parliament to the hackathon and arranging awards ceremonies for the best projects hosted by MEPs. The goal is "to build a bridge between the 'old' (*i.e.* the European Institutions and law-makers) and the 'new' world (*i.e.* the Internet and the coders/developers)".[5] Projects awarded during the 2014 edition included an information tool on current EU policy consultations, a web project that enables citizen monitoring of publicly funded projects, and an online platform to upload and discuss policy proposals.

Hacking the Law

While policy hackathons remain closely connected with the classic practices of hacking and the experience of technology-oriented hackathons, the concept of policy hacking can be expanded to other areas of policy change which are not necessarily technology-based. We are witnessing a growing range of policy reform initiatives that apply some of the characteristics of hacking and thereby differ from the classic forms of civil society interventions discussed earlier. To demonstrate this, I will take a brief journey through several localities, countries and regions.

The first stop is the city of Hamburg in Northern Germany, where the "Hamburg Transparency Law Initiative" was created in 2010. Bringing together local and national NGOs, including Transparency International, Mehr Demokratie (an initiative for participatory and direct democracy) and the Chaos Computer Club (a group of hackers and technology experts), its goal was to create a radical open data law for the city that would require, by default, to make publicly available all data held by public authorities and relevant to citizens' interests. Rather than starting a classic advocacy campaign to put pressure on the city council to create and pass a transparency law, the members of initiative decided to write the law themselves. Without significant legal knowledge, they drafted the core of a law, published it on a wiki page, and invited other civil society organisations and members of the public to comment and develop it further. After soliciting crowd-sourced input as well as expert advice, they handed over the final product to the local government and all political parties. Following a series of negotiations and revisions, the local parliament unanimously adopted the law in June 2012.

Key to the initiative's self-organised development of a new law was a review of best practices in other German states ("Laender") and other European countries. Components of information laws in, for example, Berlin and Slovakia were used for the Hamburg proposal. Since the successful conclusion of their campaign, the group has advised organisations and coalitions in other

German states and in Austria. Several of these have managed to draft laws for their specific localities, often using components from the Hamburg law, and to have them adopted by government (Lentfer, 2013).

This strategy of developing, rather than merely demanding, new policy has been applied in several countries at the national level. In Iceland, the Icelandic Modern Media Initiative (IMMI) used this approach in the wake of the financial collapse of national economy in late 2008. Set up by local social and media activists, including tech developers and internet experts, and supported by international civil society organizations, its goal was to transform the country from a safe haven for banks and financial services into a transparency haven and a favourable environment for media and investigative journalism. The initiative created a bundle of legal and regulatory proposals, with the aim to "protect and strengthen modern freedom of expression" (IMMI, 2012). These included a new Freedom of Information Act to enhance access for journalists and the public to government-held information; measures to limit libel tourism, prior restraint, and strategic law suits that serve to block legitimate information; a new law on source protection, making it illegal for media organizations to expose the identity of sources for articles, books, etc., if the source or the author requested anonymity; proposals on whistleblower and intermediary protection and on safeguarding net neutrality (Gudmundsson, 2011).

Even more than the Hamburg initiative, IMMI cherry-picked laws and regulations from other jurisdictions and created a jigsaw puzzle of tried-and-tested components, including parts of the Belgian source protection law, the Norwegian Freedom of Information Act, Swedish laws on print regulation and electronic commerce, the EU Privacy Directive, the New York Libel Terrorism Act, and the Constitution of Georgia, among others (IMMI, 2012). As the economic breakdown had affected large parts of the population and the secrecy of banks was widely debated and criticized, a significant section of the public was in favour of a radically new model, thus creating favourable conditions for policy change. With the old political class delegitimized, new social actors were temporarily swept into politics and helped advance the discussion on new laws in the national political arena. IMMI was adopted in principle by the Icelandic Parliament in 2010, and several of its components have been implemented, even though progress slowed down when the political environment changed again and traditional forces regained strength (Idir, 2016).

While the previous two examples were inspired by the possibilities of new digital media, similar practices exist in relation to a more traditional broadcast environment. The Argentinian "Coalition for Democratic Broadcasting" emerged in 2004 as a civil society coalition to campaign for a new national audiovisual media law and to replace the old law which stemmed from the times of the military dictatorship. Bringing together unions, universities, human rights groups, and community media, the coalition developed a set of key demands to serve as core pieces of a new policy framework, and a coalition member and university professor was charged by the government to draft a law based on this input. The first draft was discussed at 28 open hearings, comments by civil society groups were included in the document, and a demonstration of 20,000 people brought the final text to Parliament where it was adopted, making it a true "law of the people" (Loreti, 2011).

The resulting Law 26.522 on Audio-Visual Communication Services, adopted in 2009, broke radically with established policy traditions. In particular, it legalized community and non-profit broadcasting and assigned it a third of the radio frequency spectrum, thus becoming "one of the best references of regulatory frameworks to curtail media concentration and promote and guarantee diversity and pluralism", according to World Association of Community Radio Broadcasters AMARC (AMARC, 2010). Applying similar strategies as the initiatives in the previous two cases, the coalition did not write the new policy framework from scratch but drew heavily on existing legal frameworks and policy documents, both from international institutions (for example, the UNESCO Media Development Indicators and reports by the UN Rapporteur on Freedom of Expression) and from civil society organisations (such as a model legal framework developed by AMARC, the *Principles on Democratic Regulation of Community Broadcasting* (AMARC, 2008)). Just as the new law in Hamburg has inspired initiatives elsewhere, the Argentinian legislation has become a model law for the region. Groups and coalitions elsewhere have tried to replicate this model in their countries, thus influencing legislative development across Latin America (e.g., see Klinger, 2011).

Neighbouring Uruguay hosted similar policy innovations, created through similar processes. A new law that legalised community broadcasting was based on a draft submitted by civil society groups. As in Argentina, civil society advocates crossed the lines between inside and outside the policymaking realm: AMARC expert Gustavo Gomez became (temporarily) the National Director of Telecommunication (Light, 2011), while the new regulatory advisory body COFECOM in Argentina elected Nestor Busso, a representative of community radios, as its president (Mauersberger, 2011). The policy development processes in both countries interacted with each other as Argentine activists integrated parts of the earlier Uruguayan reform into their proposal for a new comprehensive law and Uruguayan activists, in turn, used that law as a model for further reforms.

Regional coalitions have sometimes applied the same practices as national groups. When the issue of net neutrality entered the European agenda, a network of advocacy groups picked up the idea of developing a model law. The net neutrality debate had gained momentum as the Council of Europe was developing a declaration in support of internet freedoms (Council of Europe, 2011) and the Dutch parliament passed a law in favour of net neutrality, while telecoms operators like the German Telekom announced they would start discriminating online traffic. At the Chaos Communication Congress in Germany in December 2012, members of European digital rights groups, including organisations such as the Dutch Bits of Freedom and the French La Quadrature du Net, came together to discuss what to do. Several options were discussed, including a Europe-wide coordinated public campaign, but what materialized in the aftermath of the congress was an attempt to draft a model law and promote its adoption by European governments. The network developed a comprehensive inventory[6] of (actual and proposed) legislation and soft law on net neutrality in countries such as Norway, Chile, Netherlands, Slovenia, Korea, Peru and Finland as well as a set of principles[7] for a model net neutrality law through an open process and the use of wiki pages and etherpads. Based on the legal precedents and normative principles, they drafted a framework for a new regulation. As the European activists redirected their efforts from policy development back to classic advocacy in response to proposed new net neutrality laws by the EU Commission in 2013, the development of a model law moved from the European to the global level. A new "Dynamic Coalition" on Net Neutrality emerged within the Internet Governance Forum (IGF) and picked up the project. It developed a detailed "policy blueprint" to serve as a rigorous model legal framework that would be adaptable to different jurisdictions (Belli & van Bergen, 2013).

Policy Hacking: Experimentation, Participation, and Policy Change

The two sets of case studies offer very different approaches to policy engagement and policy change, as well as different understandings of the concept of hacking. Policy hackathons focus on technical expertise and experimentation; they constitute temporary events; they often focus on policy analysis rather than policy change and on the development of tools for participation; and they are not necessarily or exclusively civil society-based but sometimes hosted or co-hosted by companies. The policy initiatives in the second set of examples, in contrast, incorporate a range of civil society actors, including classic advocacy organisations; they do not necessarily focus on technical development or even digital policy; they typically constitute sustained campaigns rather than ad-hoc events; and their goal is to develop alternative legislative proposals and change the law. So the practices, motivations and objectives, as well as the actors involved, may differ substantially.

Despite that, they share significant characteristics. To start with, all the cases highlighted here imply the claim that citizens, civil society activists and technological developers can be policy experts and are willing and able to take policy development into their own hands. Policy hackathons, according to the examples that we have seen here, either bring together people to review and improve policy or develop infrastructure to enhance public participation in policy debate. At the same time, they break down and demystify the process of policy-making and thereby serve to empower citizens to raise their concerns. As Herman Rucic, co-organiser of the EU Hackathon, notes, the purpose is not only to transmit citizen voices to policy institutions, but also to translate obscure policy debates for non-experts—to "rephrase things so that they make sense to normal people" (Rucic, 2014).

The policy change initiatives from the second set of case studies offer even stronger examples for citizen participation and empowerment as they place civil society members in the centre of policy development and lawmaking. While some of the initiatives include established civil society and legal experts, and while several combine policy development with traditional campaigning and even protest (such as the mass demonstration in Argentina), their focus lies on analysing and re-writing existing policies and developing new laws. This is significant as it surpasses civil society's traditional role as advocator and provider of information and values, and puts civil society initiatives instead in the driving seat of the policy process. Policy hacking is thus based on the prefigurative modes of action which we find in media activism and technological development, i.e. the creation of alternative standards and infrastructures. Moreover, it is often based on temporary and informal alliances and thereby relates to newer organisational forms of, e.g., "connective action" and "organised networks".

But can we really characterise both of these distinct sets of practices as "hacking"? While policy hackathons connect more easily with classic approaches to both technological and civic hacking, the inclusion of policy advocacy initiatives requires further explanation. Several points are of relevance here. First, all the examples discussed in this chapter are based in both an ethos and the practice of DIY (do-it-yourself) self-production, which lies at the very core of hacking and, more broadly, tech activism. Secondly, the specific practice of assembling policy precedents from around the world, re-packaging them, experimenting with legal "code", and manipulating, improving and upgrading it bears strong similarities with technological development. Most of the initiatives highlighted here started from an inventory of existing laws and regulations in other countries, picked out the components that seemed useful for the particular local situation, and re-assembled them to a new legislative or regulatory package. Thirdly, while technical tools do not always play the central role that they do in policy hackathons, the use of wikis and other open platforms to crowdsource policy expertise and contributions has been consistent and widespread. Fourthly, the initiatives and coalitions presented here connect legal and technical communities and combine legal and technical expertise. Those two sets of expertise are closely related, as attempts at "tinkering" with technology and with the law require similar skills and forms of reasoning (Coleman, 2009).

So in both sets of cases the "hacking" terminology connects with different practices: Policy hackathons are based more closely in the culture and experience of technological and civic hacking, whereas policy development initiatives apply practices of crowd-sourced manipulation and re-assemblage in the context of a DIY approach. Yet both strategies experiment with "the internal workings of a system" (see above) with the aim of modifying its structure, and thus they constitute two different sides of the policy hacking coin.

As policy hacking takes place in the context of policy advocacy and reform, a further question that requires our attention is whether the concept offers any innovative insights for both policy studies and social movement studies. Apart from the opening of policy processes towards wider participation, one interesting area that it addresses in the field of policy change is policy diffusion and transfer. Traditionally the spread of policies from one jurisdiction to another, and the transfer of rules and regulations across boundaries, are analysed in the context of governmental actions and institutional processes. However in the cases presented here, we can see that s They have cherry-picked regulations from other countries, and their own legal "products" have become models that came to influence legislative development elsewhere. In Hamburg, Iceland, Argentina and elsewhere, these groups have followed up their domestic policy development by advising civil society campaigns and policymakers in other countries (or, in the Hamburg case, other states) and have, in some cases, triggered policy change elsewhere. They have thus provided examples for policy transfer and diffusion that is citizen-led, not government-led.

Even though policy hacking is sometimes applied by classic advocacy organisations and has even been linked to street protest, it offers an approach to policy interventions that differs from the classic inside-outside divide that is predominant in social movement studies. With "insiders", it shares a direct and often collaborative interaction with power-holders and an (either tactical or firm) acceptance of existing policy-making structures and rules. However it does not limit itself to the provision of norms and expertise, nor to exerting pressure on policy institutions (as "outsiders" do). Instead of demanding or proposing change, its focus lies on (self-)developing concrete alternatives by re-writing policy and developing new law. As a citizen-based do-it-yourself approach to the creation of new policy environments, it displays similarities with the "beyond" strategy that was introduced earlier. While the latter focuses on the development of technological bypasses and alternative communication platforms, both are based on the ideas and practices of prefigurative action that develops new infrastructure rather than demanding change.

Finally, and returning to the question of participation, we can ask whether policy hacking is a democratic practice. As noted before, grassroots policy development of the kind described here offers an enhanced role for citizens and civil society in political processes. It may compromise established notions of representative democracy as it questions the role of elected officials and intervenes into their prescribed area of authority. However it opens up new channels for participatory engagement by citizens and for direct forms of democracy. Involving loose networks, informal coalitions and concerned individuals, it relates to more recent ideas of digital citizenship and liquid democracy (e.g., Isin & Ruppert, 2015). Moving beyond the classic focus on political parties and representative democratic processes, these democratic forms hope to enable broader participatory processes, allow for interactive and flexible decision-making, enhance active citizenship and make the public sphere more inclusive (with varying success, see Hindman, 2008).

However, even if "policy hacking" opens new avenues for participation and intervention, it contains several potentially problematic aspects that lie in its connection with the debates and action repertoires of civic and technological hacking and of digital democracy. Civic hacking has been hailed as welcome involvement of experts and creative youth to solve important problems in government, public service, community life and disaster response, but it has also been criticised as cheap outsourcing of state functions and as questionable public relations exercises (Baraniuk, 2013). Tools and practices of digital democracy have often empowered elites and small groups of experts rather than the broader citizenry (Hindman, 2008). While crowdsourced policymaking has allowed for broader citizen concerns to be heard and incorporated in law and regulation, the core of those involved with policy hacking have been technical experts and civil society advocates. Policy hacking certainly has the potential to expand beyond the focus on established actors, but it carries the risk of creating a new caste of experts rather than empowering the broader citizenry and actual policy amateurs.

The impact on policy change by policy hacking initiatives has been substantial, as several of the examples discussed here demonstrate. The broader political impact, however, is a bit less clear. The goals of hacking (both technological and civic) are typically to improve, fix and solve a discrete problem, rather than to transform power imbalances and address the political dimension of policy debates. A hack is "an elegant solution to a technological problem" (Levy, quoted in Meikle, 2002, p. 164). However, with critics like Evgeny Morozov, we may question the reductionist "solutionism" of an approach that focuses on solving discrete problems, rather than understanding (and tackling) the deeper roots and broader contexts of an issue (Morozov, 2013). The cases of policy hacking presented here follow this "solutionist" route as they focus on improving policy and fixing specific issues within a given and largely unquestioned political context. For civil society activism, policy hacking practices thus offers promising new paths for media reform but also prioritizes a problem-solving approach over a more fundamental political and social struggle.

From a critical perspective, we may interpret both the participatory and the "solutionist" character of policy hacking in the context of an increasing privatisation of media policy in which authority and decision-making power are transferred to private environments. Similar to the increased outsourcing of both policy implementation and policy development to commercial actors such as social media companies and other intermediaries (Hintz, 2015), and in line with other civic hacking initiatives (Baraniuk, 2013), policy hacking moves some aspects of policymaking out of established public and democratically-legitimated spaces and into the messier range of technological developers, civil society activists, experts, and corporate actors. This development can be situated within the broader context of neoliberal politics in which the legitimacy of the public sector has been questioned and authority has been moved away from the public realm. Operating outside of democratic control, privatised forms of policymaking have challenged—and often weakened—democratic processes and legitimacy (Couldry, 2010; Crouch, 2004) and can be regarded as part of what Freedman calls "neo-liberal media policy" (Freedman, 2008, p. 47).

As this section demonstrates, policy hacking opens up a number of debates and different, often contradictory, interpretations. As a new practice, its full implications are yet to emerge. However by addressing several shortcomings in both the study and the practice of policy change, civil society and participation, it may offer a fresh approach and new insights.

Conclusion: Policy Hacking as a New Practice?

Policy hacking expands both our understanding of hacking and established approaches to policy reform. In the widening realm of hacking, it applies the practices, ideas and cultures of technological and civic hacking to the legal and regulatory environment, and it thus opens up a field of public administration that has traditionally been rather closed and obscure. In the area of policy reform, it offers a new way for public involvement in policy change by focusing on citizen-based, do-it-yourself creation of concrete policy alternatives.

This chapter has introduced two different types of policy hacking: Policy hackathons and policy reform initiatives. They differ substantially regarding their practices, objectives, and participants. However they are both based on a DIY approach of understanding, developing, manipulating and changing a system; they reject the closed nature of that system and seek to open it up for participation; they value and apply practices of experimentation, crowd-sourcing, and dis- as well as re-assembling components; and they use digital platforms.

While policy hacking is innovative regarding both the hacking and the policy worlds, it may not be entirely new as a practice. Text work and proposals of policy language have long been part of advocates' repertoires of action. From national policy reform to lobbying at UN summits, civil society activists have invested significant time and effort into changing the wording of legislative and regulatory documents, and have developed alternative declarations (e.g., Hintz, 2009). Furthermore, policy hacking has often been embedded in a broader range of traditional action repertoires, including advocacy campaigns, protest and public pressure, and sophisticated framing exercises that communicated the need for policy change to government members, regulators and parliamentarians. Also, its successes have been context-specific. In Iceland and Argentina, for example, policy change was possible in response to economic crises, the failure of established economic models and a legitimacy crisis of the political class; and in Hamburg, a public spending scandal (as well as the steep rise of the Pirate Party in the polls) helped convince political leaders that change was necessary. Such policy windows have been crucial conditions for successful policy hacking initiatives.

Despite that, the initiatives presented in this chapter have moved significantly beyond established practices of policy advocacy as they developed complete legislative proposals, analysed and re-wrote complex policies, and facilitated public participation beyond a narrow group of civil society organisations and technical experts. While its broader political impact may be controversial, its facilitation of participation and intervention arguably supports democratic change, and the cases presented in this chapter demonstrate that it can be a viable and successful strategy for media and communications reform.

Notes

1. Some of these cases and arguments have been introduced in Hintz (2016). Here I will discuss them with a focus on hacking and add further research results. The chapter draws from ongoing research—mainly document analysis and expert interviews—on policy hacking
2. http://legalhackathon.blipclinic.org/
3. http://2013.euhackathon.eu/
4. http://2014.euhackathon.eu
5. http://2014.euhackathon.eu
6. http://www.laquadrature.net/wiki/Model_Law_on_Net_Neutrality
7. https://quadpad.lqdn.fr/XKTMzQ5Yg2

References

AMARC. (2008). *Principles on democratic regulation of community broadcasting*. Retrieved from http://legislaciones.amarc.org/Principios/Principles%20on%20Democratic%20Regulation%20of%20Community%20Broadcasting%20(eng).pdf

AMARC. (2010, January-March). *AMARC link, 13*(1). Retrieved from http://www.amarc.org/amarclink/amarc_link_AVRIL_2010_EN_final.pdf

Banaszak, L. A. (2005). Inside and outside the state: Movement insider status, tactics, and public policy achievements. In D. S. Meyer, V. Jenness, & H. Ingram (Eds.), *Routing the opposition: Social movements, public policy, and democracy* (pp. 149–176). Minneapolis, MN: University of Minnesota Press.

Baraniuk, C. (2013, 29 June). Civic hackers: Techies volunteer to rescue government. *New Scientist,* Issue 2923. Retrieved from http://www.newscientist.com/article/mg21829232.000-civic-hackers-techies-volunteer-to-rescue-government.html#.Ujxbsz_3PEc

Belli, L., & van Bergen, M. (2013). A discourse-principle approach to network neutrality: A model framework and its application. *Network Neutrality Dynamic Coalition.* Retrieved from http://networkneutrality.info/sources.html

Bennett, L., & Segerberg, A. (2012). The logic of connective action. *Information, Communication and Society, 15*(5), 739–768.

Braman, S. (2006). *Change of state. Information, policy, and power.* Cambridge, MA: MIT Press.

Chakravartty, P., & Yuezhi, Z. (2008). *Global communications: Towards a transcultural political Economy.* Lanham, MD: Rowman & Littlefield.

Coleman, G. (2009). Code is speech: Legal tinkering, expertise, and protest among free and open source software developers. *Cultural Anthropology, 24*(3), 420–454.

Couldry, N. (2010). *Why voice matters: Culture and politics after neoliberalism.* London: Sage.

Council of Europe. (2011). Internet governance principles, draft V.2.0. Council of Europe, 2011. Retrieved November 5, 2011 from http://www.coe.int/t/dghl/

standardsetting/media-dataprotection/conf-internet-freedom/Internet%20Governance%20Principles.pdf

Crouch, C. (2004). *Post-democracy.* Cambridge: Polity.

de Jong, W., Shaw, M., & Stammers, N. (2005). Introduction. In W. de Jong, M. Shaw, N. Stammers (Eds.), *Global activism, global media* (pp. 1–14). London: Pluto Press.

Delfanti, A. (2013). *Biohackers: The politics of open science.* London: Pluto Press.

DeNardis, L. (2014). *The global war for internet governance.* New Haven, CT: Yale University Press.

Freedman, D. (2008). *The politics of media policy.* London: Polity Press.

Gudmundsson, G. R. (2011). Research interview by Arne Hintz, Reykjavik, August 20, 2011.

Harold, C. (2004). Pranking rhetoric: Culture jamming as media activism. *Critical Studies in Media Communication, 21*(3), 189–211.

Held, D., & McGrew, A. C. (2003). The great globalization debate. In D. Held & A. G. McGrew (Eds.), *The global transformations reader* (pp. 1–50). Cambridge: Polity Press.

Hindman, M. (2008). *The myth of digital democracy.* Princeton, NJ: Princeton University Press.

Hintz, A. (2009). *Civil society media and global governance: Intervening into the world summit on the information society.* Münster: Lit.

Hintz, A. (2015). Social media censorship, privatized regulation, and new restrictions to protest and dissent. In L. Dencik & O. Leistert (Eds.), *Critical perspectives on social media and protest.* London: Rowman & Littlefield.

Hintz, A. (2016). Policy hacking: Citizen-based policy-making and media reform. In D. Freedman, J. Obar, C. Martens, & R. McChesney (Eds.), *Strategies for media reform.* New York City, NY: Fordham University Press.

Hintz, A., & Milan, S. (2013). Networked collective action and the institutionalized policy debate: Bringing cyberactivism to the policy arena?. *Policy & Internet, 5*(1), 7–26.

Idir, G. (2016). Research interview by Arne Hintz, Reykjavik, May 19, 2016.

IMMI. (2012). Icelandic modern media initiative. Retrieved from https://immi.is/index.php/projects/immi

Isin, E., & Ruppert, E. (2015). *Being digital citizens.* London: Rowman & Littlefield.

Keck, M. E., & Sikkink, K. (1998). *Activists beyond borders. Advocacy networks in international politics.* Ithaca, NY: Cornell University Press.

Khagram, S., Riker, J. V., & Sikkink, K. (2002). From Santiago to Seattle: Transnational advocacy groups restructuring world politics. In S. Khagram, J. V. Riker, & K. Sikkink (Eds.), *Restructuring world politics: Transnational social movements, networks, and norms* (pp. 3–23). Minneapolis, MN: University of Minnesota Press.

Kingdon, J. W. (1984). *Agendas, alternatives, and public policy.* Boston, MA: Little Brown.

Klinger, U. (2011). Democratizing media policy: Community radios in Mexico and Latin America. *Journal of Latin American Communication Research, 1*(2), 4–22.

Lentfer, D. (2013). Research interview by Arne Hintz, Hamburg, 16 May, 2013.

Light, E. (2011). From pirates to partners: The legalization of community radio in Uruguay. *Canadian Journal of Communication, 36*(1), 51–67.

Loreti, D. (2011). Research interview by Arne Hintz, Montreal, 11 February 2011.

Lovink, G., & Rossiter, N. (2005). Dawn of the organised networks. *Fibreculture Journal*, Issue 5. Retrieved from http://journal.fibreculture.org/issue5/lovink_rossiter.html

Mauersberger, C. (2011). Whose voice gets on air? The role of community radio and recent reforms to democratize media markets in Uruguay, Chile and Argentina. *Journal of Latin American Communication Research, 1*(2), 23–47.

McLaughlin, L., & Pickard, V. (2005). What is bottom-up about global internet governance? *Global Media and Communication, 1*(3), 357–374.

Meikle, G. (2002). *Future active: Media activism and the internet.* New York, NY: Routledge.

Messner, D., & Nuscheler, F. (2003). *Das Konzept Global Governance: Stand und Perspektiven*, INEF-Report 67. Duisburg: Institut für Entwicklung und Frieden.

Meyer, D. S. (2005). Social movements and public policy: Eggs, chicken, and theory. In D. S. Meyer, V. Jenness, & H. Ingram (Eds.), *Routing the opposition. Social movements, public policy, and democracy* (pp. 1–26). Minneapolis, MN: University of Minnesota Press.

Milan, S. (2013). *Social movements and their technologies: Wiring social change.* Basingstoke: Palgrave MacMillan.

Morozov, E. (2013). *To save everything, click here: The folly of technological solutionism.* New York, NY: Public Affairs.

Musiani, F. (2015). Practice, plurality, performativity, and plumbing: Internet governance research meets science and technology studies. *Science, Technology & Human Values, 40*(2), 272–286.

Network Working Group. (1993). *Request for comments: 1392. Internet users' glossary.* Retrieved from http://tools.ietf.org/html/rfc1392

Raboy, M. (2002). *Global media policy in the new millennium.* Luton: University of Luton Press.

Raboy, M., & Padovani, C. (2010). Mapping global media policy: Concepts, frameworks, methods. Retrieved from http://www.globalmediapolicy.net

Rucic, H. (2014). Research interview by Arne Hintz, 28 August 2014.

Sterling, B. (1992). *The hacker crackdown: Law and disorder on the electronic frontier.* New York, NY: Bantham Books. Retrieved from http://www.gutenberg.org/files/101/101-h/101-h.htm

Tarrow, S. (2005). *The new transnational activism.* New York, NY: Cambridge University Press.

7. *Hacking Administration—A Report From Los Angeles*

Morgan Currie
University of Edinburgh

In 2012 the City of Los Angeles began an alliance with a burgeoning civic hack scene. In this context the term "civic hacking" remains the umbrella moniker for a series of informal meet-ups and weekend-long events organized around software demos, conversations, and power-point presentations that all pose technology as a balm for civic and administrative problems. More broadly in the United States, civic hacking has prospered in several cities since Obama took office and oversaw the launch of Data.gov, a website where federal agencies publish datasets for free reuse by the public. Yet both the term and the form draw from several broader traditions: geek culture, for which hackers are deft manipulators of computer software and hardware; open source software culture, with its dedication to clever code, free speech and open licenses; and Silicon Valley, where frenzied, time-limited, overnight hackathons became a cheap means to rapid prototyping and recruiting young talent. *Civic* hacking—drawing more from these traditions than negative, shadowy depictions of hackers rooting out security breaches—captures a trend to harness the craft, ingenuity, and aesthetics of these variably outsider or industry traditions by fiscally and design-challenged governments. Says a "civic designer" in a blog post for the non-profit Code for America's website, "What began as a niche theory about the potential to improve government using technology has quickly expanded to focus more on changing the culture of government to work more effectively and creatively with its citizens." (Levitas, 2013)

This chapter analyzes civic hacking in the context of the City of Los Angeles. I make the case that civic hacking primarily offers an instrumental model of participation that engages at the level of administration. The rhetoric about civic hacking blurs citizens with experts, describes a more direct and collaborative participation, often emphasizes technological practice over debate, and is usually engaged with designing city services rather than policy or lawmaking. To analyze this kind of participation in practice, I examine civic hacking through political theories about participation in the administrative realm of government. This analysis allows me to make distinctions between types of civic hacking projects that are too often glossed over in the academic literature on the topic, particularly that between instrumental versus monitory or agonistic forms. To make these arguments, this chapter presents material from a four-year participant observation study, as well as from interviews conducted with citizens, city staff, hackathon event organizers, participants, and sponsors.

This chapter begins by defining civic hacking, then describes the civic hacking events that took place between 2013 and 2016 in Los Angeles. I next examine literature that has been critical of civic hacking to argue that many of these analyses fall short; civic hacking is an emerging phenomenon that continues to evolve and defy the claims made about it. Finally, I offer a theoretical analysis to make distinctions among civic hacking projects' goals. I draw from political theories about participation in the administrative realm of government to make the case for more monitory and agonistic forms of civic hacking that are aware of the values of the technologies being used.

What Is Civic Hacking?

Civic hacking is a type of political and civic participation through hands-on making of digital products and user-friendly design, usually by forming groups to create websites or phone apps. For example, the first White House Open Data Day Hackathon in 2013 asked participants to build web software for a new White House citizen petitioning system; at the National Day of Civic Hacking that same year, the EPA asked participants to make visualizations with data on watershed pollution. The Open Knowledge Foundation, a nonprofit dedicated to information sharing, makes the claim that civic hacking, in this regard, allows citizens to participate more directly in creating the tools of government: "This is more than transparency: it's about making a full 'read/write' society, not just about knowing what is happening in the process of governance but being able to contribute to it."[1] Yet while civic hackathons are oriented around tangible, technical products, they also create spaces of communal making that reflects the production and performance of civic desires and critiques. This aspirational place making is one key outcome of these events, not necessarily any one design product.

Civic Hacking in LA

In Los Angeles, civic hacking has taken three main forms: first, as large-scale events that attract hundreds of people and often have the structure of a prize competition; second, as one-off events that typically focus on a single issue of community concern; and third, as smaller, intimate meetups that take place weekly or monthly. The large-scale events often coincide with a National Day of Civic Hacking celebrated in Washington D.C. and by cities around the United States. Such was the case for Los Angeles' first civic hackathon in 2013, a grassroots affair that attracted around 500 people. Over 2014–2017 the City has played a larger role, hosting Hack for LA events in city administrative buildings, while helping with publicity and offering personnel support.

These large-scale events typically take shape as a competition, the form that has characterized civic hacking most vividly in literature and the press. At most Hack for LA events, organizers introduce a common set of rules: one demo per team, made from "fresh" coding that must originate during the event and that must be up and functional during the final presentations. Judges are to base their decisions on originality, a clear and focused concept, and the quality of the technology in use. The rules place an emphasis on collaboration and speed; as one organizer put it, "The hack process itself helps people to quickly problem solve."[2] Results must contend with the technical constraints of designing working demos within two-day's time.

These events also have prominent commercial sponsors who donate cash prizes. Sponsors at the December 2013 Hack for LA event, for instance, included Tapdn, a Santa Monica software company that markets to college students; Google; Livestrong, a health and fitness website; ESRI, the GIS company; and Sprint. One of Hack for LA's primary sponsors has been a non-profit, the i.am.angel Foundation, founded by the singer will.i.am from the Black Eyed Peas. The Foundation, whose mission is to support STEM in low-income neighborhoods in Los Angeles, has lent staff to organize past Hack for LA events. Officials from the City of Los Angeles also make an appearance at these events; while the industry representatives entice participants towards using their software or data, city officials prevail on attendants to draw on city data in their designs.

Civic hacking in Los Angeles also takes a second form as one-off affairs that tend to focus on particular issues. In 2015, for example, the City coordinated #techLA, a cluster of weekend hackathons held throughout the year that urged participants to design projects on the themes of water, transportation, and community. At one of these events, the July 2015 #ImmigrationHack, there were no sponsored prizes, no winners, only suggestions, "ideations", and prototypes. Non-profits and government officials talked about the very real material challenges of helping non-citizens find the correct online forms to fill out or the right desk to visit if they want to take the path towards citizenship. The end result application prototypes ranged from a user-friendly website where the undocumented could find essential government documents, to social-media websites that connect new immigrants with the settled population.

Finally, the city's broader civic hack scene encompasses more frequent meetups of programmers, data scientists, and interested citizens. In 2015 civic hack nights took the shape of monthly Hack for LA gatherings organized by volunteers and the two-person staff from Compiler LA, a Benefit Corporation that designs web apps for governments and non-profits.[3] On its Meetup page, Hack for LA defines itself more specifically as the Los Angeles "Code for America Brigade", meaning that it enjoys minimal administrative and fiscal support from the San Francisco-based civic tech non-profit Code for America. The 2015 meetups focused less on actual making and more on brainstorming how technology might address a certain issue of concern, such as homelessness and mental health. After a hiatus, beginning in June 2016, the coordinators of the monthly Hack for LA nights began weekly meetups at a space in the downtown arts district. These weekly gatherings have abandoned the social, discussion-based format of the year before and focus primarily on application design. Participants arrive to work in teams on apps and websites over the course of the three-hour meeting. Current projects as of this writing focus on public arts awareness and affordable food.

Civic hacking in Los Angeles should be viewed as a piece of a larger cultural change. One advocate I spoke to mentioned that people have been able to create livelihoods based on the City's support of civic hacking or civic technology—one example she supplied is Compiler, mentioned above, which has garnered contracts with the City, the County, and local non-profits. Other participants I spoke to told me that the ultimate goal of civic hacking isn't simply new tools—realized or unrealized—but a change in government culture towards greater transparency and data literacy by staff. Said an organizer of Hack for LA, "A lot of times in these environments we can build something that solves a problem, but even if we create a proof of concept to help officials think critically about opening data or engaging in transparent and participatory way with communities, that's a victory. We're trying to win on both of those fronts."[4]

The phenomenon of civic hacking remains an emerging phenomenon, given its shifting forms over the past three years in Los Angeles alone, making it difficult to theorize or make claims about. In the next section, I go over three important critiques that have been made about civic hackathons. However insightful these critiques, they too often gloss over distinctions among projects, the hackathon form, and how participants engage in these spaces. I move beyond these critiques, in the final sections, with a normative analysis of civic hacking that makes finer distinctions among civic hacking projects' values and goals.

Critiques of Civic Hacking

Neoliberalism and Silicon Valley

One critique of civic hacking is the influence on it of Silicon Valley, often visible in the hackathons' form and tactics, in the rhetoric deployed, and the companies explicitly involved as sponsors. According to some critics, civic hackathons are problematic because they borrow from a format exported from start-up culture in an attempt to reproduce some of the values and practices from private enterprise in public administration. This formal and conceptual transfer has been commented on in literature on civic hacking (Goldstein & Dyson, 2013; Irani, 2015; Schrock, n.d.) and by civic hackers themselves. One participant I spoke to described how the city endorses a form borrowed from private industry in order to "catch up" with it: "The city is … taking advantage of the buzz. I would say the zeitgeist."[5] Civic hackathons in this way produce particular subjects, creating "entrepreneurial" or "algorithmic citizens" that value efficiency and rapid-fire innovation (Irani, 2015; Schrock, 2016). Civic hackathons therefore encourage opportunistic, depoliticized forms of participation, such as reporting potholes, that are bereft of any impact to addresses relations of power (Morozov, 2013). The resulting institutional collaboration puts civic hackers at odds with the traditional view of hackers as dissidents or activists (Coleman, 2013).

There is also clear evidence that tropes from Silicon Valley, and the related open source software movement, have driven a shift from a discourse that valorizes centralized expertise to today's direct, crowd-sourced problem solving. In this discourse, software features provide metaphors for governance; technologies of the "free and open" become, at least for thinkers such as technologist and publisher Tim O'Reilly, a catch-all solution to collectivize social problems at large—education, publishing, architecture, and now government (Kelty, 2008). Public figures often evoke these cyberlibertarian influences when they adopt O'Reilly's slogans of "gov 2.0" and "government as platform," a phrase that describes the utility of government APIs that programmers can build software upon. O'Reilly's ideas are repeated at Los Angeles events, as when at one

Hack for LA event Mayor Garcetti cited O'Reilly to explain the government's role now that it offers open data: "We are the platform, you innovate and build on us."[6] The rhetoric of civic hacking therefore describes a direct, collaborative form of participation in government administration that collapses or inverts the distinction between citizens and experts: now it is citizen entrepreneurs who will improve policy or city bureaucracy, through technological innovation and user-friendly design. As critics point out, this rhetoric appears to put the onus of government services on citizens themselves.

Because projects often incorporate commercial software, critics have also expressed concern that civic hackathons will shift the onus of service delivery to the private sector, providing a "backdoor" to government contracts (Johnson & Robinson, 2014). Such partnerships with Silicon Valley—or Silicon Beach, the name donned by a cluster of tech companies in the coastal Los Angeles area—in this way appear to signal another example of neoliberalism as governments attempt to hand over the design and oversight of public information services to participants and private companies. Civic hacking, in this line of critique, therefore becomes another means of harnessing the efficiencies of the private sector by way of citizens who represent the skillset of nimble technology firms.

Civic involvement, as one result, becomes uncomfortably tied with corporate aims. As we have seen, the influence of start-up culture goes beyond rhetoric to incorporate public-private partnerships into the economic structure of these events, particularly in the form of sponsorships by companies that collect and monetize user data. Meanwhile, sponsors offer access to their products' API, acclimating programmers to their product. Sponsorship appears, at least in Los Angeles, endemic to civic hacking spaces; the weekly Hack for LA is sponsored by a consortium of nonprofits and private companies, including LACI and Carbon Five, a software and product design firm.

Among the participants I spoke to, some came to civic hacking events with a feeling of civic duty as much as a desire to "network" and job seek. In these spaces, the civic hacker is cast at once as a civic participant, a consumer, a potential employee, and an unpaid laborer. Indeed, the discursive and design influences from corporate Internet culture are, as mentioned, highly visible at these spaces. Google, ESRI, Intel, offer sponsorships, prize money, and booths where representatives hawk their products. Any analysis of these spaces must confront that the civic hacker has other potential gains beyond civic skills, and that this possibly weakens the civic motive. Furthermore, a general emphasis on the neutrality of data and these technical tools only means that deliberation and debate about the role of the private sector in these civic spaces are often kept at bay.

Languishing in the Speculative

Another critique is the ephemeral nature of many hackathon projects. Rather than full-fledged Deweyian publics that work together to bring political issues into focus, Lodato and DiSalvo (2016) view civic hackers as contingent "proto-publics" that simply disband after the event. Their material labor produces prototypes of imagined, better futures, but ultimately remain speculative, with no sustained presence over time. Indeed, lack of sustainability has been a problem for civic hacking in Los Angles. So far, civic hacking is better seen as a broad diagnostic of civic concerns and frustrations that are worked out through prototypes, or "demos" of possible solutions. These demos most often languish with no financial or institutional support beyond a few days' hustle of coding and design. Prize awards at civic hackathons do not seem to induce people to continue their projects once the lights are turned off and doors are closed.

In 2015 the City shifted tactics to contend with the ephemeral results of open data events. First, Los Angeles hosted a series of themed, issue-oriented hackathons (ibid.) based on concerns dictated by the Mayor's office: the drought in California, Los Angeles' ongoing transportation woes, and community engagement, with a particular focus on immigration. At Hack for LA 2015, held in the City's architecturally sublime Water and Power building, organizers announced data sets specific to those city challenges, including data on parking tickets and water use by zip code. Programmers brainstormed prototypes in the main event space, where a mural of Mulholland towered over participants who were tasked, at this particular event, with "hacking the drought" unsettling California at the time. The city encouraged participants to put their ideas into workable form through an app challenge that was to begin in September of 2015, called Challenge: LA. The Challenge promised funds and support to shepherd projects to realization. However, nothing came of Challenge: LA. According to one of the Mayoral staff, the funding never materialized, and the hackathons simply did not yield results that could move beyond prototypes. The employee explained, for example, that the City already had contracted with a private company to design water-saving information services, making the crowd-sourced approach redundant:

> from the water hackathons [hack the drought], what was interesting is people came up with ways to meter water and for individuals to get information on water use. But the Department of Water and Power had set up relationships to do that already. There was not an opportunity to use these ideas.[7]

Is participation through civic hacking ultimately better left in the speculative realm, such that the rhetoric of direct participation or collaboration with the government should be corrected and toned down? This shift in rhetoric does seem to be occurring. At an event in April 2016, the Mayor's office conceded that the City would spend less time reaching out to citizens and instead redirect their efforts towards more open data sharing across City departments and with private companies such as Google, since these use cases have proved more successful.

Confined by Solutionism

Perhaps the most consistent critique of these events centers on the civic hackathon's vision of governance, specifically its proposal that technological expertise is a way to resolve complex political problems. Lilly Irani (2015) argues that the politics of civic hackathons reside in its form more than the issues addressed: making and experimenting are privileged over debating and planning, and proponents imagine that social change can happen through small technical acts occurring "outside social movements or formal politics" (p. 17). The politics of civic hacking, therefore, are not in the various issues that projects take up, since in any case the hackathon can absorb any issue. Rather, the politics reside in an epistemological assumption about how civic concerns should be addressed. Civic hackathons, under this critique, rely on technological "solutionism," a term coined by Morozov to describe the shallow tendency to define problems narrowly through technological solutions (2013).

Lodato and DiSalvo also analyze civic hackathons in relation to literature on the role that technology and design play in politics, drawing from John Dewey's theories on publics and issue-making and from Noortje Marres' concept of "material participation" (2016). For Dewey, an "issue" is a condition of concern with immediate consequences; "publics" come into being as they cohere around and articulate an issue in the face of ongoing, collective distress. Marres refines these concepts to argue that material devices can play an important role in mediating and structuring publics and their issues; material practice offers another mode of engagement in issue-formation beyond that of discursive deliberation and debate. Civic hackathons, for Lodato and DiSalvo, present an example of material participation; they are sites where attendees give form to the conditions of political issues through tinkering and prototyping. In the authors' final analysis, however, civic hackathons reduce political issues into tractable problems that can be resolved through technical or design solutions. Complex structural issues, such as affordable housing or pollution, masquerade as technical problems that can be solved by phone apps and websites. The authors believe this mode of participation forecloses inquiry and suppresses alternate explanations of such issues beyond the narrowly technical. Since these events encourage a specific form of civic engagement, the outcomes are portrayed as a naïve view of "politics as the mechanics of government" by reducing political issues into what can be solved with technical skills alone (DiSalvo & Gregg, 2013).

The Problem of Writing About an Emerging Topic

These three critiques of civic hacking are, like any analysis of a contemporary phenomenon, based on a selection of outcomes so far, even as civic hacking continues to evolve and belie some of the claims made about it. Several critiques (Irani, 2015; Lodato & DiSalvo, 2016) are based on empirical observations of civic hackathon that took the form of a one- or two-day app contest. Yet, not all civic hacking events I attended were contests or had corporate sponsors—the Immigration Hack, for instance, was sponsored by the City and a handful of nonprofits. In fact, some civic hacking events are not based around prizes or product prototyping at all. Several civic hacking events have instead provided forums for informed discussion among participants, nonprofit representatives, and city officials, taking the shape of a more traditional public sphere. Civic hack events have offered a space where citizens can gain a better understanding of the technological infrastructures required for governance and community building—the material needs of governance and civil society crucial for public services, such as water conservation, immigration reform, preventing bullying, and bike sharing. In this way, civic hacking can offer what Carol Pateman (1970) calls the "educative dividend" of participation by helping citizens better understand how their city works.

Also, contrary to Lodato and DiSalvo, civic hacking in Los Angeles has generated a small but sustained public, since the organization of events under the banner of civic hacking has continued three years after it started. The current weekly, incubation-oriented Hack for LA meetups have by the time of this writing realized projects beyond the prototype phase; indeed, in October 2016, Hack for LA announced the first working project to come out of its group meetings, called Food Oasis LA. The website steers users to local farmer's markets, food pantries, community gardens, and grocery stores and was made in collaboration with the Los Angeles Food Policy Council.[8]

Finally, these accusations often elide key distinctions *between* types of civic hacking projects. In my fieldwork I found that civic hacking projects often narrowly focus on technological, rather than discursive, solutions. Yet I would like to modify the critique somewhat. Rather than trying to solve complex political problems, as these critiques would have it, civic hackers in Los Angeles more often focus on designing information services for city administrations: technical solutions to traditionally administrative problems. Many apps and prototypes respond to city officials' or non-profits' request for better information infrastructure for service delivery. These projects suggest the inevitable and hidden role of information infrastructure to civic organizations and government—the banal aspects of governing or organizing that are not usually open to public input or scrutiny but happen through internal IT work. For example, demos at the 2013 Boyle Heights Hack for LA event showcased a map meant to help people find retailers that take food stamps in Boyle Heights, offered data on the Los Angeles river and park information, and drew from city data on water usage so that citizens could report broken sprinklers or pipes.

In other words, the projects at civic hacking events primarily engage with the mundane trenches of the ailing nuts and bolts of administrative information services. Garcetti stated the need for technologists to help with government services in his introductory speech to 2015 Hack for LA:

> One thing I've always said is that government has the best market share out there. […] But we generally have pretty lousy products. On the flip side, we have people with great products and ideas that have no market share. So if we just kind of get married to each other, we can take the innovation that is out there and take the platform that we have, the reach that we have, to get to everybody, and we can improve the quality of life for everybody.[9]

This specific role explains why participants are often distinct from Dewey's (2012) publics, as DiSalvo & Gregg rightly argued. This distinctive kind of participation is, as Christopher Kelty (2016) puts it, neither participation in electoral politics nor by publicizing opinions—the traditional public sphere—but rather participation in "the administration of the government's practical affairs." (p. 82) These technocratic efforts are often about making administration run more smoothly in a post-recession context and to confront an "old mentality of hierarchy, bureaucratic complexity, and over-engineered, inflexible design." (Ibid, p. 79) Code for America, following this need, embeds Fellows in cities to design public information services. Fellows have built apps to help citizens in San Francisco be better serviced by SNAPs (food stamps) by using text messages rather than letters or long phone calls.[10] In Rhode Island Fellows created an online registration for a school lottery process.[11]

A more significant critique of these "citizen experts," therefore, is that participation in public information services is a very constrained view of civic activity, not because of its technical form necessarily but because, at the point at which it intervenes it has little choice but to be collaborative with government. In the characterization of the civic hacker as citizen expert, civic hacking does not challenge or take part in designing government policy, but instead aids the government in carrying out its existing priorities. Most of the solicitations by government staff at these events understand civic hacking as a form of collusion with a government interested in technological improvement. Civic hackathons, with their appeal to network technologies and metaphors of decentralization, are not actually critical of power structure, but rather instrumental and positivist in the application of technology for service delivery.

In the next section, I offer a normative, rather than purely critical, approach to civic hacking, turning to literature on administrative participation. The most important distinction that the literature has not made is the difference between two types of participation: participating in carrying out administrative tasks, as the bulk of these projects are, and participating in a public sphere to influence policy or representative politics, by provoking discussion and criticism. In the next and last sections, I point out how civic hacking can be understood in terms of theories about administrative participation, before using these theories to call for more projects that move beyond instrumental collaboration towards dialogue and even critique.

Theories of Participation in Administration

A long-standing debate among political theorists who examine the role of administration in a democracy asks this question: should democratic participation play any role in the administrative branch of government? On one side of this debate are those who believe government experts should be left alone to decide and enact government policy. Technocracy and scientific management are characterized in this classic literature on administration as anti-participatory aids against "the tyranny of the masses". One of the reasons expertise is needed in the first place, according to thinkers going back to de Toqueville, is for those instances when mass participation is an ill-advised idea (Goodnow, 1900; Wilson, 1887). To these thinkers, freeing administration from politics would be the best way to attract and reward competency.

On the other hand, public servants are not elected into office and so are not directly accountable to an electorate, making them a problematically anti-democratic aspect of governance. Mosher, a foundational scholar of modern administration theory, asks "How can we be assured that a highly differentiated body of public employees will act in the interest of all the people, will be an instrument of all the people?" (p. 19, 2016) Traditional literature in the field of public administration moreover characterizes administrative systems as top-down, rational, and authoritative, traits that purportedly conflict with expensive and inefficient values of equitable representation. (Berkley & Rouse, 2004; Kweit & Kweit, 1981; Rosenbloom, Kravchuk, & Clerkin, 2008;

Thompson, 1983) At the same time, public service is crucial for distributing public goods, and public institutions are accountable to an electorate that has some means to make demands on its formal structure through elected office.

In 20th century political theory, the debate over how power should be delegated between government experts and citizen participants is famously represented by John Dewey's response to Walter Lippmann's *Public Opinion* (1922), a treatise that calls for the need for expertise in government. According to Lippmann, experts require a place in democratic society due to the internal failings of humans who have not cultivated the habits nurtured over time by expertise, notably those of self-questioning, skepticism, and scientific inquiry. Lippmann accordingly believed both policy and its enactment should derive from experts. To give experts a more diminished role requires too utopian a vision of the non-expert citizenry. Dewey, in *The Public and Its Problems*, argues for a more limited role for experts, who are unelected and unaccountable to citizens. For Dewey, experts should only devise the means to enact policies and laws set by elected officials. Expertise still has an important role to play in politics, and Dewey advocated for scientific methods in policy setting. Yet experts should be the guides towards ends set by more democratic means.

Beginning in the 60s, as a response to a wildly ballooning federal administration, the United States began to allow more citizen participation in the area of administrative policy-making through transparency laws such as FOIA, environmental oversight laws, such as the National Environmental Policy Act, judicial litigation, public hearings, and whistleblowing protections. These examples of participation are largely adversarial—that is these policies were designed to expand the public sphere of deliberation by giving citizens access to information on which to check the power of administration. Nader's Raiders, of Ralph Nader's Center for the Study of Responsive Law, made formidable use of these new participation mechanisms to check corruption and incompetency in the Federal Trade Commission, the NEA, and the Food and Drug Administration.

Complicating this issue of administrative participation is the turn in administrative theory since the 1990s towards favoring private sector principles of efficiency and financial accountability (Kamensky, 1996). Sometimes called "New Public Management," administrative reforms during the Clinton years characterized citizens as customers or consumers with needs that public agencies respond to (Osborne & Gaebler, 1993; King & Stivers, 1998). Various theorists have made inroads into the question of whether and how government administration driven by efficiency and cost-saving can square with other ideals of democracy such as equity, citizen well-being, and environmental health. The reconciliation, according to some, comes about by the practices and attitudes of public administrators themselves. Administrators should adopt a professional ethic as guardians of the public good who are responsible for citizens' basic rights (Denhardt & Denhardt, 2000). In this perspective, the remedy to an unaccountable administrative is civil servants themselves. Yet the means for citizens to inject themselves into policy debates remain in place at the federal and local levels through, for instance, open record laws, city council hearings, and the judicial system.

Civic hacking should be seen as a relatively new method for participating in the administrative branch of government; this is also how civic hackers have described themselves to me. Claims a liaison for the Hack for LA brigade, hackers short-circuit electoral politics entirely: "There's technical stuff that hacking implies, but the basis is that we're seeking and making change outside typical channels, like voting and getting legislation passed. Hacking is making change with government, but not in ways of last 100 years."[12]

In the final section, I apply some of the theories about participation in administration to civic hacking in an attempt to distinguish weak versus strong forms—those that use making and analytics not towards instrumental ends but rather as the starting point for critical engagement in an issue.

Hacking the Administrative

Civic hacking can be a way to democratize the instrumental step of administration that both Dewey and Lippmann considered the province of experts: designing the infrastructures and technologies of service delivery. Even as civic hacking deploys the wisdom of the crowds, it still appeals to a long-standing view since the New Deal that administrative policy should be based on technical expertise and a value neutral "professional spirit" rather than through ideology and special interests (Seidenfeld, 1992, p. 1519). Many civic hacking projects evoke the rationalized, technocratic management of the bureaucratic state operating through statistics and records collection but invite citizens to design the tools of management themselves. Civic hackers, paradoxically, engage in political participation at a stage of the process that many theorists argue are beyond politics. According to a Hack for LA liaison, civic hackers design information infrastructures that are necessary *prior* to addressing more political issues:

> Our goal and mission is about access to housing, transportation and air quality, not technology. Technology is a toolset; it's the quickest way to get to those kinds of changes. That's why civic hacking is important. It's the fastest way to rebuild community and find other humans who want to do this stuff and make changes through data. To help other people build systems and get that out of the way.[13]

Yet as a result, too often civic hack projects take a weak form of participation that involve participants in administrative *tasks*, rather than *decision-making*, and as a result seek to merely reproduce administrative aims—participation at the level of delivering services per policies already set. In these projects, citizens and administrators often share an epistemological orientation of administrative problem solving that begins at the same starting point: government and citizen are in alignment, whether on the need to save water, service immigrants, have more efficient fire services, or encourage biking. In this sense, this kind of material practice does not start from a place of deliberation about societal problems but already at the point of their status as settled matters—at the point of policy enactment, not policy making. This form of participation has instrumental, pragmatic aims, rather than the goals of contestation, oversight, or structural critique.

While this type of civic hacking project predominates, civic hacking projects can engage in monitory and agonistic forms of participation within and beyond the administrative, either by calling for greater transparency, in the case of monitory forms, or, in the case of agonistic forms, by criticizing and challenging policy and administrative goals for gross power asymmetries, particularly between governments and citizens and between private industry and consumers. The best examples of these types of civic hacking projects do not use technology or government data to provide a service alone, but to make an argument or provoke debate on an issue.

The weakest form of this type are simple transparency projects. Govtrack.us, one of the earliest civic hacking projects to consider itself as such, publishes data on federal legislation and bills as well as information about Congressional representatives. In Los Angeles, an example of this work was a prototype of a website that displayed pie chart visualizations revealing city expenditures and salaries. Yet transparency alone, as some have argued (Gray, 2014), is not automatically tied to the safeguard of public well-being and human rights; it can also be used to support technical innovation, government efficiency, and economic growth (such as lucrative government contracts for open data software providers ESRI and Socrata).

More successful examples go beyond transparency to publicize a controversy. Chicago's Chi Hack Night, for instance, produced a text message alert system that sends a text when wind blows 15 miles per hour or more in Chicago, with the words "Wind Alert! Avoid petcoke exposure by limiting outdoor activity," and a link to learn more. Petcoke, short for "petroleum coke", are air contaminates known to be released by area facilities owned by Koch industries. The alert was part of a wider campaign to tighten government regulations on these facilities. In another example, Chicago's Million

Dollar Blocks project drew from the work done by Laura Kurgan's Spatial Information Lab at Columbia to design a map of the costs of incarceration by zipcode across the city. In these examples, software and data visualizations not only offer a service but also prompt questions and raise awareness about pressing issues of political concern.

In addition to the distinction between civic hack projects that produce information services versus transparency and issue publicizing, projects must be analyzed by how much they engage with the values embedded in the technologies they use. The Petcoke alerts and the Million Dollar Blocks projects, for instance, make their code available under free licenses. Million Dollar Blocks relies on open source software, using Open Street maps data rather than Google Maps. These projects avoid commercial software that engages in data-collection and convert the citizen-user into a consumer.

To conclude this analysis, imagine a quadrant. On one axis you have a scale moving from instrumental/collusive on one side and monitory/agonistic on the other, with transparency projects somewhere in the middle. With the other axis you consider value-awareness, on a scale of no awareness to fully aware, such that all technologies used exhibit some consideration towards their relationship to the user.

Beyond Civic Hacking

Thomas Lodato (2014), another researcher of civic hackathons, writes that he hopes to see "hacked civics", not civic hackathons, because

> hacked civics are beyond user-friendliness, beyond vowel-less ventures, and beyond end-user license agreements. These hacked civics rethink the state; they cobble together various citizenries; they break and reassemble civic life; they don't agree that the answer is technology; and, most of all, they don't agree on civics. (n.p.)

Following Lodato's provocation, perhaps civic hacking as a term has too much baggage, and we need another term to understand how people can use their technical skills to contribute to the political and civic sphere. The Million Dollar Blocks Project, for instance, does not advertise itself a civic hacking project, and it derived primarily from an academic setting. To one of my interviewees at a Los Angeles civic hack night, Million Dollar Projects counts as an important example of civic hacking. Yet perhaps we need new terms for these projects that use data and software towards humanistic, agonistic, and pointedly political—not only instrumental—ends.

Notes

1. http://opengovernmentdata.org/#sthash.YYinvpoQ.dpuf
2. Interview conducted December 8, 2013.
3. B-corporations is a for-profit corporate entity that includes positive impact on society and the environment as one of its defined goals, along with profit.
4. Interview conducted June 22, 2016.
5. Ibid.
6. Heard at #Tech LA, May 31, 2014.
7. Interview conducted February 8, 2016.
8. https://foodoasis.la/
9. Heard at Hack for LA on June 6, 2013.
10. The app is called Promptly: https://github.com/codeforamerica/promptly/wiki/How-to-Promptly.
11. The website is called Golden Ticket. https://github.com/codeforamerica/golden-ticket-console
12. Interview conducted July 9, 2015.
13. Interview conducted July 9, 2015.

References

Berkley, G. E., & Rouse, J. E. (2004). *The craft of public administration.* Boston, MA: McGraw-Hill.

Coleman, E. G. (2013). *Coding freedom: The ethics and aesthetics of hacking.* Princeton, NJ: Princeton University Press.

Denhardt, R., & Denhardt, J. (2000). The new public service: Serving rather than steering. *Public Administration Review, 60*(6), 549–559.

Dewey, J. (2012). *The public and its problems: An essay in political inquiry.* University Park, PA: Penn State University Press.

DiSalvo, C., & Gregg, M. (2013, November 21). The trouble with white hats. *The New Inquiry.* Retrieved July 23, 2015, from http://thenewinquiry.com/essays/the-trouble-with-white-hats/

Goldstein, B., & Dyson, L. (Eds.). (2013). *Beyond transparency: Open data and the future of civic innovation* (1st ed.). San Francisco, CA: Code for America Press.

Goodnow, F. (1900). *Politics and administration.* Berkeley, CA: University of California Libraries.

Gray, J. (2014). *Towards a genealogy of open data* (SSRN Scholarly Paper No. ID 2605828). Rochester, NY: Social Science Research Network. Retrieved from http://papers.ssrn.com/abstract=2605828

Irani, L. (2015). Hackathons and the making of entrepreneurial citizenship. *Science, Technology & Human Values*, 0162243915578486. DOI:10.1177/0162243915578486.

Johnson, P., & Robinson, P. (2014). Civic hackathons: Innovation, procurement, or civic engagement? *Review of Policy Research, 31*(4), 349–357. DOI:10.1111/ropr.12074.

Kamensky, J. (1996). Role of reinventing government movement in federal management reform. *Public Administration Review, 56*(3), 247–256.

Kelty, C. M. (2008). *Two bits: The cultural significance of free software*. Durham, NC: Duke University Press Books.

Kelty, C. M. (2016). Too much democracy in all the wrong places: Towards a grammar of participation. *Current Anthropology, 58*(S15), S77–S90.

King, C. S., & Stivers, C. (1998). *Government is us: Public administration in an antigovernment era*. Thousand Oaks, CA: Sage Publications.

Kweit, M. G., & Kweit, R. W. (1981). *Implementing citizen participation in a bureaucratic society: A contingency approach*. New York, NY: Praeger Publishers.

Levitas, J. (2013). Defining civic hacking. *Medium*. Retrieved August 9, 2016, from https://medium.com/civic-innovation/defining-civic-hacking-16844fc161cd

Lippmann, W. (1922). *Public opinion*. San Diego, CA: Harcourt, Brace.

Lodato, T. J. (2014). Three positions on civic hacking. Retrieved August 9, 2016, from http://thomaslodato.info/thoughts/three-positions-on-civic-hacking/

Lodato, T. J., & DiSalvo, C. (2016). Issue-oriented hackathons as material participation. *New Media & Society, 18*(4), 539–557. DOI:10.1177/1461444816629467.

Marres, N. (2016). *Material participation: Technology, the environment, and everyday publics*. New York: Springer.

Morozov, E. (2013). The Meme Hustler. Tim O'Reilly's crazy talk. *The Baffler*, No. 22.

Mosher, F. C. (2016). Democracy and public service. In J. Dolan & D. H. Rosenbloom (Eds.), *Representative bureaucracy: Classic readings and continuing controversies: Classic readings and continuing controversies*. London: Routledge.

Osborne, D., & Gaebler, T. (1993). *Reinventing government: How the entrepreneurial spirit is transforming the public sector*. New York, NY: Plume.

Pateman, C. (1970). *Participation and democratic theory*. Cambridge: Cambridge University Press.

Rosenbloom, D., Kravchuk, R., & Clerkin, R. (2008). *Public administration: Understanding management, politics, and law in the public sector*. New York, NY: McGraw-Hill Education.

Schrock, A. (n.d.). Hotspot 4 – turf wars: What is a civic hacker? *Civic Paths*. Retrieved July 24, 2015, from http://civicpaths.uscannenberg.org/hotspot-4-turf-wars-what-is-a-civic-hacker/

Schrock, A. R. (2016). Civic hacking as data activism and advocacy: A history from publicity to open government data. *New Media & Society, 18*(4), 581–599. DOI:10.1177/1461444816629469

Seidenfeld, M. (1992). A civic republican justification for the bureaucratic state. *Harvard Law Review, 105*(7), 1511–1576. DOI:10.2307/1341745.

Thompson, D. (1983). *Bureaucracy and democracy. In democratic theory and practice*. Cambridge, MA: Cambridge University Press.

Wilson, W. (1887). The study of administration. *Political Science Quarterly, 2*, 197–222, esp. p. 214.

8. *Why Locality and Presence (Still) Matter for Political Activism*

Sebastian Kubitschko
Aperto

Introduction

Ever since the first invention of digital machinery, and even more so since the mass distribution of computing in the early 1980s, practices related to media technologies and infrastructures (MTI) become an ever more consistent part of political arrangements. Hackers have accompanied, influenced, co-determined and critically reflected on the role technological "innovations" play for social, cultural, economic and political constellations from its early days. Thanks to remarkable research on the cultural significance of free and open software (Kelty, 2008) as well as political protest and disruption (Coleman, 2014) it is understood that hackers not only rely on digital media but also act on the very politics of technology. What we do know much less about is the relevance that hackers attribute to physical locations and presence. Contemporary research on hackers, hacking and hacktivism does point to the overall significance of shared locality and of face-to-face interactions. Yet, due to the specific aim and focus of many writings, they understandably do not narrow in on this question (see Schrock, 2014). There is need, in other words, to ask: does locality and presence (still) matter for hackers in particular and political activism more generally? And, if so, why and how? In this chapter I want to make a small step forward to deepen understandings of hacker cultures by zeroing in on hackers' (continuing) appreciation of and reliance on locality and presence to bring their political projects into being. By doing so this chapter shifts the focus of hacker studies away from loosely networked activism towards more organization-based engagement. This, I want to argue, is of particular relevance as putting our attention solely on distributed networks of political action might not only miss critical facets of hacker cultures but of contemporary engagement in general. Considering the vast means of networked communication today, one could, theoretically, sit in a dark cellar in the middle of nowhere and still be part of a collective engaging with politics. Yet, as will be shown in the coming sections, robust and meaningful political activism in many cases still very much relates to presence and is grounded in concrete locales.

The continuing importance of locality and presence might at first hand sound surprising in the context of hackers who people tend think of as "the quintessential digital subjects" whose agency "unfolds in the ethereal space of bits and bytes" (Coleman, 2010, pp. 47, 48). Yet, as will be detailed in the following, locality and presence still matter above all for two interrelated reasons: the social dimension and the political dimension of hacking. The case study at hand on which this line of reasoning rests is the Chaos Computer Club (CCC). Officially founded in 1984 in Germany the CCC has seen a constant rise of members that today figure around 5500, which makes it Europe's largest and one of the world's oldest hacker organizations. Founded at a time when the communication landscape was changing drastically—home computing was thriving, online worlds emerging—the Club has since then not only accompanied but also critically questioned the political consequences of emerging media technologies and infrastructures. Within its more than thirty year long history, the CCC has developed from a more or less loose association of individual "cyber anarchists" into a civil society organization that is recognized as one of the most influential digital rights organizations in Germany. For organizations like the CCC who practice a constructive vision of politics that brings together direct and indirect modes of collective action, locality and presence are relevant building blocks for their political endeavors. Instead of presenting an in-depth analysis of a few selected examples, the aim here is to give insights to the wide spectrum of concrete instances where locality and presence matter for the hacker organization.

The data set presented in this chapter is composed of 40 face-to-face interviews (with new and old members, co-founders and spokespersons of the Club) and participant observation (at hackerspaces, during public gatherings and hacker conventions across Germany, meetings with journalists and private get-togethers) over a three-year period (2011–2014). In addition, the data set includes a media analysis (media coverage, different forms and styles of media access as well as the hackers' own media outputs) dating back to the early 1980s.

Underworld, Government, and Something Civic in Between

Along with the multiplication and diversification of hacking over the past decades, the discourse around hackers has seen constant metamorphosis—depicted as nerds in the early days, heroes of the computer age in the 1970s and 1980s, and as people associated with computer intrusion or as actors involved in cybercrime and cyberwarfare in the beginning of the 21st century (Nissenbaum, 2004). It is from this point of view that networked collectives like *Anonymous* and other politically-driven initiatives are often demonized as criminals without further questioning their motivations or investigating their reasoning and political aims (Coleman, 2014). It is without doubt the case that a vast number of individuals, organized syndicates and terrorist networks with relevant computing competences are involved in illegal activities that often take place in the so-called dark or deep web—a world that remains untouched by popular online platforms and search engines. Here it is interestingly to note that the roots of "cybercrime" are often cultural rather than scientific, and frame both the way people view online deviance as well as legal and policy responses to cybercrime (see Wall, 2008). Either way, what the focus on cybercrime tends to leave aside is that a vast number of hackers work for government agencies and have become an essential part of the global labor market as their positions range from system administration to security research, from information architecture to programmers and developers.

Seen from this perspective one is tempted to picture the current scenario as a world of two extremes: on the one hand there is the underworld and on the other hand there is the corporate and governmental domain. This, however, is still not the whole picture. There is a third domain—if that is what we want to call it—that plays a fundamental role for contemporary social transformations and political constellations. As could be witnessed in recent years, hackers increasingly play a vital role for the foundation and functioning of social movements and progressive political action (Postill, 2014). Explicitly acknowledging the relevance of such research, the ambition of this chapter is to contribute to ongoing discussions on the role hacker organizations play for democratic arrangements at large. Accordingly, this chapter—similar to the other chapters in this volume—aims to tell a story that avoids stereotyping and deepens understandings of how exactly hackers (and makers) contribute to the formation of society.

Let me start my line of reasoning by making a general statement that hacking appears to be a location-specific phenomenon. This is not least the case, as Edward Snowden's exposures have shown strikingly, because legal and cultural conditions can vary drastically from state to state, sometimes from city to city or even from building to building, as Julian Assange's refuge in the Ecuadorian embassy demonstrates. That is to say, despite the communication and information networks that span around the globe there are strong national, regional and local differences when it comes to hacking. The ongoing commercialization and diffusion of MTI might in fact perpetuate this diversification as they are "bent in specific ways according to local power dynamics, levels of expertise, cultural negotiations, and social interactions" (Rodríguez, Ferron, & Shamas, 2014, pp. 153–154). Here we only have to think of the triumphal march of locative media and geo-coded data; which, admittedly, is seldom contextualized in political terms (see De Souza e Silva & Frith, 2012). While the hacker scene was more homogenous in its early days over the past 60 years, hackers have turned into businessmen (and, to a lesser degree, businesswomen). They have turned into greatly feared criminals, and hackers were and continue to be heavily involved in major technological inventions. As a consequence, today it seems appropriate to talk of hacker cultures in the plural and to avoid generalizations about "the hacker".

The case of the CCC underlines the notion that there is no such thing as the typical hacker or the typical hacker collective. The Club's organizational structure is best described as polycentric, consisting of multiple, interconnected nodes across Germany, Austria and Switzerland, and several less formal affiliated national and international hacker spaces. This decentralized formation does not rule out certain formal structures. Represented by an executive board and exclusively sustained by voluntary work, membership fees and donations the Club acts as a civil society organization that is registered as both a not-for-profit and a lobby organization. In contrast to many other hacker collectives the CCC decided from early on to take a political route that would take them through institutionalized centers of power. As a consequence, Club members consider interactions with politicians, legislators, judges and media representatives as one set of needful solutions to problematize and thematize the politics of technology.

It is understood that locality was critical in times before the digitalization and pervasiveness of media technologies and infrastructures. A telling case here is, for example, the relevance of community print shops for political activism in London in the 1970s (Baines, 2016/in print). Along with the quasi-omnipresence of digital media "locality seems to have lost its ontological moorings" in an apparently "dramatically delocalized world" (Appadurai, 1995, p. 208). This idea, however, only holds partially true. To start with, from a more materialist perspective one can say that along with the pervasiveness of what has been referred to as "algorithmic culture" (Striphas, 2015) locality very much matters when it comes to the positioning of relevant media infrastructures. This has predominantly three reasons: stability, speed and resources. First, data centers, for example, have to be situated at stable locations—stability here includes both geographical and socio-political dimensions. Second, to stay with the example of data centers, media infrastructures tend to be resource-intensive, which makes it much more cost-efficient to position them either at cooler locations in Nordic regions or embed them in existing (urban) infrastructures like heating systems (see Velkova, 2016/forthcoming). Finally, with the spread of high-speed processes relying on "real time" data transmission (e.g. autonomous driving, surgical robots, financial transactions, etc.) there is need for locally proximate servers.

From a socio-political point of view, the first assumption one might derive from the quasi-omnipresence of media technologies and infrastructures is that the lives of people who are extensively engaging with media must be dominated by indirect social relationships. Interestingly enough, Howard Rheingold wrote that even the prototype "virtual community" the WELL was grounded in the everyday physical world as "WELLites who don't live within driving distance of the San Francisco Bay are constrained in their ability to participate in the local networks of face-to-face acquaintances" (Rheingold, 1993, p. 2). In his article on communications technology and the transformation of the urban public sphere Craig Calhoun (1998) engages in a debate that seems obsolete in some way—namely, the explicit distinction between "online" and "offline" communities. What remains highly relevant, however, is the larger reference he makes towards the relevance of local confrontation throughout history and the ongoing relevance of local solidarity for dealing effectively with political power structures. Locality, as it is understood in this chapter, is primarily relational, contextual and historically grounded rather than scalar or spatial (Appadurai, 1995, pp. 208, 212); which takes into account the term's etymology in *localité* (French) and *localitas* (late Latin) from *localis* "relating to a place".

With this in mind, it becomes clear that locality in many ways relates to the notion of presence. One can trace the study of how social behavior is related to the physical presence of others back to Goffmann (1967), who builds his understanding on the former's understanding of co-presence. This chapter in many ways resonates with these classical approaches as it conflates presence with sharing the same physical location. That said, this line of reasoning is not based on a normative distinction between technologically mediated and direct face-to-face communication. I would not deny that one could also theorize presence in the context of mediated environments, as recent writings have in fact done (Campos-Castillo & Hitlin, 2013). There are indeed situations where mediated co-presence creates intense awareness of distant others

(Madianou, 2016). Anthony Giddens rightfully pointed out that, "a person may be on the telephone to someone twelve thousand miles away and for the duration of the conversation be more closely bound up with the responses of that distant individual than with others sitting in the same room" (Giddens, 1991, p. 189). Perhaps even more eloquently David Harvey stressed that, "in modern mass urban society, the multiple mediated relations that constitute that society across time and space are just as important and as 'authentic' as unmediated face-to-face relations" (Harvey, 1993, p. 55). These are important views to take into account because they avoid idealizing presence as the one and only way to establish personalized connectivity, emotionality and solidarity (see also Miller, 2016). At the same time, overstressing the affordances of ubiquitous media environments might entail the danger of underestimating (or even ignoring) the ongoing significance of sharing the same physical location at the same time when it comes to political activism.

In the context of the CCC, I am not particularly interested in presence at the micro-level and its consequence for macrostructures, mainly because I consider drawing these kinds of causal relationships as problematic. My focus is on the role presence plays for the CCC as a politically engaged civil society organization. This, of course, relates back to larger social constellation, but it does so in rather irregular and multifaceted ways. While co-presence takes into account the psychological and emotional side of interaction (Collins, 2004), there is also a more rationalized and more banal dimension that I want to highlight. There are moments and occasions—particularly in formalized and institutionalized settings—where you simply have to be physically present to be part of a specific situation and the decisions that result from it.

The Social Dimension of Locality

To start with it might be necessary to state the obvious. Hackers are "social animals" just like all of us. Along with the human need for sociality evolves the need for creating a common locale where one can talk, learn from one another, drink mate lemonade, build stuff, and be amongst people on the same page. This, in fact, is how the hacker organization got going in the first place: The Club has its starts as a group of people sitting together at a round table in Berlin in 1981, discussing the importance digital technology would have in the near future for the construction of social realities at large and political arrangements in particular. Out of this meeting emerged what would some years later be named the Chaos Computer Club. In parallel with the spread of digitally networked communication technologies in the early 1980s (e.g. Bulletin Board Systems) more and more local and regional CCC meet ups—so called Erfa-Kreise (short for information exchange circles)—started to arise in German speaking countries. They were not yet referred to as hackerspaces but, after all, that's what they were.

Beyond extensively using and creating early collaborative media, previously rather isolated individuals interested in technology would now meet in person with more or less like-minded people, establishing a sense of community. As Coleman put it, "the advent of networked hacking should not be thought of as a displacement or replacement of physical interaction" (Coleman, 2010, p. 49). Put in more concrete terms, the CCC has its organizational roots in bringing together mediated communication and local face-to-face meetings. One can interpret this dynamic as an early example of bringing together logics of networking and aggregating (see Juris, 2012). The ability to merge offline and online communication showed the initial Club activists that new modes of sociality and collective action were possible that would combine existing and emerging modes of engagement. The CCC was (and still is) at the forefront of building, using and sustaining decentralized communication networks that heavily rely on MTI. At the same time, Club members felt (and continue to feel) a strong need for meeting up face-to-face, to establish physical meeting points, to create a common space for exchanging information and experiences. In 2015 the Club had 26 localized "information exchange circles" across German speaking countries.

The relevance of locality also became recognizable during the mid and late 1980s when the CCC was dominated by what one might refer to as technological escapades. Following the initially prosperous emergence and public reception of the Club the second half of the 1980s saw a loss of collectivity and a fragmentation of the collective's constructive political ambitions. This, to a large degree, was due to individual members' involvement in activities that ranged from dubious to straightforward illegal. Amongst the list of operations that were attributed to CCC-affiliated hackers the NASA and KGB hacks were the most sensational ones. The hacks were followed by extensive investigations by police units from several European countries. Following the immense public pressure, the Club was dominated by internal disputes, distrust and accusations. Leading members and some of the co-founders took a backseat or even discontinued their affiliation entirely. The Club's destabilization resulted in a serious de-legitimation of the collective and, ultimately, almost led to its disbanding. The potential collapse was eluded by the commitment of a collective of old and new members that moved the Club's epicenter from Hamburg to Berlin, where the fall of the Berlin Wall had opened up new possibilities to bring people together in a largely unpredetermined urban environment. Following the formation of a new core group the

CCC recovered from its turbulent times and saw a reformation that led to its remarkable revival from the mid 1990s onwards.

Over recent years, the ongoing or even growing importance of establishing localities for hacker cultures can be witnessed in the global spread of hackerspaces (Maxigas, 2012). Broadly speaking, hackerspaces are physical infrastructures for exchange, collaboration and creating where people meet to build and do stuff together. Hackerspaces that commonly include workbenches, desks and fabrication space, tools like 3D workstation and laser cutter have become a global phenomenon over the past years. They tend to be an urban phenomenon; which does not mean that all of the hundreds of hackerspaces are located in cities but the majority is located in metropolitan areas. This is far from saying that all hackerspaces are the same. As Lindtner and Li highlighted in their research on XinCheJian, the first hacker space in China that opened its doors in Shanghai in 2010, "technologies and values are sites of negotiation, remaking, and constant appropriation as they are translated into particular local settings" (Lindtner & Li, 2012, p. 19). CCC members highly value their local hackerspaces and often refer to them as their "living room" where they meet other active members face-to-face, where they exchange news, sit down for hour-long conversations after work, where they can experiment with new tools. In that sense hackerspaces provide its members a base for engagement with broader political discourses. In addition, there was also a more practical part to the story: to tinker with technology often requires a whole lot of equipment and fabrication space that not everyone is able to afford or accommodate. The activities that take place in CCC-affiliated hackerspaces include non-commercial tech-collaborations, artistic projects, and workshops on cryptography.

Historically, the CCC has been heavily involved in establishing and spreading the idea of localities that are commonly labeled hackerspaces today. Many of the early hackerspaces were more or less directly inspired and influenced by the c-base in Berlin, which emerged out of the CCC milieu in 1995 as one of the first independent, stand-alone hacker spaces in the world. In 1998 another major CCC-affiliated hackerspace called C4 opened its doors in Cologne in. More recently, the "Hackerspace Design Patterns", written by long-term CCC-members Jens Ohlig and Lars Weiler, received large attention and acknowledgment throughout hacker communities—in particular by US-based hackers. The way ideas traveled and knowledge was exchanged in this case further underlines the relevance of meeting people face-to-face for bringing projects into being. A group of hackers chartered a private plane from the 2007 Defcon hacker convention in Las Vegas that flew them directly to Germany. The "Hackers on a Plane" trip brought the visitors to the CCC Camp and included a tour of hackerspaces in Germany and Austria where they would get a first-hand experience of how to set up and maintain a hackerspace.

The reference to Defcon suggests that, besides long-standing local information exchange circles as day-to-day meeting points, large-scale events that take place at regular intervals are also critical for hackers' sociality. "More than ever, hackers participate in and rely on a physical space common to many types of social groups (such as academics, professional groups, hobbyists, activists, and consumer groups): the conference" (Coleman, 2010, p. 49). Since 1984, the Club hosts the annual Chaos Communication Congress. What started as a rather small get-together in Hamburg gathered more than 10,000 participants from around the globe in 2015 and is considered the world's longest-running annual hacker conference today. Gatherings, like the annual Congress, "take on an extraordinary and, for some hackers, a deeply meaningful aspect to their lifeworld" as they allow hackers "to collectively enact, make visible, and subsequently celebrate many elements of their quotidian technological lifeworld" (Coleman, 2010, p. 50). The Club's grand annual convention is complemented by the Chaos Communication Camp that takes place very four years and various smaller events organized by local CCC meet ups across Germany.

These events are locales that form parts of the organization's "interaction ritual" (Collins, 2004). For new members they are excellent occasions to get to know other associates. For longer-term Club members these events are a great opportunity to meet old and new acquaintances face-to-face at least once a year. Gatherings where people come together "in the physical world in the form of real people, with faces, bodies, and voices" (Rheingold, 1993, p. 3) are highly significant for the hacker organization as they initiate and deepen social ties, which, ultimately, are the base for collaboratively working on concrete projects. In addition, the conventions are also critical as they gain vast media attention and coverage; not necessarily exponentially, but along with the growing number of visitors, the media interest increased. It is the locale where hackers become visible for larger publics outside the context of hacks. As a consequence, the events themselves, as well as the associated media coverage, regularly attract people to join the hacker organization. Taking into account that the Club is exclusively funded through donations and membership fees, the physical get-togethers are not only highly relevant for members to strengthen solidarity and to exchange information but they are also critical for the continuing existence of the Club per se.

Despite the ability to build a sense of community since at least the mid-1980s when mediated communication became part of the daily communication routines of CCC members the relevance of nurturing sites where "locality is materially produced" (Appadurai, 1995, p. 209) has never been replaced. As the increase of local meet ups over the past years and the ongoing growth of the Club's events demonstrate, "face-to-face interactions work in concert with digital interactivity to constitute social worlds" (Coleman, 2010, p. 47). Establishing and fostering stable localities for embodied interaction—ranging from joint hackerspaces to periodical conventions—enable members of the hacker organization to create "dense, multiplex, and systematic networks of relationships" (Calhoun, 1998, p. 373). Here, we need to take into account that different localities offer different social and political potentialities. As a methodological side note, this is also one reason why qualitative research in general and ethnographic accounts in particular continue to be of great importance for our understanding of the world (see, for example, chapters by Kumar, LaDue and Sartoretto in Kubitschko & Kaun, 2016/in print). In addition to creating "social worlds" the above-mentioned localities are central for active Club members to piece together relevant information, knowledge and experiences as well as to share skills that build the fundament for coordinating as well as executing political projects.

The Political Dimension of Presence

A considerable part of the interactions between civil society organizations and political authorities take place through mediated encounters—either directly via email, instant messaging and telephony or indirectly via popular online platforms and mainstream media (Fenton, 2010; Tufte, 2014). At the same time, contentious action like protest and demonstrations continue to strongly rely on localized initiatives and activities—the Occupy movement as well as many less-covered initiatives were based on attending local gatherings, assemblies and occupying strategically chosen locations with large symbolic capital (see Kavada, 2015). Despite the richness of mediated communication, as my research of the CCC shows, physical proximity to the centers of power continues to matter. This in fact might not be particular to the hacker organization or any other civil society organization but a more general aspect of actors who aim to gain influence and wish to co-determine political arrangements.

Activists' engagement often remains grounded in very specific issues and is focused on their localities even in cases where they operate as part of larger networks (Sassen, 2004). In the context of low-power radio activism, for example, it has been shown that locality (or localism) can serve as an anchor for policy goals (Dunbar-Hester, 2013). In the context of the CCC the case of the Transparency Law in Hamburg is a telling example at hand. In 2011 the Club, together with Mehr Demokratie ("More Democracy") and Transparency International, filed a people's initiative for a new transparency law in Hamburg. As a consequence, Hamburg was the first federal state in Germany to introduce a law that gave citizens the right to information from their government and administration free of charge and anonymously—including senate resolutions, building permits, contracts concerning public services and expert appraisal. The successful transparency initiative demonstrates that for political action to be executed effectually there is need for actors who are accustomed to and hold relevant knowledge of the locality; which includes, legal issues, local contacts, access to resources, and so on.

Much of decision-making continues to be played out in face-to-face interactions and is based on personal relationships that require co-presence. Take the case of lobbying in the US. It is no coincidence that all of the large lobby and pressure groups have offices in Washington, DC. In fact, large tech-corporations like Google regularly increase their workforce and expand their offices in the US capital (see Hamburger & Gold, 2014). Put bluntly, lobbying requires presence. Going to a hearing requires presence. Attending a court case requires presence. Meeting politicians for lunch requires presence. One simply can't do these things online as a decentralized, computer-mediated network. When you act as an expert for the constitutional court—as the Club did on five occasions—or you contribute your knowledge and arguments to politicians and legislators—as the CCC regularly does in committees and backdoor meetings—most of the time you do so in person.

When it comes to hacker organizations the appreciation of the political importance of presence might be less obvious, but it is not less important for those hackers who aim to co-determine the political landscape by interacting with institutionalized politics and relevant media representatives. Policymakers, political institutions and journalists process the flood of information disproportionately as they allocate attention to some problems rather than others (Jones & Baumgartner, 2005; Powers, 2015). Being able to meet journalists for a personal conversation, being located in walking distance to the offices of legislators, being based in a city where international meetings and workshops take place regularly, residing in an environment where other relevant actors and possible collaborators are present does not guarantee political efficacy. Yet, for the hacker organization, these actions are fundamental for facilitating existing and new embodied interactions with "traditional" centers of power. This, again, is far from saying that mediated communication does not matter, but to stress that bringing together mediated and face-to-face communication plays an imminent part for the Club's political work. In this regard, it is relevant to note that one kind of co-presence is not like another. It can be public, inclusive and on large scale (e.g. Congress and Camp). It can just as well be more private, exclusive and on a smaller scale (e.g. task groups and hackerspace). Similarly, it does, of course, make a difference whether Club members meet amongst each other or with “outsiders”.

For civil society organizations like the CCC who rely on voluntary work and donations, locality and presence also matter because they do not have unlimited resources. Being physically close to relevant actors means that one can interact with them in resource-efficient ways. Also, in an apparently fast-paced world politicians and journalist are often looking for an informed voice in a promptly manner. Physical proximity therefore also means to be more or less ready for delivery on demand, being able for spontaneous meetings, gatherings and interactions. The spectrum of face-to-face interactions ranged from invite-only soirées at politicians’ private residencies to public panel discussions, from informal conversations with journalists during lunch to speaking at the Bundestag’s committee of inquiry related to the National Security Agency (NSA) scandal, from sitting in political talk shows to meeting legislators behind closed doors. Some of these meetings might be regularly, others only sporadically or even onetime. These are not necessarily occasions where one executes political influence in a straightforward way. But being in the position to meet relevant actors face-to-face makes possible, or at least simplifies, longer-term relationships and to build up trust. When it comes to the relationship with media outlets, the majority of press inquiries—several thousands every year—unavoidably remain “indirect relations mediated by technology” (Calhoun, 1998, p. 379). All the same, numerous interactions with journalists are embodied and “directly interpersonal relations” (Calhoun, 1998, p. 379). Some of these interactions even lead to official employment relationships.

A good way to concretize the relevance of locality and presence to execute political work is the so-called Staatstrojaner (“Federal Trojan”) hack. In 2011 the CCC got hold of a hard drive apparently containing governmental surveillance software. Considering the fact that the members who were best qualified to solve the issue were spread around Germany, the first step was to form a core group of people who would solve both the technical challenges and the publicity issues. This first organizational step largely relied on mediated communication. Yet, instead of solving the issue solely via online chats or digital collaborative tools the group decided to also meet up in person to sit down, discuss the issue at hand, to think through the practical procedure of solving the technical facets and planning the public campaign strategy. The case of the Staatstrojaner hack is also telling from a legal perspective. Hamburg is one of the states in Germany that does not prohibit reverse engineering of software. Accordingly, to actually prove that the surveillance software violated the terms set by the German constitutional court parts of the technical tinkering had to be done in Hamburg. Finally, before making the findings public, the CCC sent a personal messenger (German politician and civil liberties advocate Burkhard Hirsch) to the Department of the Interior to give them lead-time to withdraw the software from any current investigations. Following the week-long planning and technical executing the hack was publicized and initiated a heated political debate about governmental surveillance tactics in Germany—two years before the issue of surveillance gained global currency owing to Snowden's revelations.

Organizational Capacity

It is understood that the world we are living in is getting more and more complex. This is true for many aspects of life, but in particular for technological "innovations". Take, for example, the development from the first (mechanical) typewriter to the first electronic keyboard to touch-sensitive displays. Deconstructing and explaining the functioning of a typewriter might be a manageable task. Doing the same with a touch-sensitive displays is out of the question for any layperson. Along with the pervasive computerization of societies the possibilities to deconstruct and explain the functioning and effects of technological artifacts has diminished. Consequently, the ability to uncover and understand the potential political consequences particular technical developments might entail decreases. This also means that the ability to engage politically with particular technological developments becomes rather difficult, if not even impossible to some degree. At the same time, it is ever more apparent that almost any form of political engagement in form of protest, demonstration, boycott, direct action, and so on in one way or another relies on media technologies and infrastructures for the sake of publicizing, mobilizing and organizing. Following this paradox one central question that comes to mind is: who is in the position to act on the politics of MTI? Hackers like the CCC are best considered actors who aim and are able to make the complex phenomena that are part of people's everyday experience intelligible and comprehensible. In times when technology is increasingly the subject of policy-making and jurisdiction, this is a highly political, perhaps even meta-political endeavor.

The aim of this chapter so far has not been to argue that establishing common locations and spending time in the physical presence of others is sufficient for bringing about political change. Rather, to deepen understandings of hackers' contemporary political engagement, we need to take into account their need to build adequate locales and to assemble outside of digitally mediated environments. This is particularly the case because despite the fact that there is "a multiplicity of ways in which conversational practices interface with technological devices" (Hutchby, 2001, p. 1), situated interactions remain vital sources for establishing and cultivating collectivity as well as for enabling collective action. At the same time it has been shown that hackers' engagement is not always related to "global" issues but often materializes in very concrete locatable contexts. This not only demonstrates how locality and presence overlap when it comes to contemporary activism but also that in practice it would artificial to detach social from political dimensions—as has been underlined, for example, in the case of the Staatstrojaner hack.

Bringing the above, one can note two main facets: On the one hand, the Club's robustness is rooted in the mutual reinforcement of networked communication, physical interaction and adequate localities since its early days. The fact that many people know each other for some time, and they do so not only from "virtual" encounters but from numerous regular meetings and conversations that took place in co-presence stabilizes the CCC as a hacker organization. On the other hand, most of the Club's political activism is based on collaborative work amongst committed members who bring together relevant skills, information, knowledge and experiences. For the Club socializing face-to-face in concrete locations is an absolutely essential part for making (new) things happen. Activities that involve face-to-face engagement on site not only deepen the experience of the individual members but also contribute to the organization's overall ability to bring its political aims into being. All in all, locality and presence enhance the Club's organizational capacity, which in turn enables the CCC to establish, execute and sustain meaningful political action over time. From this one can conclude that the standing that allows the hacker organization to embed its voice in public and political discourse to a large degree relies on physical proximity, direct exchange and face-to-face meetings.

As should be clear by now, it appears to be impossible to pigeonhole the Club as an anti-state or collaborationist collective. The hackers take a middle-way that aims to bring into being a constructive, yet critical, mode of engagement. While specific forms of hacking like digital piracy can be considered a direct challenge to the authority of the state (Beyer, 2014), hacking in the case of the CCC is more of a critique, observation and confrontation of particular forms of state authority. CCC's activism has strong confrontational elements (e.g. hacking governmental surveillance software or suing the German government for their involvement in NSA surveillance practices), but it has also collaborative elements (e.g. partaking in government initiatives like the German Bundestag's "Internet and Society" committee) and the belief that to execute power and influence actors need to go through institutions (at least to some degree). The CCC is a civil society organization that brings together multi-socialized and multi-determined individuals that have considerable amount of expertise related to media technologies and infrastructures. Hacking in this context refers to the critical, creative, reflective and subversive engagement of technology that allows creating new meanings. It is due to this form of engagement that locality and presence are of such high relevance.

If you want to act as a registered not-for profit association that collaborates with existing institutions like governmental agencies and media outlets, you need an address for other to reach you. For legal and for practical reasons, that address cannot be an e-mail address. It needs to be a concrete building with an actual mailbox and a door you can knock at and walk into. Ideally, that door is not in the middle of nowhere but somewhere close to people who are in the position to make important decisions. Despite the media manifold that has become a routine part of actors' daily lives, physical proximity, approachability and sociality matter when it comes to politics. This is also the case when it comes to acting on the politics of media technologies and infrastructures. In the case of the CCC this is particularly important, as they tend to engage with issues on the "national", "regional" and "local" level. In other words, the Club's engagement is concerned with subjects that are global phenomena (e.g. privacy, surveillance, information freedom), but their concrete struggle is inevitably contextualized in concrete settings—what has fittingly been referred to as "grounded, practiced connectedness" (Massey, 2005, pp. 187–188). To put it in more concrete terms, locality and presence are also critical components for the CCC's endeavors because their struggles and agency often occur in particular, localized environments. Its major incentive is to co-determine the very (social and political) environment they are part of; which does not exclude other forms of activities, like financially supporting Snowden's lawyer team or helping to sustain decentralized infrastructures like The Onion Router (Tor). The Chaos Computer Club has managed to stabilize its role as a prominent actor in the constitution of political arrangements in Germany in general and the advocacy of digital rights in particular by continuously grounding its political activism in the realm of physical locations and face-to-face encounters.

References

Appadurai, A. (1995). The production of locality. In R. Fardon (Ed.), *Counterworks: Managing the diversity of knowledge* (pp. 208–229). London: Routledge.

Baines, J. (2016/in print). Engaging (past) participants: The case of radicalprintshops.org. In S. Kubitschko & A. Kaun (Eds.), *Innovative methods in media and communication research*. London: Palgrave Macmillan.

Beyer, J. (2014). *Expect us: Online communities and political mobilization.* Oxford: Oxford University Press.
Calhoun, C. (1998). Community without propinquity revisited: Communications technology and the transformation of the urban public sphere. *Sociological Inquiry, 68*(3), 373–397.
Campos-Castillo, C., & Hitlin, S. (2013). Copresence: Revisiting a building block for social interaction theories. *Sociological Theory, 31*(2), 168–192.
Coleman, G. (2010). The hacker conference: A ritual condensation and celebration of a lifeworld. *Anthropological Quarterly, 83*(1), 47–72.
Coleman, G. (2014). *Hacker, hoaxer, whistleblower, spy: The many faces of anonymous.* London: Verso.
Collins, R. (2004). *Interaction ritual chains.* Princeton, NJ: Princeton University Press.
De Souza e Silva, A., & Frith, J. (2012). *Mobile interfaces in public spaces: Locational privacy, control, and urban sociality.* New York, NY: Routledge.
Dunbar-Hester, C. (2013). What's local? Localism as a discursive boundary object in low-power radio policymaking. *Communication, Culture & Critique, 6*(4), 502–524.
Fenton, N. (2010). NGOs, new media and the mainstream news. In *New media, old news* (pp. 153–168). London: Sage.
Giddens, A. (1991). *Modernity and self-identity: Self and society in the late modern age.* Cambridge: Polity.
Goffman, E. (1967). *Interaction Ritual – Essays on Face-to-Face Behavior.* Garden City, NJ: Doubleday.
Hamburger, T., & Gold, M. (2014). Google, once disdainful of lobbying, now a master of Washington influence. *The Washington Post,* 12 April.
Harvey, D. (1993). Class relations, social justice and the politics of difference. In M. Keith & S. Pile (Eds.), *Place and the politics of identity* (pp. 41–65). London: Routledge.
Hutchby, I. (2001). *Conversation and technology: From the telephone to the internet.* Cambridge: Polity.
Jones, B., & Baumgartner, F. (2005). *The politics of attention: How government prioritizes problems.* Chicago: University of Chicago Press.
Juris, J. (2012). Reflections on #occupy everywhere: Social media, public space, and emerging logics of aggregation. *American Ethnologist, 39*(2), 259–279.
Kavada, A. (2015). Creating the collective: social media, the occupy movement and its constitution as a collective actor. *Information, Communication & Society, 18*(8), 872–886.
Kelty, C. (2008). *Two bits: The cultural significance of free software.* Durham, NC: Duke University Press.
Kubitschko, S., & Kaun, A. (2016/in print). *Innovative methods in media and communication research.* London: Palgrave Macmillan.
Lindtner, S., & Li, D. (2012). Created in China: The makings of China's hackerspace community. *Interactions, 19*(6), 18–22.

Madianou, M. (2016). Ambient co-presence: Transnational family practices in polymedia environments. *Global Networks, 16*(2), 183–201.
Massey, D. (2005). *For space.* London: Sage.
Maxigas. (2012). Hacklabs and hackerspaces—tracing two genealogies. *Journal of Peer Production, 2.* Retrieved May 10, 2016, from http://peerproduction.net/issues/issue-2/peer-reviewed-papers/hacklabs-and-hackerspaces/
Miller, V. (2016). *The crisis of presence in contemporary culture: Ethics, privacy and speech in mediated social life.* London: Sage.
Nissenbaum, H. (2004). Hackers and the contested ontology of cyberspace. *New Media & Society, 6*(2), 195–217.
Postill, J. (2014). Freedom technologists and the new protest movements: A theory of protest formulas. *Convergence, 20*(4), 402–418.
Powers, M. (2015). Contemporary NGO–journalist relations: Reviewing and evaluating an emergent area of research. *Sociology Compass, 9*(6), 427–437.
Rheingold, H. (1993). *The Virtual Community.* Reading, MA: Addison Wesley.
Rodríguez, C., Ferron, B., & Shamas, K. (2014). Four challenges in the field of alternative, radical and citizens' media research. *Media, Culture & Society, 36*(2), 150–166.
Sassen, S. (2004). Local actors in global politics. *Current Sociology, 52*(4), 649–670.
Schrock, A. (2014). "Education in disguise": Culture of a hacker and maker space. *Interactions, 10*(1). Retrieved from https://escholarship.org/uc/item/0js1n1qg
Striphas, T. (2015). Algorithmic culture. *European Journal of Cultural Studies, 18*(4–5), 395–412.
Tufte, T. (2014). Civil society sphericules: Emerging communication platforms for civic engagement in Tanzania. *Ethnography, 15*(1), 32–50.
Velkova, J. (2016). Data that warms: Waste heat, infrastructural convergence and the computation traffic commodity. *Big Data & Society, 3*(2).
Wall, D. (2008). Cybercrime and the culture of fear: Social science fiction(s) and the production of knowledge about cybercrime. *Information, Communications and Society, 11*(6), 861–884.

Section III. Organizing Introduction

JEREMY HUNSINGER
Wilfrid Laurier University

This section of the book deals with organizing. To speak of organizing is to speak of a will to arrange things, usually with an eye to future events and collaborations. Organizing is a political act in that it arranges the contemporary or past in in order to provide for the future. It is an act of not only building the future, but planning for it; which engages with the nature of hacking and making directly. As much as these are political considerations, they are also ethical considerations. Organizing is not merely done to provide goods and order the world for people, but also with an eye to what is best for those people and what will enable those people to flourish. Organizing hacking and making is an ethical project, much like the hacking and making are inherently ethical, even if occasionally the ethics aren't universally good. The papers in this section each deal with organizing by directly engaging in ethical or political questions of organizing or illustrating elements of those processes.

Historical contingencies are important to organizing, as they provide the basis for decisions. Von Lunen explores historical contingences to situate hacking and making as bricolage and tinkering. Starting with terms bastein, bricolage, and tinkering from the 19th and 20th centuries. Von Lunen first consider the cultural and linguistic differences that found in those words and their uses in particular reference to radios and the activities that hobbyists perform around those technics. Later he builds on that analysis with an analysis of historically contextualized activity in aviation. His analysis also relates basteln, bricolage, and tinkering to their relationship with economic circumstance and gender relations. By considering the historical context, Von Lunen provides us several considerations for organizing the modes of hacking and making.

Jennifer H. Maher's work engages feminisms and free/libre/open source software (F/LOSS). This mode of software development and distribution is central to various traditions in hacking and making. Her work specifically uses the notion of the "poison gift" giving to trouble easy assumptions of universal good. She establishes a reflexive understanding of F/LOSS in context in the start of her essay by asserting that the gift of F/LOSS is both open and closed, both free and unfree. Drawing from the women's Linux user group *LinuxChix* and contemporary analyses of women's writing and women's labor, Maher argues that the masculine economy of F/LOSS enables and perhaps constructs the poison gift in a reputational economy. The poison gift is a choice made in organizing, and one to be recognized and avoided. Her problematization of hacker culture demonstrates why reflexivity needs to be built into organizational processes.

Alison Vogelaar makes a case for Occupy Wall Street to be considered part of the maker movement. She starts her analysis by drawing on amateur/expert practices and knowledges of organizing, with the understanding that the maker movement is a resurgent phenomena. She then establishes similarities between Occupy and the maker movement to highlight the possibilities for democratic action. She centers these questions around organizing by emphasizing how Occupy changed in relation to the development of knowledge. Her work is particularly interesting because it infuses questions of organizing into discussions of hacking and marking from an associated movement.

Light provides the last chapter in the section. Her work engages Spivak's concept of strategic essentialism in the analysis of divergent values in the self-organizing groups of the maker sphere. Values are central to organizing because we derive the evaluation of the ends from those values. As such, Light's analysis of the discourses and practices of design activities emphasizes important ways to understand the phenomena. Her research uses several studies to illustrate the challenges of negotiating values in the organizations of the maker sphere.

Values, contexts, reflexivity, and knowledge all play significant roles in the analysis of the organization of hacker and maker communities. These chapters engage organizing as individual parts of a broader story, without needing to have the grand narratives uniting them. Instead, these chapters find their value both on their own but also in their relation to the rest of the material in the book with which they are in conversation. They extend the possibilities of organizing not just into the future, but into the past. These understandings provide new analytical and theoretical tools for understanding hacking and making in our world.

9. *Basteln, Tinkering, and Bricolage: A Cultural History of Hacking*

ALEXANDER VON LÜNEN
University of Huddersfield

Introduction

Despite—or because of?—the plethora of literature on hacking, the terms "hacker" and "hacking" appear fuzzier than ever before. Revolving around the (usually male) computer nerd and his tendency for criminal exploits, publications differ on what "hacking" actually implies, and whether there is a culture and history of hacking beyond the computer world. While it is not the ambition to provide a definitive answer to the question what "hacking" actually is, this chapter attempts to look at it from a specifically non-IT angle to broaden the scope in which hacking is usually discussed in.

This chapter will look at this culture and history of hacking and, to some extent, of making. As case studies, technology in the interwar years will be discussed, particularly radio and aviator equipment. After these historic examples of hacking and making, this chapter will discuss cultural connotations and concepts of hacking and making by examining the words "basteln" (German, verb; noun: bastler) and "bricolage" (French, verb; noun: bricoleur), both of which are translated into English as "tinkering". It will be pointed out, however, that "basteln" and "bricolage" have slightly different cultural connotations. The word "basteln" does not always have a negative or derogative tone to it when it comes to technology, as the English "tinkering" may have. While "basteln" often refers to arts and crafts activities, German engineers would use the word to describe what many nowadays would call a "hack": a quick, imperfect technological feat that has nonetheless some ingenuity to it. Furthermore, as I will point out, tinkering with radios (German:

"Radiobasteln") in Weimar Germany had a certain political connotation to it that resembles today's *hacktivism*.

Tinkering With Radios

In the early days of radio, tinkering was inevitable—both for mere radio listeners, as well as amateur radio operators. The technology was still in its infancy, with broadcasting stations and the radio apparatus industry still learning how to respond to an increased consumer demand. Even casual radio listeners would try to enhance their listening experience by tinkering with their receiver. This would only cease in the late 1920s when radio became more mature.[1]

Radio in Britain

The situation in the USA was used by British policy-makers to exemplify the need for tight regulations on amateur radio and no free market for broadcasters (Stoller, 2010, p. 13). The British authorities regarded the situation across the pond as utter chaos (i.e. too unregulated) and were keen to prevent the same cenario in the UK. After World War One, the Post Office started negotiations with industrialists to set up a monopolised broadcasting company. Especially negotiations with Guglielmo Marconi, an Italian engineer who made some path-breaking inventions and discoveries in the field of radio transmissions, proved tedious. Marconi, who had set up a company in Britain and was a resident in the country, tried to safeguard his 152 patents related to radio and had to be given substantial leeway in the newly created British Broadcasting Company (BBC).[2] While it was the government's intention to prevent a monopoly by Marconi, six out of of the eight stations of the BBC that went live in 1922 and 1923 were equipped with Marconi transmitters. Listeners had to pay for an annual radio license (10s) and could only operate licensed receivers from companies involved in running the BBC (Geddes, 1991, pp. 14–15).

Book series like *Every Boy's Hobby Annual* or serials such as *Hobbies* had a broad range of topics for tinkerers, ranging from cabinet making to building a device to code and transmit images. Obviously, it would feature articles on how to built a radio (e.g. in February 1932) as well, and magazines more specifically dedicated to radio, such as Wireless Constructor, or the magazine of the Wireless Society of London, would frequently feature circuits of radios and discussions of its components.

There was a particular *raison d'etre* in the early 1920s for tinkering: a loophole in the BBC license regulation allowed an "experimenter license".

The regulation determined for a fee to be paid for each radio set by manufacturers; which they would simply include into the retailing price. "Experimenters" were exempt from this tariff on devices. Offerings of construction kits from magazines and commercial vendors therefore became quite popular to avoid the surcharge on receivers. The Sykes Committee, a parliamentary commission set up to oversee the BBC, recommended in August 1923 that the tariff (and monopolies) imposed on receivers should be dropped, and the manufacturing opened up. Resistance from the BBC meant this would delayed until the end of 1924, but eventually it was implemented (Geddes, 1991, pp. 34–38).

In both the USA and the UK in those years, there was little political impetus around radio—neither in ham radio nor in commercial broadcasting. Amateur radio organizations showed a keen compliance to the law and the political system in order to avoid tighter regulations, and the labour movement in the Anglo-American sphere did not seem to be too interested in radio politics (cf. Beers, 2010).

The situation in Germany in the 1920s proved to be somewhat different. Not only seemed tinkering more popular for a time, the political situation gave way to something resembling the hacktivism encountered later with computers.

Weimar Germany

Contrary to popular belief, Weimar Germany wasn't the fancy and footloose society many historians have characterized it as. Especially the climate for individual inventors became increasingly hostile compared to Imperial Germany prior to World War One (Gispen, 1992, p. 396). Weimar Germany saw a pronounced effort towards concentration of R&D in big companies, leaving little room for entrepreneurs (Gispen, 1992, p. 400). Radio was no exception to this.

In the Weimar Republic, ownership of a radio was first restricted to the *Detektor*, a simple crystal set receiver with no amplification (only headphones). Radio broadcasting was controlled by the state and organized regionally, in line with Weimar Germany's federal structure. All valve-based radios required an operator license and membership in a state-approved radio club. Part of the reason to be so restrictive on radio licenses was the political turmoil that erupted in Germany after World War One and the Treaty of Versailles. Several attempted coups and local rebellions from both left-wing and right-wing extremist led to a climate in which the young, fledgling democratic government of Germany tried to curb political propaganda. Politicians feared that sophisticated radio receivers could be easily converted into transmitters and thus be used for illicit communication (Klingsporn, 1988, p. 34). This regulation was lifted on November 1, 1925, after political life in Germany became a bit more settled, but prices and programming guaranteed that radio was largely an urban, middle-class affair (Führer, 1997, pp. 736–737).

The magazine *Der Radio-Amateur* (The Radio Amateur) divided listeners into two categories in an editorial in October 1923: Enthusiasts and Tinkerers. The radio enthusiasts are "friends of entertainment who want to have some fun for themselves or their guests. This can become a trend on a larger scale, but individually it is usually short-lived." Tinkerers, on the other hand, "want to create wireless apparatuses according to plans made by others or themselves that 'at least work' or may even introduce a novelty. [...] The tinkerer possesses a great degree of ambition and in this strive for always greater goals results the continuity of his passion" (my translation).

Indeed, a closer look at the figures for radio subscriptions in Weimar Germany gives some credibility to the short-lived nature of radio enthusiasm. The number of subscriptions continually rose, yet on closer inspection it is revealed that this was just a net growth. While a high number of new subscribers was registered, an almost equally high number of existing subscribers would drop out after a short while, a trend obviously vindicated by the economic crisis of 1929, but due to technical issues as well (such as poor reception; Führer, 1997, p. 740). Economic reasons are also to be blamed for a low subscription rate among blue-collar workers in the 1920s, but perhaps attitudes of the elite played an equally strong role. Programming was geared towards classical music and intellectual lectures. Führer (1997, pp. 752–753) attributes this to the authoritarian and conservative nature of the Weimar elites, which regarded radio as a means to police leisure time, i.e. to make sure the listeners would not spend time on lowly pursuits and be educated instead.

Radiobasteln became thus also a political issue. Blue-collar workers who were willing to pay the license fee every month were indeed radio buffs rather than casual listeners. While middle-class radio owners were increasingly offered stylish devices that looked like furniture pieces to fit in with the rest of interior decoration, the simple Detektor with its low price remained the receiver of choice for those on low incomes. Left-wing organisations thus started the "Arbeiter-Radio-Klubs" (Engl: Worker Radio Clubs) as the equivalent of today's *maker spaces*: working class radio buffs would meet to discuss technical improvements to their radio sets and they would find the equipment needed for tinkering in the clubhouse they otherwise would not be able to afford; but they would also discuss the political implications of the advent of radio.

These working-class radio clubs were denied membership in the national federation of radio clubs in 1924, mainly because the federation feared the politicisation of radio and thus a conflict with authorities. This, likewise, radicalised the blue-collar radio clubs and their tinkerers (Klingsporn, 1988, pp. 47–48).[3]

The "Workers' Radio Club of Germany" (German: Arbeiter-Radio-Klub Deutschland, ARKD) was founded April 10, 1924, initially to overcome the limited accessibility to radios mentioned above. In 1928 there were approximately 210 local chapters with c. 10,000 members (Dussel, 2004, pp. 45–48). The agenda of the ARKD was not only technical support for its members, but also to critique existing programming and demand more broadcasts relevant for the working class.[4] Radio was supposed to be democratic, i.e. the listeners should have a greater say, if not produce their own programme. Said left-wing playwright Berthold Brecht in 1929 (quoted from Silberman, 2001, p. 39): "The increasing concentration of mechanical means and the increasingly specialized education [...] call for a kind of rebellion by the listener, for his mobilization and redeployment as producer." In 1928, however, the ARKD split after voting in an entirely Social Democratic committee, renaming itself to "Workers' Radio Federation of Germany" (German: Arbeiter-Radio-Bund Deutschland, ARBD). Among the first actions of the new committee was to drop the demand for a working-class radio station, which led to the Communists leaving the ARBD and forming their own radio club in 1929.

The Worker Radio Clubs had thus a political angle similar to hacker organizations/publications such as 2600 in the US or the Chaos Computer Club in Germany. Rather than just tinkering with technology out of joy or economic necessity—both of which were crucial elements—these radio clubs also sought to democratize radio broadcasting by demanding the end of top-down programming—just as hacker organizations demand democratized access to information.

Hacking in Aviation

In the interwar years, aviation became a major public spectacle with records—such as speed, distance or height—not simply being a showcase for technical feats, but also epitomizing the technical prowess of the record-holder's nation.[5] There was yet no institutional framework in place between the military and private engineers—as a matter of fact, most of military aviation was hopelessly behind in technical terms and relied heavily on the industry to bring forward innovations (see Hughes, 1989, p. 99ff. for a discussion of the American case).

This on the other hand, meant that some areas got less attention, such as pressure suit design for high-altitude flight. Military directors had decided that pressure cabins were favourable, as the suits of those days were rather clunky and hampered the pilot's ability to move. For the daredevils of 1930s aviation, on the other hand, they were a necessity. Pressure cabins would make the airplane much heavier and thus counteract the intention to break the height record. Thus, pressure suits were considered to be the only option, and both private aviators as well as military air forces had to find industrial partners for this. Remarkably, these contacts were fairly unbureaucratic.

In Britain, the initiative was driven by Oxford physiologist John Scott Haldane and London based diving suit manufacturer Siebe & Gorman, represented by their chief engineer and managing director Robert Davis, after being contacted by American Mark Ridge in 1933, who wanted to take an open basket balloon into the stratosphere. Ridge was turned down in the USA by both government agencies and private companies, first and foremost because he was very unprofessional in both background and appearance.[6] While he convinced Haldane and Davis, officials in Britain were equally unimpressed as their US counterparts by Ridge's attitude. Not least because the RAF had been contemplating their own height record flight; and while happy to cooperate with Haldane and Davis to use the pressure suit, RAF officials had no appetite to risk a major PR disaster with Ridge's ill-minded stunt flight.

Similar efforts were under way in other countries. The Italian Royal Air Force started its own high altitude flying group at Guidonia air base for the sole purpose of breaking the height record in aviation. While set up in a highly regulated environment—the military in fascist Italy—the approach taken within this group has stronger resemblance to hacker culture than one might expect (more on this below).

In Nazi Germany, the government had not much interest in pressure suits, mainly since German aircraft engines weren't powerful enough to have any chance of breaking the height record—and some companies, like Junkers, had working prototypes of pressure cabins. The company Dräger was contracted in 1936 to develop a suit for the Luftwaffe, but the results were regarded as unsatisfactory.

Gerhard Klanke, a meteorologist, entered the *Reichswetterdienst* (RWD, National Weather Service) in 1931 as weather pilot. Being part of first the Ministry for Transport and then relocated to the Air Ministry in 1935, the RWD was a government agency and Klanke not supposed to design and develop new technology. Klanke, on the other hand, had a strong technical interest, filing for his first patent in 1925 (an indicator device for bicycles).

Weather pilots, at a time when no satellite surveys could be utilized for weather forecasting, flew to considerable heights on a regular basis, usually between 5000 and 6000 m. Physiologists had already discovered around 1900 that above 5000 m, supplemental oxygen is required to counteract the decreased pressure. Above c. 12,000m a device to maintain an ambient pressure equivalent to lower heights would be required, as even breathing pure oxygen could not alleviate the loss of pressure necessary for the oxygen to enter the lung cells. Hence, the idea for pressure cabins or pressure suits. As Klanke intended only to go up to between 5000 and 6000 m, an oxygen supply would have sufficed. However, in the early 1930s, oxygen devices were imperfect and especially the breathing masks caused more problems than they solved. Klanke thus had the idea to construct a suit that would completely enclose the pilot and to compress the inflowing air to levels equivalent to tolerable heights (i.e. below 5000 m). Not only with this unconventional idea, but first and foremost with its implementation, Klanke demonstrated true *hacker power*.

The suit was constructed in 1933, according to Klanke's own account, and consisted of two parts and a helmet. The jacket was made out of a rubber-cotton fabric by the Klepper company (sewn together by Klanke's wife), which produced this cloth for water-tight garments and tents. The trousers were standard off-the-shelf protective trousers by the Dräger company, produced for fire-fighters and workers in the chemical industry. The helmet was custom-made out of brass, held together by two stove rings. Suitcase locks were used as fasteners for the helmet, the visor made of plexi-glass. The two parts (jacket and trousers) were joined and sealed by a gas-tight belt at the waist.

As so often with the pressure suits of those days, the major technical problem was introduced by the fabric of the suit. The suits need to be both flexible and inelastic, something that was hard to achieve with the materials available in those days. In the inflated suit the expansion became prohibitive, severely restricting the ability to move, hence Klanke used rubber bands for the chest and arms to limit the expansion. Another issue: the expansion of the suit would lift the helmet above the pilot's head, completely inhibiting vision. In order to prevent this Klanke attached the helmet to leather belts, similar to braces.

Dräger's design from 1936 was equally crude, with chief engineer Hermann Tietze systematically trying different methods to contain the overt inflation of the suit, first leather belts and then a steel mesh. So while it could be argued that Dräger had a solution of equal quality to Klanke's suit, the one by Dräger was never actually used, not even for test flights.[7] The reason for this is a discrepancy in approaches to engineering, as I will discuss in the remainder of this chapter.

Bricolage

Looking at the two German pressure suit constructors, Gerhard Klanke and Hermann Tietze, the resemblance to the bricoleur vs. the engineer juxtaposition by French anthropologist Claude Lévi-Strauss is striking. The French word "bricoleur" is difficult to translate, but is usually given as "jack-of-all-trades" or "handyman" in dictionaries. The traditional meaning of the word "bricolage" in French refers to deflections in ball games or game movements in hunting (Lévi-Strauss, 1968, p. 16); modern dictionaries translate it as "tinkering about" or "makeshift repair" (Brick, 2002, p. 100).[8] A bricoleur is someone who "uses devious means compared to those of a craftsman" (Lévi-Strauss, 1968, p. 17). According to Lévi-Strauss, the engineer approaches a technical problem in a systematic manner, creating the tools or other technological devices as necessary. The bricoleur, however, takes tools and materials readily available and utilizes them the best way possible (Lévi-Strauss, 1968, p. 18). He (or she) is therefore creative in the use of existing technology, applying it outside their intended designation, as his or her tools do not "have only one definite and determinate use" (Lévi-Strauss, 1968, p. 18).

The engineer analyses a problem, unravels the difficulties and deliberates the tools and materials required. Prior to implementation the engineer has made a detailed plan about tackling the problem. The bricoleur, on the other hand, picks the tools as the problems come. He or she gets into a "dialogue" with his or her "set of tools" and "before choosing between them [has] to index the possible answers which the whole set can offer to this problem" (Lévi-Strauss, 1968, p. 18).[9]

Looking back at Klanke and Tietze, it is obvious who the bricoleur and who the engineer is. Tietze tried to find or create new materials, methodically trying them in test labs; while Klanke appropriated existing materials—such as the off-the-shelf gas-tight trousers—to create his pressure suit. On the other hand, Tietze's first design had a bricolage element to it, since it used existing components, such as a gas mask as part of the helmet. Unhappy with the result—for Tietze sought to find the optimal technical solution rather than having something imperfect that has practical value—he set out to solve the problem by methodically and meticulously finding and testing materials. For Tietze, it was true what Resnick and Rosenbaum (2013, p. 164) have noted: "Many people think of tinkering in opposition to planning—and they often view planning as an inherently superior approach." Tietze was certainly in the latter camp.

Hacking and Making

The concept of bricolage by Lévi-Strauss would lend itself to be the equivalent of hacking and making, and the similarities are apparent. However, bricolage—as an anthropological concept—usually refers to make-shift technology in pre-modern or non-Western contexts. Bricolage is often done out of scarcity rather than recreation. While a good number of elements are found in both bricolage and hacking—such as curiosity or experimenting—hacking and making are often closer to artistic expression than economic necessity.

Attempt of a Definition

As Taylor (1999, p. 13) rightly remarks, there seems to be a host of concurrent, if not contradictory, definitions of the term "hacker". Levy (1984) in his hallmark publication on the first-generation hackers at MIT is biased to both the academic and the IT world by defining hacking as "highly skilled but largely playful activity of academic computer programmers searching for the most elegant and concise programming solution to any given problem." As this chapter has already alluded to, hacking is older than IT and consequently far broader.

Obviously, there is still a prevailing definition of hacking as computer-based crime—such as "Hacking is accessing a computer-based system without appropriate authorization, which can be unlawful." (Bell, Loader, Pleace, & Schuler, 2004, p. 104). While narrowing down the definition of hacking down to criminal activities is somewhat unjust, hackers have long revelled in this "rebel" and "renegade" attitude that comes with it. Being deviant and defiant is part and parcel of the very idea of hackerism; this very idea being to take control of technology rather than just being a mere user. The definition by Taylor (1999, p. 15) sums up these characteristics quite nicely:

The main characteristics of a hack are thus:

1. Simplicity: the act has to be simple but impressive.
2. Mastery: the act involves sophisticated technical knowledge.
3. Illicitness: the act is "against the rules".

The last point, "against the rules", does not necessarily refer to an illegal act, but rather breaking a convention or a way of using technology as it was intended by its designers. In the words of Webber (2014, p. 93):

"Controlling a device through technical mastery and making it do things the inventor never intended is the hallmark of the hacker." A *hack*, according to Raymond, is: "Originally, a quick job that produces what is needed, but not well", although nowadays is affiliated with technology "in a playful and exploratory rather than goal directed way." A person acting in this spirit, a *hacker*, is consequently "[o]ne who enjoys the intellectual challenge of creatively overcoming or circumventing limitations" (Raymond, 1991, pp. 189, 191). Ciborra (2002, p. 49) re-iterates this by stating that hacking is "making use of any technology in an original, unorthodox, and often playful way."

Escaping the Iron Cage

Philosophers of technology and other scholars have coined a variety of terms and concepts to describe the limiting nature of technology and technological systems. Sociologist Max Weber used the metaphor of the "iron cage of rationalization" to warn of the dangers of overreaching bureaucracies in his landmark publication *The Protestant Ethic and the Spirit of Capitalism* in 1905. Philosopher Martin Heidegger introduced the concept of "enframing" to point out that technology both enables and restricts how we access the world; his critique—as a reactionary who idolized the hardship of the peasant—was that technology would convert nature into a mere resource, a mere commodity (cf. Heidegger, 1977). Based on Heidegger's work, French philosopher Michel Foucault would coin the term *apparatus* to criticize technological systems that increasingly proscribe human actions, i.e. the only and best way to act, like an unwritten manual (cf. Agamben, 2009). All these critiques aim at the dichotomy of technology as being both a liberator and a captor at the same time. Aviation, to stay in the framework of the above example, liberated humanity in many ways. It made flying possible, a dream of humanity since antiquity. Yet, aviation as technological system also restricts, like any other transport technology. It requires a complex infrastructure (international treaties, air control agencies, airports, fuel refineries etc.) and strict travel regimes (time management, security checks, etc.). The freedom to travel is thus a lack of control over travelling at the same time.

On a more individual level, industrialization brought with it increasingly standardized—i.e. monotonous—work procedures. Karl Marx already described this in his "alienation theory" from 1844 in which he described the factory worker as being disconnected—if not disenfranchised—from the product of his labour. The division of labour, according to Marx, separates the worker from the product, makes labour "external" to the worker; "in his work, therefore, he does not affirm himself but denies himself." (Marx,

1974, p. 66) Authors discussing hackerdom and makerdom have made similar points—although without linking it to Marx. Turkle (2005, p. 188) refers to early open source projects as "a mode of production built on a passionate involvement with the object being produced."

These two points raised—passion and control—reverberate with studies on the culture of playing and risk-taking. Historian Johan Huizinga wrote the path-breaking study on the history of play, *Homo Ludens*, in 1938 and he outlines both the characteristics and function of it in human culture. He pointed out, for example, that *play* is the deliberate defiance of structures: "Play to order is no longer play" (Huizinga, 1949, p. 7). To play means to break out of the ordinary routine (Huizinga, 1949, p. 9). However, play is not—or at least not always and entirely—anarchism. Historically and culturally, as Huizinga (1949, p. 46ff.) discusses, play was a way to contain the usually hot-tempered adolescents in society, often in the form of sports. Both play and contest had thus a "civilizing function", according to Huizinga, that would channel the aggression of male youngsters into something more constructive.

Hacking can be looked at through the prism of *homo ludens* as well—certainly when defining play as breaking out of daily routines and defying traditional patterns of learning and production. Using the two strands from above—alienation of the worker and play to defy routine—one can discern a pattern: play to take back control. Hacking and making both offer this on an individual level, i.e. for individuals to take back control of the way they work; but there is also an angle here for alternative modes of production on a larger scale—namely the Open Source movement with its roots in hacktivism, and which is also a defining factor in the Maker movement.

Picking up Turkle's statement on Open Source from above and putting it into context of alienating work, the next logical step would be to examine how Raymond (2001) discussed what he called the "Cathedral vs. the Bazaar" taxonomy of software development. In the *cathedral* (i.e. big, monolithic companies with strict hierarchies) software developers and others follow orders and are limited in their creativity. In the *bazaar* approach many enthusiastic individuals cooperate on a project without an absolute end, i.e. a well-defined, static and standardized product. Himanen (2001) regards this as a challenge to what Max Weber has coined "protestant working ethics": Weber characterized Western capitalism as driven by the Protestant ethics of hard labor as a spiritual calling. To Himanen "the radical nature of general hackerism consists of its proposing an alternative spirit [...] that finally questions the dominant Protestant ethic" (Himanen, 2001, pp. 12–13). The *hacker culture*—as expressed in the Open Source movement—thus epitomizes the playful and joyful side of engineering.

In this regard the Italian high-altitude flying group in the 1930s worked along the same lines. The working culture within the group was rather personal and joyful. Although embedded in a military hierarchy, the structure within the group was comparably flat. Mario Pezzi, for example, was both the commander of the group and its chief test pilot, i.e. he could make decisions more oriented towards actual problems, rather than dealing with military and bureaucratic procedures. The group at Guidonia thus, too, was organized more like a bazaar than a cathedral. Engineers, scientists and others were working enthusiastically on the project—or: were *hacking* a height record. The same spirit could be found in the Royal Aircraft Establishment at Farnborough (cf. Lünen, 2010).

In Germany the government simply *ordered* Dräger to produce such a suit, and the parties involved—the private company Dräger, various departments of the Air Ministry and affiliated research labs—only communicated via official reports and specifications. Dräger had no commercial incentive to delve into the topic, and no one felt personally committed to dedicate his resources to this issue beyond ordinary measures. As Schimank (2005, p. 47) mentioned, rationalized production processes lead to "rationalized acting": "ready-made programs of actions that releases the actor from time-consuming reflexion" (my translation). This also expresses what Schimank coined as the *fiction of rationality*: In rationalized—a.k.a. industrialized—working environments, the worker is more and more alienated from his or her work, as Marx had pointed out. Thus, identification with the work is decreased and leads to irrational behavior, since the worker no longer actively pursues a common goal, but just follows orders. Hence the German setup in the 1930s was such an *actorless social body* Schimank spoke about.[10]

With Klanke and his suit the situation was different, obviously. Klanke had his heart in the matter; it is thus no surprise that Klanke would be willing to use this otherwise hair-raising contraption. But herein lies the crucial point why Dräger failed and Klanke succeeded, in terms of using their respective suit in actual flights: Klanke was more than willing to take the risk of using the highly imperfect suit, whereas Dräger—being a commercially successful, internationally renowned company—would not take such a risk. Looking at Dräger's strategy reveals a very systematic and top-down approach, in which new materials were sought and thoroughly tested before used, rather than taking the bricoleur-style attitude of a risk-affine Klanke. As mentioned, risk-taking is also an important narrative in hacking. It should also be clear that this is an attitude more in line with stereotypes of masculinity.[11]

De Palma (2001) points out that computer science courses are stereotypically associated with a playful (i.e. irrational) accesses to technology. Meaning:

boys choose to study computer science (or engineering) because they played computer games (or with Lego or Brix in the case of engineering) before. Holth (2014, p. 101) expands on this, pointing out that "most of the men, but few of the women" among her interviewees "expressed enthusiasm and passion" when talking about their relationship with technology. Indeed, as was pointed out in relation to Huizinga's *homo ludens* above, play was historically seen as a male domain, as contest to both contain and prove manhood.[12]

It is thus hard to ignore the gender issues in hacking as a continuation of gender issues in the general interaction with technology. Motivations such as "play" may be characteristic for male engineers; however, it seems as though "control" —in the sense of hubris—seems to be the stronger incentive. The examples in aviation hacking given above certainly point in that direction.

Conclusions

In conclusion, it can be said that hacking, and to some extent making as well, has the motives of *play* and *control* in its centre. These motives articulate themselves in a variety of ways, be it as risk-taking or rebelling against rationalized forms of production.

Reading through narratives by hackers, whether through interviews done by scholars or self-confessions in publications such as *2600*, these often start with curiosity about how technical artefacts work; often resulting in a disassembly of the artefact and tinkering with it. The desire to use technology in an unconventional way—unintended by its designers—is about this very notion, and is often part of the hacker's rhetoric: taking (back) control over technology. This is not restricted to hobbyists, however, as Kleif and Faulkner (2003, p. 319) point out. This motivation of taking control of technology is present in professional engineers, as well as hobbyists. As a matter of fact, as the authors elucidate, the boundary between these two groups—professionals vs hobbyists—is quite fuzzy. The crucial distinction between the two groups perhaps being that hobbyists have an option to leave the technology alone should they no longer desire to engage with it; something that professionals usually can not do without fearing pecuniary repercussions (Kleif & Faulkner, 2003, p. 313). Here, the *ludic* dimension of the hacker comes to the forefront again which was mentioned above: one cannot be ordered to play.

As seen in the example of blue-collar worker radio clubs in Weimar Germany, participating in the technology as consumer quickly led to calls for participation on a higher level, i.e. greater influence on radio programmes, if not a demand for independent radio stations altogether. While amateur radio has some claim to be a hacker legacy, I would argue that the hams are closer to making than hacking. Amateur radio, from its early beginnings on, was in a privileged position: it had frequencies reserved just for itself. Or as Campbell (2003, p. 64) put it: "Radio amateurs are in an unusual position within society. While closely regulated and supervised by the federal government, on the one hand, they are protected by both national and international laws, on the other hand." This is surely no position hackers and hacktivists find themselves in. Yet, it must be conceded that computer hacking came only into existence because of big government and big business. Were it not for the Cold War and its multi-billion R&D budgets by government agencies such as ARPA, hacking would still be restricted to tinkering with radios.

On the other hand, this may well generally characterize hacking and making: that they are inevitably a response to political, national and institutional frameworks. The *radiobastler* of Weimar Germany responded to a specific political set-up; the radio licensing situation was not very different to, say, the situation in the UK. Yet, while there was a tinkerer scene in the UK, it was far less politicised than the one in Germany in the 1920s. The *radiobastler* were consequently not only a response to a socio-economic situation, but to a political one as well. Likewise, the hacking in aviation described above was a negotiation between bricoleurs and institutional frameworks present in the interwar years. The desire to break height records in aviation to show off national prowess created both incentives and space for hacking.

Furthermore, as Oldenziel and Hård (2013) point out, consumers hack as well. Users of technical products have always tinkered with the technology and its application, customizing and appropriating it to their needs and desires. Tinkering with radios in the interwar years was not restricted to building and modifying radio receivers, it also accommodated more mundane hacking and making tasks, such as putting the headphone of your crystal set into a jar to create a make-shift speaker so that the whole family could listen in.

Tinkering, basteln, bricolage are thus not limited to engineers or otherwise tech-savvy people; they are about appropriating—if not expropriating—technology from its designers, just like hacking. Users have always hacked their technical devices to customize them to their needs or just to find out about their internals, and professional engineers often use some form of hacking for prototype development. The only difference maybe being that bricolage is often done in a context of necessity, i.e. economic hardship or shortage, whereas hacking is often a ludic choice. One could therefore define "hacking" as "bricolage plus homo ludens".

Notes

1. One could argue that amateur radio had the most obvious connection to hacking and making. As I will briefly outline at the end of this chapter, however, this is not the case.
2. The BBC was indeed a "company" first, a joint venture of c. 200 single private businesses; "a curious hybrid of commercial interest and government responsibility" as Street (2002, p. 28) calls it. It was turned into the publicly owned British Broadcasting Cooperation on January 1, 1927.
3. All radio clubs had Radiobastler in their ranks, but only the Workers' Radio Clubs would politicise the tinkering with radios.
4. This negative stance towards the radio establishment was facilitated by the decision of broadcasters to exclude Ernst Thälmann, communist candidate for the presidential elections in 1925, from speaking on the radio when all other candidates were given airtime (Klingsporn, 1988, p. 52).
5. I have written about aviation records—particularly height records and related technology—extensively in my doctoral dissertation (Lünen, 2010). All the information discussed in this section is from that publication, unless noted otherwise.
6. B.F. Goodrich had manufactured pressure suits for renowned aviator Wiley Post in 1933/4. Post, however, had some wealth and good contacts to the aviation industry due to his celebrity status as record aviator.
7. Klanke claims to have made c. 300 flights with his suit.
8. More recently, *bricolage* has also been used in visual communcation theory, specifically in semiotics, to describe the mixture of media and styles to form hybrid media; also, youth cultures that mix, say, vintage clothing and modern culture have been characterized as cultural bricolage (cf. Barker, 2004, p. 17).
9. Someone has characterized the TV character McGyver, who creates bewildering tools out of everyday items, as bricoleur.
10. Schimank (2002, p. 53) defined organizations in which not much use is made of individuals and their ideas, but in which decisions are rather based on pre-configured, formalized procedures as an "actorless social body" (my translation).
11. I will only briefly discuss gender issues here, for X in chapter Y is dealing with this in much more detail.
12. Again, computing technology was not the starting point of this gender stereotype. Haring (2003, p. 735) states about radio amateurs that "an estimated 95 or 99 percent were male."

References

Agamben, G. (2009). *What is an apparatus? And other essays.* Stanford, CA: Stanford University Press.

Barker, C. (2004). *The SAGE dictionary of cultural studies.* London: SAGE.

Beers, L. (2010). *Your Britain: Media and the making of the labour party.* Cambridge, MA: Harvard University Press.

Bell, D., Loader, B. D., Pleace, N., & Schuler, D. (2004). *Cyberculture: The key concepts.* London; New York, NY: Routledge.

Brick, M. (Ed.). (2002). *Collins/Robert French-English English-French Dictionary. Unabridged.* Glasgow; Paris: Harper Collins/Dictionaires Le Robert.

Campbell, B. (2003). Compromising technologies: Government, the radio hobby, and the discourse of catastrophe in the twentieth century. In S. M. Squier (Ed.), *Communities of the air: Radio century, radio culture* (pp. 63–75). Durham, NC; London: Duke University Press.

Ciborra, C. (2002). *The labyrinths of information: Challenging the wisdom of systems.* Oxford: Oxford University Press.

De Palma, P. (2001). Why women avoid computer science. *Communications of the ACM, 44*(6), 27–29.

Dussel, K. (2004). *Deutsche Rundfunkgeschichte.* Konstanz: UVK.

Führer, K. C. (1997). A medium of modernity? Broadcasting in Weimar Germany, 1923–1932. *The Journal of Modern History, 69*(4), 722–753.

Geddes, K. (1991). *The setmakers: A history of the radio and television industry.* London: BREMA.

Gispen, K. (1992). National socialism and the technological culture of the Weimar Republic. *Central European History, 25*(4), 387–406.

Haring, K. (2003). The 'freer men' of ham radio: How a technical hobby provided social and spatial distance. *Technology & Culture, 44*(4), 734–761.

Heidegger, M. (1977). *The question concerning technology and other essays.* New York, NY: Harper Torchbooks.

Himanen, P. (2001). *The hacker ethic and the spirit of the information age.* London: Secker & Warburg.

Holth, L. (2014). Passionate men and rational women: Gender contradictions in engineering. *NORMA: International Journal for Masculinity Studies, 9*(2), 97–110.

Hughes, T. P. (1989). *American genesis: A century of invention and technological enthusiasm, 1870–1970.* Chicago: University of Chicago Press.

Huizinga, J. (1949). *Homo Ludens: A study of the play element in culture.* London; Boston, MA and Henley: Routledge & Kegan Paul.

Kleif, T., & Faulkner, W. (2003). "I'm no athlete [but] I can make this thing dance!" – Men's pleasures in technology. *Science, Technology & Human Values, 28*(2), 296–325.

Klingsporn, K.-M. (1988). *Die Einführung des Rundfunks in Deutschland und die Reaktion der organisierten Arbeiterradiobewegung auf das neue Medium.* Master's thesis, Freie Universität Berlin.

Lévi-Strauss, C. (1968). *The savage mind* (2nd impression ed.). London: Weidenfeld and Nicolson.

Levy, S. (2010 [1984]). *Hackers: Heroes of the computer revolution* (25th anniversary ed.). Sebastopol: O'Reilly.

Lünen, A. V. (2010). Under the waves, above the clouds: A history of the pressure suit. Online publication (PDF). Retrieved from http://tuprints.ulb.tu-darmstadt.de/2103

Marx, K. (1974). *Economic and philosophic manuscripts of 1844* (4th revised ed.) Moscow: Progress Publishers.

Oldenziel, R., & Hård, M. (2013). *Consumers, tinkerers, rebels: The people who shaped Europe*. London: Palgrave Macmillan.

Raymond, E. S. (Ed.). (1991). *The new hacker's dictionary*. Cambridge, MA; London: MIT Press.

Raymond, E. S. (2001). *The cathedral and the bazaar: Musings on Linux and open source by an accidental revolutionary*. Cambridge, MA: O'Reilly.

Resnick, M., & Rosenbaum, E. (2013). Designing for tinkerability. In M. Honey & D. E. Kanter (Eds.), *Design, make, play: Growing the next generation of STEM innovators* (pp. 163–181). New York, NY; London: Routledge.

Schimank, U. (2002). Organisationen: Akteurkonstellationen – Korporative Akteure – Sozialsysteme. In J. Allmendinger & T. Hinz (Eds.), *Soziologie der Organisation* (pp. 29–54). Wiesbaden: Westdeutscher Verlag.

Schimank, U. (2005). *Die Entscheidungsgesellschaft. Komplexität und Rationalität der Moderne*. Wiesbaden: Verlag für Sozialwissenschaften.

Silberman, M. (Ed.). (2001). *Bertolt Brecht on film & radio*. London: Methuen.

Stoller, T. (2010). *Sounds of your life: The history of independent radio in the UK*. New Barnet: John Libbey Publ.

Street, S. (2002). *A concise history of British Radio 1922–2002*. Tiverton: Kelly.

Taylor, P. A. (1999). *Hackers: Crime in the digital sublime*. London; New York, NY: Routledge.

Turkle, S. (2005). Hackers: Loving the machine for itself. In *Second self: Computers and human spirit* (pp. 183–218). Cambridge, MA: MIT Press.

Webber, C. (2014). Hackers and cybercrime. In R. Atkinson (Ed.), *Shades of deviance: A primer on crime, deviance and social harm* (pp. 95–98). London; New York, NY: Routledge.

10. *Women's Hacking of the Poison Gift of Free/Libre/Open Source Software*

Jennifer Maher
University of Maryland Baltimore County

A revolutionary approach to software development for the micro- or personal computer began in September 1983 with an announcement from Richard Stallman. In his post "new UNIX implementation" made to the Arpanet newgroup net.unix-wizards, a bulletin board for discussion of the proprietary UNIX operating system, Stallman wrote of his desire to counter what had become the dominant proprietary software model that restricted users ability to share, modify, and redistribute software. Rather than a closed-model rooted in the basic tenets of a capitalist exchange of software as a commodity, free software was to be rooted in the principle of the Golden Rule. As Stallman explained in his post, "I consider that the golden rule requires that if I like a program I must share it with other people who like it ... So that I can continue to use computers without violating my principles, I have decided to put together a sufficient body of free software so that I will be able to get along without any software that is not free" (1983). Using the recursive acronym "GNU's not UNIX," the GNU Project would expand over the years not only through the work of other free software developers but also open source developers. "Open source," a term created in 1998 to shake off the negative connotations of "free" that struck some as antithetical to the profit motive of the corporate world, still used the same development model as free software and thereby helped to expand the use of freely shareable, modifiable, and redistributable software through projects such as the web browser Firefox, the word-processing program Open Office, the repository GitHub, and server distributions such as Apache and Red Hat.

The free and open software communities, in spite of their differences (Chopra & Dexter, 2008; Coleman & Golub, 2004), share a fundamental principle known as the hacker ethic. As defined in Eric Raymond's (1996) *The New Hacker's Dictionary*, the hacker ethic is a "belief that information-sharing is a powerful positive good, and that it is the ethical duty of hackers to share their expertise by writing free software and facilitating access to information and to computing resources wherever possible" (1996, p. 234). As a result, the software hacker ethic relies upon the economy of the gift rather than exchange (Söderberg, 2008; Weber, 2004; Zeitlyn, 2003). Unlike proprietary software that piggybacks on the notion of Fordist scarcity in order to facilitate monetary value in a product that, unlike a car or a house, is endlessly reproducible (Kinsella, 2001; Lemley, 2014; Merges, 2011; Vaidhyanathan, 2001), free/libre/open source software (F/LOSS) facilitates abundance through a recursive process of giving and regiving code. But to tighten access to and use of software that would otherwise have no material limitations except those created out of artificial scarcity would be to deny not only the good that results from such giving but also the joy. As Levy (2001) illustrated in his seminal work *Hackers: Heroes of the Computer Revolution*, the hacker ethic predates the F/LOSS communities that developed with the advent of the personal computer. Writing of hackers working on the PDP-1 mainframe computer in the early 1960s at the Massachusetts Institute of Technology (MIT), Levy explained, "The question of royalties never came up … using the computer was such a joy that they would have paid to do it … Wasn't software more like a gift to the world, something that was reward in itself?" (p. 38). The possibility for joy through giving increased exponentially with the processing possibilities ushered in with the microcomputer, so much so that free software evangelist Stallman would compare it to air, invoking the idea that to put a monetary price on software was akin to charging for one of the most elemental necessities of life (1987).

In spite of its enormous benefits, the gift of F/LOSS is not without serious problems, most notably, for my purposes here, is the sexism and gender discrimination that occurs in the development of freely shareable software. The challenge that F/LOSS development offers to traditional economies of exchange might seem likely to challenge also the masculinist domination embedded and reproduced through the market economy of capitalism, where, as Irigaray (1985) explained, "The work force is thus always assumed to be masculine, and 'products' are objects to be used, objects of transaction among men alone" (p. 175). Although this conception of the work may seem antiquated to some, especially given that, in the U.S., for example, women account for 46.8% of the total labor force as of 2015 (Bureau of Labor Statistics, 2016), we need only look to the lack of women in corporate leadership roles worldwide (Noland, Moran, & Kotschwar, 2016), as well as the gendered history of computing and the disproportionately low number of women who code (Abbate, 2012; Chun, 2004; Clegg & Trayhurn, 2000; Cohoon & Aspray, 2006; Engsmenger, 2003; Henn, 2014; Misa, 2010) as evidence of a masculinist force at play. Because F/LOSS is freed from being an object through which economic value is located, it might stand to reason that F/LOSS development culture might also be freed from that force that perpetuates the artificial construction of computing as a masculine endeavor. Yet, this is not the case. According to a 2014 report by the Bureau of Labor and Statistics, women make up almost 20% of software developers in the United States. The most recent F/LOSS Survey (2013) shows, however, that only 11% of those involved in F/LOSS are women. Although the percentage of women programming, rather than doing documentation, artwork, etc., is unclear in the 2014 survey, previous estimates in 2009 pointed out that as few as 1.5% of F/LOSS developers were women (James, 2010), which was up from 1.1% in 2001 (Ghosh, Glott, Krieger, & Robles, 2002). Regardless of the likely increase in the number of women in F/LOSS, simply increasing that number, while certainly desirable, does not address the fact that hacker code creation is very much rooted in an artificial logic that constructs women coding, and coding F/LOSS, in particular, as the exception rather than rule. To both illuminate and combat this phenomenon, I argue that hacker gift-giving must be understood as both open and closed, both free and unfree. More specifically, I employ discussions of the dual nature of the gift (Bourdieu, 1997; Cixous, 1976; Mauss, 2002) to examine how freely shareable software is both "present" and "poison" and also identify how what Cixous calls *écriture féminine*, or women's writing, reveals an opportunity to rebel against F/LOSS as the "gift-that-takes" and transform it into the "gift-that-gives."

F/LOSS as Gift

To a great degree, F/LOSS development has succeeded because of evangelists such as Stallman, Raymond, and Linus Torvalds, all of whom, in their varying ways, have spread the good news of the philosophical and practical benefits of hacking (Maher, 2015). Open source developer and evangelist Raymond, the open source community's self-described "resident ethnographer," has had a significant role in teasing out of the idea of freely, shareable, modifiable, and redistributable software as hacker gift. As he explained in the *How to Become a Hacker*:

> There is a community, a shared culture, of expert programmers and networking wizards that traces its history back through decades to the first time-sharing minicomputers and the earliest ARPAnet experiments. The members of this

> culture originated the term "hacker". Hackers built the Internet. Hackers made the UNIX operating system what it is today. Hackers run Usenet. Hackers make the World Wide Web work. (2001)

Unlike "black hat" hackers or "crackers," who engage in criminal coding activities for personal gain, software hackers are literally responsible for much of the infrastructure that supports everyday, digital life. What is more, their source code (i.e., the human-readable instructions in which code is written) is made available to read, modify, and redistribute. This is the gift of F/LOSS. Even as the proprietary model came to dominate software development in the 1980s (Ceruzzi, 2003), hackers resisted the increasing demand that software and its source code be treated as the "crown jewels," used only under guard but otherwise locked away. Hackers did so not only by continuing to produce freely shareable software but also fighting off attempts to squash F/LOSS. Paramount in this fight was F/LOSS evangelism that spread the good news of an alternative to proprietary software. The gift of F/LOSS then was not only the development of freely shareable code but also a technological evangelism that worked to ensure the availability of that code, even to the most non-technical end user.

To give, according to the *Oxford English Dictionary*, is "the action of giving, an instance of the same; a giving, bestowal" and that the giving of a gift occurs "gratuitously, for nothing." This definition serves to imbue the giver with a sense of selflessness and good will, interested only in presenting to the receiver something desirable or useful. But, Mauss (2002) argued, in his examination of archaic societies, that "exchanges and contracts take place in the form of presents; in theory they are voluntary, in reality they are given and reciprocated obligatorily" (p. 3). While certainly different than those transactions that occur in what is often identified as the restrictive economy of exchange in Fordist capitalism, the gift appears to be antithetical to market self-interest. As Gell (1992) sums up: "gift-reciprocity-Good/market-exchange-Bad" (p. 142). In the context of F/LOSS, the early evangelism of free software appeared to reproduce this very dichotomy, with the result that Stallman and the Free Software Foundation (2002)—"a nonprofit with a worldwide mission to promote computer user freedom"—later attempted to clarify that the "free" in "free software" is "*free* as in *free speech*, not as in *free beer*." But, by then, F/LOSS had already been compared by Microsoft to communism, and the rebranding of "free software" as "open source" was undertaken to throw off the appearance of anti-capitalist sentiment. But as Söderberg (2008) notes, F/LOSS does not challenge capitalism but instead "sets the contours of the capitalist relation in relief" (p. 187). Thus, hacking and the creation of F/LOSS is in many ways once removed from the kind of transaction that would place it squarely at the center of the restrictive economy of exchange, which proprietary software is located. F/LOSS is instead rooted in exchange as service, which is the model by which F/LOSS-based businesses accrue capital (Fuchs, 2010).

If the gift brings with it certain constraints for and expectations of reciprocity that shape the giver as interested in some way other than capital accumulation, what then is the payoff of the gift for the giver? Besides the obvious benefit of code, it is social status (Himanen, 1999; Stewart, 2005). As Raymond explained, "Like most cultures without a money economy, Hackerdom runs on reputation. You're trying to solve interesting problems, by how interesting they are, and whether your solutions are really good, is something that only your technical peers or superiors are normally equipped to judge" (2001). With the shift from an emphasis on working for money to coding for pleasure, reputation takes on even greater significance. Himanen tied this emphasis on status among software hackers to the possibility of a post-Protestant work ethic. Rather than toil as punishment, as was the understanding of laboring in the pre-Protestant age, the Protestant work ethic, originally described by Max Weber, identified work as moral duty to God. But as society became increasingly secular, this moral impetus was co-opted by capitalism itself with the result that work is elevated "to the status of the most important thing in life" (Himanen, 1999, p. 12). But, with the mutually affirming emphasis on giving and joy, F/LOSS hacking, especially given its successes over the last 30 years, challenges the dichotomy between work and pleasure. This shift has allowed for a moral dimension in technological development (Hunsinger, 2011), as evident in the moral encoding of freely shareable code through software evangelism. With the hacker emphasis on reputation, it is little wonder then that heavy-handed ethos moves sometimes come into play, as in Stallman's announcement of the GNU Project more than three decades ago: "I am Richard Stallman, inventor of the original much-imitated EMACS editor, now at the Artificial Intelligence Lab at MIT. I have worked extensively on compilers, editors, debuggers, command interpreters, the Incompatible Timesharing System and the Lisp Machine operating system. I pioneered terminal-independent display support in ITS." (1983). In establishing his hacker credibility, Stallman relied upon his status as a free software programmer who had made important contributions that were both useful to others and reified the hacker ethic. Although the quality and creativity of code are obviously subject to hacker evaluation and judgment, these measures are not the only avenues by which respect can be garnered or lost in F/LOSS culture.

The Gift as Poison

As commonly conceived, F/LOSS is the gift of freely shareable, modifiable, and redistributable software to all digital developers and users. But drawing upon the word's double meaning in Germanic languages, Mauss explained that there is also a dangerous element to the gift. Not simply a "present," the gift also function as "poison." Using the example of the giving and receiving of animals, food, and land, Mauss described a complex relationship that forms among the giver, the recipient, and the gift itself to affect complex social relations that do not always lead to the feel-good culture that is sometimes erroneously attributed to gift economies. One extreme form of such relations is the potlatch, which has often been compared to F/LOSS development practices (Bergquist & Ljungberg, 2001; Ghosh, 2005; Raymond, 2000; Wayner, 2000). Meaning literally "to feed" or "to consume," the potlatch took shape as an on-going festival of gift-giving rooted in a competition for honor and prestige (Mauss, 2002, pp. 7–8). But often overlooked in discussions of F/LOSS is Mauss' description of the potlatch of some indigenous tribes in the Pacific Northwest of the United States as nothing less than destructive, as "one must expend all that one has, keeping nothing back" in a "competition to see who is the richest and almost the most madly extravagant. Everything is based upon the principles of antagonism and rivalry" (p. 48). Because of its economy, a gift-giving culture is easily and ideally reduced to an, if not pure, then certainly purer state by which to engage in social relations (see Raymond's *The Cathedral and the Bazaar*, 1999). But this improvement in social relations is seldom realized for two reasons. First, the gift of F/LOSS is really more a case of *quid pro quo*, as Opp (2010) surmises: "You participate. You solve shared problems. Others do the same. In many ways, you give to get." Second, as the potlatch evidences, this gift can be just as much a source of ill as good.

For F/LOSS hackers, the potlatch occurs through the constant giving and getting of code that is obviously positive and productive, at least from this particular conception of gift-giving. However, the competition for reputation is judged not only on the quality of the code one produces but also on the manner of participation on the forums and mailing lists that are so integral to the distributed manner by which F/LOSS projects are developed. For instance, in a how-to entitled *How To Question the Smart Way*, first written in 2001 and last revised in 2014, Raymond remarked, "In the world of hackers, the kind of answers you get to your technical questions depends as much on the way you ask the questions as on the difficulty of developing the answer" (2014). While including basic writing tips—"Write in clear, grammatical, correctly-spelled language." and "Be precise and informative about your problem."—his advice is ultimately intended to keep forum participants from being a "loser," a word that appears throughout the document. From its earliest incarnation at MIT, the hack was considered a feat "imbued with innovation, style and technical virtuosity" (Levy, 2001, p. 23). Consequently, the love a good challenge is valued just as much by hackers as the ethic of information freedom. And what better way to demonstrate one's programing prowess than sharing the code for all to see. But to fall short in reifying these values, including at the level of question asking, can be met with derision, as Raymond explained:

> Hackers have a reputation for meeting simple questions with what looks like hostility or arrogance. It sometimes looks like we're reflexively rude to newbies and the ignorant. But this isn't really true.
>
> What we are, unapologetically, is hostile to people who seem to be unwilling to think or to do their own homework before asking questions. People like that are time sinks—they take without giving back, and they waste time we could have spent on another question more interesting and another person more worthy of an answer. We call people like this "losers" (and for historical reasons we sometimes spell it "lusers").
>
> We realize that there are many people who just want to use the software we write, and who have no interest in learning technical details. For most people, a computer is merely a tool, a means to an end; they have more important things to do and lives to live. We acknowledge that, and don't expect everyone to take an interest in the technical matters that fascinate us. Nevertheless, our style of answering questions is tuned for people who *do* take such an interest and are willing to be active participants in problem-solving. That's not going to change. Nor should it; if it did, we would become less effective at the things we do best. (2014)

Simply put, hacking and hostility often go hand-in-hand. To believe otherwise would suggest a compromised commitment to hacking and its ethic. It is easy to understand how, in the gifting of freely shareable code and, at the same time, imparting digital freedom to a world that would otherwise be enslaved (yes, enslaved, according to Stallman) by closed, proprietary software, protecting those methods that ensure hacker gift-giving is of the utmost importance. Consequently, to be put off by the hostile rhetoric sometimes emitted in the course of code talk can itself constitute being a "loser," concerned less with code than social etiquette, hence Raymond's advice "On Not Reacting Like a Loser." In this way, the hacker culture of F/LOSS reifies "the male dominated, highly impersonal and individualistic culture of STEM fields" (Wuhib & Dotger, 2014) that, in turn, forward values where the individual, not larger systemic inequalities, is to blame for lack of inclusion or failure.

In much the same way that technical endeavor is typically thought to be cocooned within a bubble of instrumental neutrality that protects it from any traditional moral or political judgments (Ellul, 1980; Winner, 1988), the creation of F/LOSS not only excuses all matter of what might otherwise be considered sins but justifies them because, in the end, code is the only end that matters. Take, for example, a December 2012 exchange on the Linux Kernel Mailing List, the primary means of communicative exchange about the ongoing development of Linux, the heart of a free/open operating system that, together with work from the GNU Project, first garnered major success for the F/LOSS communities. In response to kernel maintainer Mauro Carvalho Chehab's assertion that the location of a bug must be in the audio application PulseAudio and not in the Linux code itself, Torvalds, the creator of the Linux kernel and its "benevolent dictator" who has overseen its development since 1991, wrote the following:

> Mauro, SHUT THE FUCK UP!
>
> It's a bug alright—in the kernel. How long have you been a maintainer? And you *still* haven't learnt the first rule of kernel maintenance?
>
> If a change results in the user programs breaking, it's a bug in the kernel. We never EVER blame the user programs. How hard can this be to understand? (2012)

Then quoting from Chehab's email in which he asserted, "[I]t looks that pulseaudio/tumbleweed has some serious bugs and/or regressions," Torvalds went on to reemphasize his point by chastising Chehab further:

> Shut up, Mauro! And I don't _ever_ want to hear that kind of obvious garbage and idiocy from a kernel maintainer again. Seriously.
>
> I'd wait for Rafael's patch to go through you, but I have another error report in my mailbox of all KDE media applications being broken by v3.8-rcl, and I bet it's the same kernel bug. And you've shown yourself to not be competent in this issue, so I'll apply it directly and immediately myself.
>
> WE DO NOT BREAK USERSPACE!
>
> Seriously. How hard is this rule to understand? We particularly don't break user space with TOTAL CRAP. I'm angry, because your whole email was so _horribly_ wrong, and the patch that broke things was so obviously crap. The whole patch is incredibly broken shit. It adds an inane error code (ENOENT), and then because it's so inane, it adds a few places to fix it up ("ret == —ENOENT? —EINVAL: ret").
>
> The fact that you then try to make *excuses* for breaking user space, and blaming some external program that *use* to work, is just shameful. It's not how we work.

> Fix your f*cking "compliance too", because it is obviously broken. And fix your approach to kernel programming. (2012)

Leaving little doubt that the potlatch is not simply a matter of an expert antagonistically chiding a newbie, Torvalds' response and the degree to which it schools Chehab and, at the same time, reifies Torvalds' own social standing, clearly illustrates a destructive element, often misrecognized as positive, in F/LOSS gift-giving culture.

The hostility supposedly justified, even measured, by commitment to the hacker ethic constitutes the poison of freely shareable software. Yet, an emphasis on the creativity and joy that arises from hacking allows for the appearance of a gender-neutral meritocracy where everyone with interest and skill is free to participate. In fact, judged purely on the prowess of one's code and participation in the culture, *The New Hacker's Dictionary* claimed that a "female hacker" need not suffer the same kinds of oppression—quotidian or otherwise—that other women might out in the technical world: "Hackerdom is still predominately male. However, the percentage of women is clearly higher than the low-single-digit range typical for technical professions, and female hackers are generally respected and dealt with as equals" (Raymond, 1996). The reason for such egalitarianism was attributed to the "positive effect of text-only network channels," in which code creation often occurs virtually and distributively. These channels allow for a "gender- and colorblindness" that can, if not thoroughly erase, then certainly minimize the cultural significance of otherness. With gender rendered essentially meaningless in a network where life is for the consumption and production of code, not the encoding of and by the body, hackerdom purports to offer the instantiation of an essentially genderless world composed of 1s and 0s. But as the statistics of the dismally low participation rate of women in F/LOSS evidences, not only is hacker participation by women incredibly low, it suffers a worse rate than proprietary software industry, even though, as James (2010) pointed out, 75% of F/LOSS software is now being written by people paid to do so rather than simply for the love of hacking.

Critiques of the F/LOSS communities often point to practices other than coding itself as a force that identifies hacking as a decidedly masculine activity that thereby ensures the exclusion of women whether purposeful or not. In one of the earliest analyses of this intersection between language and gender in the F/LOSS communities, Adam (2003) highlighted the effect of frontier metaphors in the conceptualization of hacking "where not only is the hacker's mastery over savage nature (the computer and the program) celebrated but so too is the hacker's disdain for the prevailing norms of society." The result is a certain kind of social asceticism among hackers, one that rejects unnecessarily constraining norms and serves as point of pride at the hacker ability to put code above all else. Leach (2009) also noted that in spite of an explicit rhetoric of ethics in F/LOSS development, these ethics are too often limited to code as innate and ultimate goodness: "The fact that certain aspects of the constitution of the social form of f/loss do not come under scrutiny internally is a direct effect of the power of the imagination in relation to software's potential contribution to a better world" (p. 67). As a consequence, "An ethos of expanding the boundaries of knowledge and making code as functional as possible stands as its own moral good (p. 61)." Respectively, Coleman (2004, 2013) and Kelty (2008) described this phenomenon in terms of a hacker agnosticism that frees F/LOSS from the constraints of traditional politics and morality. But, like Leach, Naufus (2011) argued that the rhetoric of F/LOSS only fosters a misrecognition of what hacking is and therefore exacerbates the gender problem: "'Open' is believed to mean that technology circulates on its own accord, set free to have its own impacts on the world" (p. 682). To this end, "In order to open the doors to 'anyone,' the possibility that there are social loops creating knowledge and passing it along—the very mechanism that both excludes women and could serve to include them—threatens the basis on which it is possible to claim that the door is open." To admit otherwise threatens to shatter the illusion that freely shareable software is a gift without harm. In fact, the kind of rhetoric demonstrated by Torvalds and sanctioned by Raymond interestedly perpetuates this illusion. As Naufus explained, "[Hackers] must create the truthfulness of code, and thus technology's edge, through highly masculinized, aggressive online talking" (p. 679). Importantly, through her team's observations, interviews, and surveys, Naufus deduced that such aggression has its own pay off: "We found cases where people who do contribute a great deal of code, but do not get involved in flame wars about the worth of code, are perceived as less knowledgeable." Writing, in the constitution of both words and code, is therefore integral to the reputation of hackers.

For these reasons, to identify the gift of F/LOSS as without beneficial return for the giver is little more than what Bourdieu (1997) identified as "collective self-deception" (p. 192) that pervades ideas about the gift-giving economy of F/LOSS economy. In what Schrift (1997) described as the "ungendered" discussions of gift-giving embodied by Mauss (p. 11) and evidenced in most analyses of F/LOSS, the claim to genderlessness, much like the gender blindness often attributed to hacker culture, must be understand as something decidedly masculine. Because of the self-deception regarding both the gift and the giver, the difference between the exchange economy, exemplified by proprietary software, and the gift economy, exemplified by F/LOSS, only appear to be a difference in kind rather than degree. But the gift of F/LOSS and proprietary software are not so far removed from another as both are rooted in the same kind of economy that is not limited to the economic. After all, in F/LOSS development, "code was just as readily given away as a way of cutting ties" (Nafaus, 2011, p. 670). But to recognize that there is a problem with F/LOSS as gift through recognition and redress of the gender problem in software hacking threatens to upset the foundational way in which gift-giving is conceived of by hackers. And yet, by rebelling against the poison gift of F/LOSS, some hackers are seeking to do just this.

Women's Writing

As Balsamo (1996), Haraway (2001), Harding (1986), and Wajcman (1991) have respectively argued, modern science and technology are grounded in dichotomies that ultimately serve certain interests at the expense of others. Haraway explained, "In the traditions of 'Western' science and politics—the tradition of racist, male-dominant capitalism; the tradition of progress; the tradition of the appropriation of nature as resource for the productions of culture; the tradition of reproduction of the self from the reflections of the other—the relation between organism and machine has been a border war." (p. 292). Not unlike in Fordist production, hacking, in its various forms (see Dunbar-Hester's 2008 work on radio tinkering, for example), often reifies borders between the masculine and feminine to the point of the latter's exclusion. In order to move beyond such borders, Haraway argued that we must come to see ourselves as cyborgs, "creatures in a post-gender world" (p. 294). In contrast to early utopian visions of that Internet as a technology that would free us from gender altogether, the cyborg skips the origin stories of the masculine and the feminine and instead embraces a kind of partiality, even taking "pleasure in the confusing of boundaries and ... taking responsibility in their construction" (p. 295). In the context of F/LOSS, this partiality might initially be located in the participation of women hackers. But while important, this kind of presence alone does not necessarily catalyze the kind of boundary blurring which Haraway and others have called for and instead too often reifies the myth of meritocracy that permeates hacker culture.

Instead, I want to suggest that the cyborg begins to be made manifest in F/LOSS hacking culture through what Cixous (1986) called *écriture feminine* or women's writing. Cixous see writing as a way for women to write themselves into the world in such a way that it challenges "the subordination of the feminine to the masculine order" (p. 65). Recognizing that "men and women are caught up in a web of age-old cultural determinations that are almost unanalyzable in their complexity," including that of "woman" and "man" (p. 83), Cixous pointed to a subversive way by which women, in particular, can find a way to rebel against their subordination and invisibility that occurs as a result the different ways in which the gift is materialized in the different and, certainly not limited to economic, economies through which "a whole huge system of cultural inscription" is made "legible as masculine or feminine" (1986, pp. 80–81). Through women's writing, women are able to blur and even cross the boundaries set for them, no longer "kept out of the way, on the edge of the stage, on the kitchen side, the bedside" (p. 70). Of course, men also write. But the difference for Cixous lies in the gift of women's writing, rather than masculine writing. Not limited to woman alone, women's writing escapes the limitations of masculine "*rapport*," meaning return. For men, "loss and expense are stuck in the commercial deal that always turns the gift into a gift-that-takes. The gift brings in a return" (p. 87). As I have suggested, hacking is often misrecognized as an escape from this limitation because freely shareable software is not limited to a commodity in the market of exchange. But even if the gift of F/LOSS was not limited to *quid pro quo*, the pay off that comes from the aggressive masculinity in F/LOSS rhetoric thoroughly compromises the gift, meaning that it exacts, even if unknowingly, a great price from those to whom it is gifted. Cixous asked, "In return for his gift, what does 'the traditional man' want?" She answered: "At first what *he* wants, whether on the level cultural or of personal exchanges, whether it is a question of capital or of affectivity … is that he gain more masculinity: plus-value of virility, authority, power, money, or pleasure …" (p. 87). In spite of placing information freedom before, but not necessarily exclusive of, monetary profit, hackers still desire reputation, which is the aim of the potlatch. In Cixous' terms, freely shareable code is a gift-that-takes because it is meant to come back to the giver in form of not just code but increased social status. In the context of F/LOSS, the challenge to the market economy offered by dismantling code as commodity does not therefore translate to absence of gain by some other measure via poisonous means. Thus, whether through exchange or gift, software development is built upon a masculine economy whereby "masculinity/femininity are opposed in such a way that it is male privilege that is affirmed in a movement of conflict played out in advance" (p. 80).

But what of the gift made possible by women's writing? Is it not also subject to such return? While women may, of course, desire the same kind of return, it is in a particular kind of women's writing that an escape from the "law of return" is to be found by those who wish to rebel against the gift-that-takes. Cixous acknowledged, "There is no 'free gift'. You never give something for nothing" (1986, p. 87). But the difference in women's writing "lies in the why and how of the gift, in the values that the gesture of giving affirms, causes to circulate." Evidenced not only in the kind of rhetoric that justifies the use of name-calling like "loser" but also in the masculine culture common not just to hacking but to tech generally (Cave, 2014; Spiegel, 2008; Thomson, 2008), the poison gift too often comes at the expense of another, as "a man is always proving something; he has to 'show off,' show up others" (p. 87). Women's writing, in contrast, seeks to break free of this dynamic "by willing the togetherness of one-another, infinitely charged with a ceaseless exchange of one with another." But this kind of exchange (not to be confused with exchange economy) that occurs in a feminine economy is different than the exchange of code that occurs in hacker culture, for example, because, in women's writing, she does not seek return, to "try to 'recover her expenses'." As such, rather than the gift-that-takes, women's writing constitutes the "gift-that gives" (Cixous, 1976, p. 888). For Cixous, a feminine economy is one of openness, premised upon the gift given without the expectation of return: it "is an economy which has a more supple relation to property, which can stand separation and detachment, which signifies that it can also stand freedom—for instance, the other's freedom" (as quoted in Schrift, 1997, p. 12). In a masculine economy, the absence of return creates an uncomfortable debt or obligation that relegates social alliances always uneasy to varying degrees (Cixous, 1981). For, "to refuse to give" in fulfillment of the obligation created by having been gifted "is tantamount to declaring war; it is to reject the bond of alliance and commonality" (Mauss, 2002, p. 17). And even in more subtle ways, such as how one asks or answers a question, the lack of appropriate return can suggest a weakness constituted by a lack in hacker belief and practice.

In a feminine economy, however, these negative affects are transformed to demonstrate that the gift-that-gives does not demand return because it is based on generosity (Diprose, 2002; Joy, 2013; Schrift, 2001). For women to write "toward women" is to create "the very possibility of change, the place that can serve as a springboard for subversive thought, the precursory movement of a transformation of social and cultural structures" (Cixous, 1976, p. 879). But this gift neither ensures that change nor obligates others to help catalyze it. In fact, women's writing put out in the world is likely to receive rebuke meant to deter those who may wish now or in the future to take their "turn to speak." Thus, rather than a sign of weakness, the absence of return in the gift of women's writing denotes a strength that comes from generosity, especially in regards to the other. In theorizing this disposition, Diprose (2002) argued, "Generosity operates at the level of sensibility … The openness to otherness that characterizes generosity is … carnal and affective, and the production of identity and difference that results is a material production" (p. 9). To engage in women's writing therefore is to break down the artificial boundaries and exclusionary tactics of masculine economy, while also demonstrating that, unlike in a gift-giving culture centered upon the potlatch, it is "not impossible or pathetic to be generous" (Cixous, 1986, p. 72).

The gift-giving of code is easily misrecognized as stemming from generosity. But because of the return expected from the gift of F/LOSS, as well as the hostility manifest in its development, F/LOSS's hacker culture builds boundaries often in quotidian ways meant to lift up some at the expense or exclusion of others. And even if generosity of code did exist, it certainly would not extend to generosity with one another. In truth, a generous disposition is likely to be considered a sign of weakness in the masculine economy of hackers, the kind of thing pounced upon as a fragility born of care about trivial social norms that have nothing to do with code. To argue otherwise is likely to be reduced to the loser reaction of whining or some other pathetic annoyance (Raymond, 2014). What is more, to care about anything other than code is to reveal oneself as an outsider, whose hackerness is less than it ought to be. For, in the masculine economy, it is better for hackers and for code to be as Torvalds has described: "I'm not a nice person, and I don't care about you. I care about the technology and the kernel—that's what's important to me" (Machkovech, 2015). Yet, in spite of the potlatch always already taking place in hacker culture, women's writing has sought to challenge and change the closed, exclusionary gift-giving of the masculine economy of F/LOSS. Threatening to unveil the tenuous solidarity constituted through both the philosophy and practice of the hacker ethic, women's writing, as cyborg practice, fights back against the gift-that-takes and offers an alternative, a remedy to the hacker potlatch and its particular kind of poison.

The Gift Reimagined

In "Open Source is Way Whiter and Maler than Propriety Software," Matthew Asay (2013), Vice President of Mobile at Adobe and emeritus board member of the advocacy organization the Open Source Initiative, noted an inherent problem with software hacking: "In the open source world, there is no safety net and no real on-ramp to participation. Communities self-police, and not always very well. Stepping into a overwhelmingly white, male community could be difficult, even if the community isn't overtly exclusionary." But what constitutes overt exclusion? And, more to the point, is overtness even needed if the conditions are made to exclude in more subtly pervasive ways?

Ashe Dryden (2013), a programmer with 15 years experience who now works as a diversity advocate and consultant, has explained, "I know many women that either don't contribute to [open source software] because they've been dismissed for being women—being too pretty, not pretty enough, being forced to prove their competence more than their male counterparts because they're women." And herein lies the dangerous duplicity. Even as it evangelizes openness and freedom, hacker culture excludes through a whole range of writing practices that, while of little matter to most, matters quite a bit to some.

To disrupt the law of return that constructs freely shareable, modifiable, and redistributable software as a "gift-that-takes" and to combat the potlatch that renders F/LOSS a poison for many, hackers who acknowledge and wish to change this exclusionary culture have employed women's writing. In doing so, F/LOSS as the gift-that-takes is challenged by the gift-that-gives. Women's hacking has thus far taken form in a number of ways that undermine the belief that F/LOSS development necessitates the potlatch. For instance, F/LOSS hackerspaces constituted by women's writing, as well as women's coding, have created communities that reject the hostility and aggression trumpeted as natural and necessary to hacking. The first meeting of LinuxChix, for example, was held in September 1995 in a San Francisco coffeehouse where women were able to geek out comfortably about Linux and also share stories about the discrimination that they often experienced (Bowman, 1999). Founder Deb Richardson (1999) later explained on the online site of the LinuxChix community that the impetus for the organization arose not only from a desire to have fun but also from the fact that "I got tired of seeing new users being browbeaten for asking stupid questions. I got tired of seeing people respond to perfectly valid question with RTFM [read the fucking manual], or we're not a Linux help channel, and other such not-terribly-useful things. I got tired of the lock-room mentality of the more popular online Linux forums." Through women's writing, LinuxChix was created out of a generosity toward others to provide an alternative to the poison gift of F/LOSS. In opening up a space to those all too often othered in F/LOSS development, LinuxChix offered inclusivity without the expectation of the kind of return demanded in masculine economy. Rather than a space comparable to the Linux Kernel Mailing List, Richardson explained that LinuxChix was meant for "new users" who might otherwise be scared off by those who are "brash, harsh, and less tolerant of people who know less than they." Consequently, LinuxChix was not premised upon the promise of giving or receiving new patches of code, let alone building the kind of hacker reputation traditionally valued in F/LOSS development.

Hackerspaces constituted by women's writing aim to provide an alternative to the masculine economy that undergirds traditional hackerspaces. As defined by the informational site of the same name, "Hackerspaces are community-operated physical places, where people share their interest in tinkering with technology, meet and work on their projects, and learn from each other." To facilitate these objectives, these volunteer networks often utilize public mailing lists to coordinate meet-ups and to talk hacking and tinkering in between meets. In sum, hackerspaces, according to Maxigas (2012), were "set up by hackers for hackers with the principle mission of supporting hacking." Hacking is therefore rife with strict gender boundaries that privilege the masculine at the expense of the feminine. The seemingly open process of hacking, whereby one need only be interested in hacking to participate in hackerspaces, too often leads to what Nagbot (2016) identified as a closing-off that manifests in an exclusion most obviously identified in a lack of diversity in hacking. In a discussion of her local Linux User group, Amber Wu (2014), for example, highlighted how rhetorics of openness too often accomplish just the opposite:

> "Free as in Freedom" applied to the language, culture, gender, and identity in general just as much as source code. Prejudiced in-groups such as the worst of [hacker] meetups, however, hide behind the same belief. Events being "open to everyone" was a token political statement. If they didn't recognize someone as being part of their in-group, all they had to do was make them uncomfortable. After all, that person is always "free" to leave.

But through women's writing, there is a means to call out and rebel against the collective self-deception that pervades the hacker sensibility that allows closure to exist as openness. But, in hackerspaces such as LinuxChix, women's coding, as a form of women's writing, is just as important in this gift-giving. Borrowing from Kelty's work on recursive publics of free software, Toupin (2014) has remarked that feminist hackers are "hard at work shaping the very conditions which underwrite their debates and projects. In other words, a concern for the kind of hackerspaces that they want to hack in—and how such spaces can be actualized—is at the heart of their association." In this way, women's writing, in both natural and programming languages, stands to constitute a feminine economy of hacking.

Governed by the principle—"Be Polite. Be Helpful."—LinuxChix is a feminist hackerspace that serves as "a community for women who like Linux and anyone who wants to support women in computing." Created through the generosity of women's writing by those who want to hack but are put off by the poison gift of F/LOSS, the virtual and physical manifestation of this desire provides a space where women in the Open Source Community, as well as the Free/Libre community, are able to make themselves visible without the threat of repercussion that would otherwise come with the "gift-that-takes." With LinuxChix, as well as its Spanish-language offshoot ChicasLinux, the emphasis is on having a space "where everyone is and feels welcome." Consequently, in meetings and on the mailing list, "sexist remarks" and "all sweeping generalizations" are to be avoided, as "sexism can be targeted at men as well, which is also significantly uncool." And to keep the space as open and welcoming as possible, "All the silly—isms are discouraged and will not be tolerated: sexism, ageism, racism, etc." Explicit guidelines such as this are not completely unheard of in F/LOSS development, of course. Raymond advised that to answer question in a helpful way—"*Be gentle ... If you don't know for sure, say so! ... If you're going to answer the question at all, give good value ... Help your community learn from the question*" (2014). But the problem with such directives is that they are issued in a space where hacker norms so often conflict with politeness. For instance, Raymond reaffirmed in the same document that "rudeness" is just a natural part of hacker culture where "much of what looks like rudeness ... is not intended to give offense. Rather it's the product of direct, cut-through-the-bullshit communications style that is nature to people who are more concerned about solving problems than making others feel warm and fuzzy" (2014.) In hackerspaces such as LinuxChix, however, the desire to "create a more hospitable community" that "encourages participation" (Richardson, 1999) is borne out in practice as a productive force, not fragility.

The desire to create hackerspaces that are open to all does at time necessitate action that helps to ensure the actualization of the "gift-that-gives." In order that generosity and its materializations not be abused, some of the LinuxChix chapters have had to restrict membership to women due to threats that compromise the mission of the space. Women's writing necessarily speaks to the destructive force of the poison gift and is integral to normalizing generosity and the feminine economy. To speak with this purpose is not meant to be provocation, but, in the words of Cixous, "it means that woman admits there is an other" (1986, p. 101). As architect and maintainer of the Linux USB 3.0 host controller driver, Intel employee Sarah Sharp has played an active role on Linux Kernel Mailing List. However, in addition to work on the driver, Sharp has also addressed on an ongoing basis the hostility that too often undercut list messages. Responding to an exchange that, while playful in many ways, nonetheless included Torvalds writing things like—"Have you guys *seen* Greg? The guy is a freakish giant. He *should* scare you."—while also encouraging Greg, to stop being a "door-mat" and "to learn to shout at" contributors (2013). In response, Sharp wrote:

> Seriously, guys? Is this what we need in order to get improve -stable?
> Linus Torvalds is advocating for physical intimidation and violence.
> Ingo Molnar and Linus are advocating for verbal abuse.
>
> Not *fucking* cool. Violence, whether it be physical intimidation,
> verbal threats or verbal abuse is not acceptable. Keep it professional
> on the mailing lists.
>
> Let's discuss this at Kernel Summit where we can at least yell at each
> other in person. Yeah, just try yelling at me about this. I'll roar
> right back, louder, for all the people who lose their voice when they
> get yelled at by top maintainers. I won't be the nice girl anymore. (2013)

For Torvalds, Sharp's response, which in many ways adheres to the style demands of masculine hacker rhetoric, is a step in the right direction:

> That's the spirit.
>
> Greg has taught you well. You have controlled your fear. Now, release your anger. Only your hatred can destroy me.
>
> Come to the dark side, Sarah. We have cookies.

While seemingly tongue-in-cheek, Torvalds' message is only an exaggeration of an insidious truth materialized by the poison gift. But in her women's writing, Sharp appropriated a style in order not to receive return but to speak, in part, for all those who have been bullied into not taking their turn to speak. Sharp made clear, in spite of the joking back and forth, "I'm serious about this. Linus, you're one of the worst offenders when it comes to verbally abusing people and publicly tearing their emotions apart." Offering links to two of his messages in the PulseAudio thread, Sharp concluded, "I'm not going to put up with that shit any more." What followed was a response by Torvalds in which he wrote of how he despised subtlety and niceness; how, being Finnish, he was a "minority" from a culture that "includes cursing" and Sharp was therefore being culturally insensitive in asking him to not curse at community members; and how Sharp needed to be made "aware that your 'high horse' isn't necessarily all that high." Rebukes continued, with Sharp receiving a private email accusing her of "playing the victim card," to which, she publicly responded, "I will repeat: this is not just about me, or other minorities. I should not have to ask for professional behavior on the mailing lists. Professional behavior should be the default." But for Torvalds, the "victim card" was exactly what was at play as Sharp was "trying to enforce your particular expectations on others, and trying to do so in a very particular way." Although certainly not asking for the same materialization of generosity demonstrated in the women's writing and coding for LinuxChix, Sharp nonetheless argued for a baseline professionalism that would induce some measure of generosity currently missing on the Linux Kernel Mailing List and open a pathway whereby inclusivity might be encouraged rather than thwarted.

Although Torvalds wrote in this exchange that he "really fundamentally believes that being honest and open about your emotions about core/process is good," his conception of openness is predicated upon the hacking potlatch of the masculinist economy where hostile rhetoric is not just an appropriate but beneficial means for return. Because Torvalds functions as what Bourdieu describes as a spokesperson whose "speech concentrates within it the accumulated symbolic capital of the group which has delegated him and of which he is the *authorized* representative" (1991, p. 24), he is not simply exercising what he claims is a nation-based cultural style of writing but instead is sanctioning others to embrace the worst effects of the gift-that-takes under the guise of doing good. In doing so, Torvalds encourages the border war made possible, even expected, by masculinist economy. Free is only free for some; openness only open to some. But through the women's writing of Sharp and others like Kees Cook, who commented during the Linux Kernel Mailing List (lkml) exchange, "People's mistakes can be pointed out without the kind of abuse I've read on lkml," the possibility for change, rooted in a generosity that expects no return except recognition and respect of otherness, is recognized and makes possible the materialization of a feminine economy in F/LOSS hacking culture.

Conclusion

Women's writing by hackers has documented acts of poison gift-giving, testified to the worth of diversity, and worked to bring about change to the masculine economy of F/LOSS development, specifically, and the tech world, generally. And the call for increased inclusivity and openness that comes from the "gift that gives" has not gone unheard. In 2014, for instance, the open source company Red Hat established the Women in Open Source Award to recognize the contributions of those so often marginalized in F/LOSS. In 2015, Sharp became the first winner in the Community category "for her efforts in improving communications and inviting women into open source community" (Red Hat, 2014). Through her work coordinating Outreachy, an organization that "helps people from groups underrepresented in free and open source software get involved," Sharp has helped to cultivate opportunities like internships that "are open internationally to women (cis and trans), trans men, and genderqueer; to residents and national ... residents and nationals of the United States of any gender who are Black/African American, Hispanic/Latin@, American Indian, Alaska Native, Native Hawaiian, or Pacific Islander." But because notice of the need for inclusivity and diversity can also be met with hostility, it is not surprising that Sharp's award, which she celebrated on her blog in an entry entitled "I won Red Hat's Women in Open Source Award!" was met with much rebuke. Posted on Reddit, Sharp's blog entry garnered a great deal of attention, although much of the support centered on Sharp's work on Linux USB 3.0 rather than with Outreachy. Yet, the nature of the award also raised attention to the work Sharp had done on the Linux Kernel Mailing List in calling for inclusivity (Reddit, 2015). That being said, other Reddit users questioned the need for the Women in Open Source Award. User 9729 commented that, in regards to Outreachy, "[I]t sounds pretty biased on who gets those internships. I mean the authors [*sic*] mindset seems geared toward giving women a leg up. It's difficult getting into the industry with little to no experience for any gender" (Reddit, 2015). But as user send-me-to-hell responded, "If there's a group you're a part of that's conspicuously absent from a larger community, it makes it hard to feel comfortable in that community and generally harder for people in that community to have an awareness of how to deal with people from your particular group. It's in both group's interests to have diversity."

Although there has been progress in working toward a feminist economy of F/LOSS, where generosity is the gift-that-gives without expectation of return, the poisonous gift-that-takes not surprisingly still dominates. As Byfield (2014) wrote in a *Datamation* article about the myths concerning feminism and free/open source, "In the last five years, the role of women has grown and feminism has become a recognized part of the free and open source ... community. However, this change has not gone unchallenged." The depth of these challenges is in many ways crystallized in Sharp's decision to leave the Linux kernel community. In her October 5, 2015 blog post, "Closing a Door," which she noted had been sitting in her drafts folder for a year because "I have always been worried about the backlash," Sharp explained her decision to leave:

> Here's the deal: I'm not a Linux kernel developer any more. I quietly transferred the maintainership of the USB 3.0 host controller driver in May 2014. In January 2015, I stepped down from being the Linux kernel coordinator for the FOSS Outreach Program for Women (OPW), and moved up to help coordinate the overall Outreachy program. As of December 6, 2014, I gave what I hope is my

> last presentation on Linux kernel development. I was asked to help coordinate the Linux Plumbers Conference in Seattle in August 2015, and I said no. My Linux Foundation Technical Advisory Board (TAB) term is soon over, and I will not be running for re-election.
>
> Given the choice, I would never send another patch, bug report, or suggestion to a Linux kernel mailing list again. My personal boxes have oopsed with recent kernels, and I ignore it. My current work on userspace graphics enabling may require me to send an occasional quirks kernel patch, but I know I will spend at least a day dreading the potential toxic background radiation of interacting with the kernel community before I send anything. (2015)

While Sharp praised the unmatched technical abilities of the kernel community, the poison gift-that-takes had finally taken too much. Due to the continued support given to maintainers "to spew whatever vile words they needed to in order to maintain radical emotional honesty," Sharp lamented that, although "technically respected" in the Linux kernel community, she was not given, nor could she ask for, "personal respect." The "communication style" of this hacker community had created an "awful power dynamic there that favors the established maintainer over basic human decency." Therefore, Sharp, who had for so long taken a turn to speak out against the masculinist economy in this context and its embrace of the "gift that takes" noted that she had no choice but to leave the community. But in giving up the possibility of taking addition turns to speak out on the Linux Kernel Mailing List, Sharp remarked that she felt "sad every time someone thanks me for standing up for a better community norms, because I have essentially given up trying to change the Linux kernel community." Weary from the border war between masculine and feminine economies in the context of Linux development, Sharp's generosity had apparently been exhausted. And yet, in Sharp's writing of her decision to leave the community, we see, possibly in spite of herself, her continued embrace of women's writing.

The example of Sarah Sharp illustrates just how closed the openness, just how unfree the freedom of F/LOSS development can be because of the poison gift. To engage in language and coding as women's writing that rebels against the gift-that-takes and instead offers the gift-that-gives is no easy matter, and, as in the case of Sharp, sometimes necessitates weighing where one's efforts are best spent. But through a generosity of spirit that wants for others, as well as oneself, the kind of freedom that encourages not just F/LOSS development but the technological world to be inclusive, there is a path toward a feminine economy that sees the strength in such a dream. But to be achieved, we must, even in the midst of despair, heed Cixous and "Write, let no one hold you back, let nothing stop you."

References

Abbate, J. (2012). *Recoding gender: Women's changing participation in computing.* Cambridge: MIT Press.

Adam, A. E. (2003). Hacking into hacking: Gender and the hacker phenomenon. *ACM SIGCAS Computers and Society, 32*(7). p. 3.

Arjona-Reina, L. L., Robles, G., & Dueñas, S. (2014, January). *The FLOSS2013 Free/Libre/Open Source Survey* (2013). Retrieved from http://floss2013.libresoft.es/results.en.html

Asay, M. (2013, December 11). Open source is way whiter and maler than proprietary software. *Readwrite.* Retrieved from http://readwrite.com/2013/12/11/open-source-diversity

Balsamo, A. (1996). *Technologies of the gendered body: Reading cyborg women.* Durham, NC: Duke University Press.

Bergquist, M., & Ljungberg, J. (2001). The power of gifts: Organizing social relationships in open source communities. *Information Systems Journal, 11*(4), 305–320.

Bourdieu, P. (1993). *Language and symbolic power.* Cambridge, MA: Harvard University Press.

Bourdieu, P. (1997). *Pascalian meditations* (R. Nice, Trans.). Stanford, MA: Stanford University Press.

Bowman, L. (1999, September 15). She-geeks confess love for Linux. *ZDNet.* Retrieved from http://news.zdnet.com/2100-9595_22-515695.html

Bureau of Labor Statistics. (2016). *Statistics from the current population survey.* Retrieved from http://www.bls.gov/cps/cpsaat03.htm

Byfield, B. (2014, January 16). 9 myths about free/open source software and feminism. *Datamation.* Retrieved from http://www.datamation.com/open-source/9-myths-about-freeopen-source-software-and-feminism-1.html

Cave, K. (2014). IDG: Workplace bullying in technology companies. *Workplace Bullying Institute.* Retrieved from http://www.workplacebullying.org/idg

Ceruzzi P. E. (2003). *A history of modern computing.* Cambridge, MA: MIT Press.

Chopra, S., & Dexter, S. D. (2008). *Decoding liberation: The promise of free and open source software.* New York, NY: Routledge.

Chun, W. H. K. (2004). On software, or the persistence of visual knowledge. *Grey Room, 18*, 26–51.

Cixous, H. (1976). The laugh of the medusa (K. Cohen and P. Cohen, Trans.). *Signs, 1*(4), 875–893.

Cixous, H. (1981). Castration or decapitation (A. Kuhn, Trans.). *Signs: A Journal of Women in Culture and Society, 7*(1), 41–55.

Cixous, H. (1986). Sorties: Out and out: Attacks/ways out/forays (B. Wing, Trans.). In H. Cixous & C. Clément (Eds.), *Newly born woman* (pp. 63–134). Minneapolis, MN: University of Minnesota Press.

Clegg, S., & Trayhurn, D. (2000). Gender and computing: Not the same old problem. *British Educational Research Journal, 26*(1), 75–89.

Cohoon, J. M., & Aspray, W. (Eds.) (2006). *Women and information technology: Research on underrepresentation.* Cambridge, MA: MIT Press.

Coleman, G. E. (2004). The political agnosticism of free and open source software and the inadvertent politics of contrast. *Anthropological Quarterly, 77*(3), 507–519.

Coleman, G. E. (2013). *Coding freedom: The ethics and aesthetics of hacking.* Princeton, NJ: Princeton University Press.

Cooke, K. (2013, July 15). Re: [00/19] 3.10.1-stable review. *Linux Kernel Mailing List.* Retrieved from http://lkml.iu.edu/hypermail/linux/kernel/1307.1/03705.html

Diprose, R. (2002). *Corporeal generosity. On giving with Nietzsche, Merleau-Ponty, and Levinas.* Albany, NY: SUNY Press.

Dryden, A. (2013, November 12). The ethics of unpaid labor and the OSS community. Retrieved from https://www.ashedryden.com/blog/the-ethics-of-unpaid-laor-and-the-oss-community

Dunbar-Hester, C. (2008). Geeks, meta-geeks, and gender trouble: Activism, identify, and low-power FM radio. *Social Studies of Science, 38*(2), 201–232.

Ellul, J. (1980). *The technological system.* New York, NY: Continuum.

Engsmenger, N. (2003). Letting the "computer boys" take over: Technology and the politics of organizational transformation. *International Review of Social History, 48*(11), 153–188.

Free Software Foundation. (2000). Why we must fight UCITA. *GNU Operating System.* Retrieved from http://www.gnu.org/philosophy/ucita.en.html

Free Software Foundation. (2002). What is free software? *GNU Operating System.* Retrieved from https://www.gnu.org/philosophy/free-sw.en.html

Fuchs, C. (2010). Labor in informational capitalism and on the internet. *The Information Society, 26,* 179–196.

Gates, B. (1976). Open letter to hobbyists. *Homebrew Computer Club Newsletter.* Retrieved from http://www.digibarn.com/collections/newsletters/homebrew/V2_01/gatesletter.html

Gell, A. (1992). Inter-tribal commodity barter and reproductive gift-exchange in old Melanesia. In C. Humphreys & S. Hugh-Jones (Eds.), *Barter, exchange and value: An anthropological approach* (pp. 142–168). Cambridge: Cambridge University Press.

Ghosh, R. A. (2005). Cooking-pot markets and balanced value flows. In *Code: Collaborative ownership and the digital economy* (pp. 153–168). Cambridge, MA: MIT Press.

Ghosh, R. A., Glott, R., Krieger, B., & Robles, G. (2002). Free/libre and open source software: Survey and study. Retrieved from http://www.flossproject.org/report

Hackerspaces. (2006). Hackerspaces. Retrieved from http://www.hackerspaces.org

Haraway, D. (2001). A cyborg manifesto: Science, technology and socialist-feminism in the late twentieth century. In D. Bell (Ed.), *An introduction to cybercultures* (pp. 291–324). New York, NY: Routledge.

Harding, S. (1986). *The science question in feminism.* Ithaca, NY: Cornell University Press.

Henn, S. (2014, October 1). When women stopped coding. *National Public Radio.* Retrieved from http://www.npr.org/sections/money/2016/07/22/487069271/episode-576-when-women-stopped-coding

Henry, L. (2014, February 3). The rise of feminist hackerspaces and how to make your own. *Model View Culture.* Retrieved from https://modelviewculture.com/pieces/the-rise-of-feminist-hacker spaces-and-how-to-make-your-own

Himanen, P. (1999). *The hacker ethic and the spirit of the information age.* New York, NY: Random House.

Hunsinger, J. (2011). From hacklabs to hacker markets: Becoming complicit in self-exploitation. Retrieved from https://www.researchgate.net/publication/283123285

Irigaray, L. (1985). *This sex which is not one* (C. Porter with C. Burke, Trans.). Ithaca, NY: Cornell University Press.

James, J. (2010, April 6). IT gender gap: Where are the female programmers? *TechRepublic.* Retrieved from http://www.techrepublic.com/blog/software-engineer/it-gender-gap-where-are-the-female-programmers/

Joy, M. (Ed.). (2013). *Women and the gift: Beyond the given and the all-giving.* Bloomington, IN: Indiana University Press.

Kelty, C. (2008). *Two bits: The cultural significance of free software.* Raleigh: Duke University Press.

Kinsella, N. S. (2001). Against intellectual property. *Journal of Libertarian Studies, 15*(2), 1–53.

Leach, J. (2009). Freedom imagined: Morality and aesthetics in open source software design. *Ethnos, 74*(1), 51–71.

Lemley, M. A. (2014). IP in a world without scarcity. *Stanford Public Law Working Paper No. 2413974.* Retrieved from http://papers.ssrn.com/sol3/papers.cfm?abstract_id=2413974

Levy, S. (2001). *Hackers: Heroes of the computer revolution.* Sebastopol, CA: O'Reilly.

LinuxChix. About LinuxChix. Retrieved from http://www.linuxchix.org/content/about-linuxchix

Machkovech, S. (2015, January 15). Linus Torvalds on why he isn't nice: "I don't care about you." *Ars Technica.* Retrieved from http://arstechnica.com/business/2015/01/linus-torvalds-on-why-he-isnt-nice-i-dont-care-about-you/

Maher, J. H. (2015). *Software evangelism and the rhetoric of morality: Coding justice in digital democracy.* New York, NY: Routledge.

Mauss, M. (2002). *The gift: The form and reason for exchange in archaic societies.* New York, NY: Routledge.

Maxigas. (2012). Hacklabs and hackerspaces: Tracing two genealogies. *The Journal of Peer Production, 2.* Retrieved from http://peerproduction.net/issues/issue-2/peer-reviewed-papers/hacklabs-and-hackerspaces/

Merges, R. P. (2011). *Justifying intellectual property.* Cambridge, MA: Harvard University Press.

Misa, T. J. (Ed.). (2010). *Gender codes: Why women are leaving computing.* Hoboken, NJ: Wiley.

Nagbot, S. (2016). Feminist hacking/making: Exploring new gender horizons of possibility. *Journal of Peer Production, 8.* Retrieved from http://peerproduction.net/issues/issue-8-feminism-and-unhacking/

Naufus, D. (2011). 'Patches don't have gender': What is not open in open source. *New Media & Society, 14*(4), 669–683.

Noland, M., Moran, T., & Kotschwar, B. (2016). Is gender diversity profitable? Evidence from a global survey. *Peterson Institute for International Economics*. Retrieved from https://piie.com/publications/wp/wp16-3.pdf

Opp, J. (2010, July 13). Participating in a gift economy: Are you giving enough? *Opensource.com*. Retrieved from https://opensource.com/business/10/7/participating-gift-economy-are-you-giving-enough

Outreachy. Retrieved from https://www.gnome.org/outreachy/

Powell, W. E., Hunsinger, D. S., & Medlin, B. D. (2010). Gender differences within the open source community: An exploratory study. *Journal of Information Technology Management, 21*(4). http://jitm.ubalt.edu/XXI-4/article3.pdf

Raymond, E. S. (1996). *The new hacker's dictionary* (3rd ed.). Cambridge, MA: MIT Press.

Raymond, E. S. (1999). *The cathedral and the bazaar: Musings on Linux and open source by an accidental revolutionary*. Sebastopol, CA: O'Reilly Media, Inc.

Raymond, E. S. (2000). Homesteading the noosphere. Retrieved from http://www.catb.org/esr/writings/homesteading/homesteading/index.html

Raymond, E. S. (2001). *How to become a hacker*. Retrieved from http://www.catb.org/esr/faqs/hacker-howto.html

Raymond, E. S. (2014). *How to ask questions the smart way*. Retrieved from http://www.catb.org/faqs/smart-questions.html

Reagle, J. (2013). 'Free as in sexist?': Free culture and the gender gap. *First Monday, 18*(1). Retrieved from http://firstmonday.org/article/view/4291/3381

Reddit. (2015). I won Red Hat's Women in Open Source Award. Retrieved from https://www.reddit.com/r/linux/comments/3cpus9/i_won_red_hats_women_in_open_source_award_the/?limit=500

Red Hat. (2014). Presenting the women in open source award. *Red Hat*. Retrieved from http://www.redhat.com/en/about/women-in-open-source

Richardson, D. (1999). General LinuxChix FAQ. Retrieved from http://www.linuxchix.org/docs/faq.html

Schrift, A. D. (Ed.). (1997). *The logic of the gift: Toward an ethic of generosity*. New York, NY: Routledge.

Schrift, A. D. (2001). Logics of the gift in Cixous and Nietzsche: Can we still be generous? *Angelaki, 6*(2), 113–123.

Sharp, S. (2013, July 15). Re: [00/19] 3.10.1-stable review. *Linux Kernel Mailing List*. Retrieved from http://lkml.iu.edu/hypermail/linux/kernel/1307.1/03517.html

Sharp, S. (2015, June 25). I won Red Hat's Women in Open Source Award! *The Geekess*. Retrieved from http://sarah.thesharps.us/2015/06/25/i-won-red-hats-women-in-open-source-award/

Söderberg, J. (2008). *Hacking capitalism: The free and open source software movement*. New York, NY: Routledge.

Spiegel, E. (2008, April 14). Dealing with an IT bully. *Datamation*. Retrieved from http://www.datamation.com/columns/smit/article.phb.3740601/Dealing-with-An-IT-Bully.htm.

Stallman, R. (1983, September 27). New UNIX implementation. Retrieved from http://www.gnu.orggnu/initial-announcement.en.html

Stallman, R. (1987). *GNU Manifesto. GNU Operating System*. Retrieved from http://www.gnu.org/gnu/manifesto.en.html

Stewart, D. (2005). Social status in an open source community. *American Sociological Review, 70*(5), 823–842.

Taylor, P. A. (2003). Mestros or misogynists? Gender and the social construction of hacking. In Y. Jewkes (Ed.), *Dot.cons: Crime, deviance and identity on the Internet* (pp. 126–146). Portland, OR: Willan Publishing.

Thomson, R. (2008, March 24). IT workers being bullied, says union. *Computer Weekly*. Retrieved from http://www.computerweekly.com/news/2240085434/IT-workers-being-bullied-says-union

Torvalds, L. (2012, December 23). Re: [regression w/patch] Media commit causes user space to misbahave [sic] (was: Re: Linux 3.8-rcl). Retrieved from https://lkml.org/lkml/2012/12/23/75

Toupin, S. (2014). Feminist hackerspaces: The synthesis of feminist and hacker cultures. *Journal of Peer Production, 5*. Retrieved from http://peerproduction.net/issues/issue-5-shared-machine-shops/peer-reviewed-articles/feminist-hackerspaces-the-synthesis-of-feminist-and-hacker-cultures/

Toupin, S. Feminist hackerspaces as safer spaces. *dpi: Feminist Journal of Art and Digital Culture, 27*. Retrieved from http://dpi.studioxx.org/en/feminist-hackerspaces-safer-spaces

Vaidhyanathan, S. (2001). *Copyrights and copywrongs: The rise of intellectual property and how it threatens creativity*. New York, NY: New York University Press.

Wajcman, J. (1991). *Feminism confronts technology*. University Park, PA: The Pennsylvania State University Press.

Wayner, Peter. (2000). *Free for all—How Linux and the free software movement undercut the high-tech titans*. New York, NY: Harper Business.

Weber, S. (2004). *The success of open source*. Cambridge, MA: Harvard University Press.

Winner, L. (1988). *The whale and the reactor*. Chicago, IL: University of Chicago Press.

Wu, A. (2014, July 21). 'Females' in open source. *Model View Culture*. Retrieved from https://modelviewculture.com/pieces/females-in-open-source

Wuhib, F. W., & Dotger, S. (2014). Why so few women in STEM: The role of social coping. 4th IEEE Integrated STEM Education Conference. *Institute of Electrical and Electronics Engineers*. DOI:10.1109/ISECon.2014.6891055.

Zeitlyn, D. (2003). Gift economies in the development of open source software: Anthropological Reflections. *FLOSShubs*. Retrieved from http://flosshub.org/system/files/rp-zeitlyn.pdf

11. *Making Space for a Revolution: Occupy Wall Street as a Maker Movement*

ALISON E. VOGELAAR
Franklin University Switzerland

CHARLOTTE M. MCKERNAN
University of New Mexico School of Architecture and Planning

What history *makes* of Operation Occupy Wall Street is yet to be told. Much of its material and symbolic effect will be the product what we—scholars, teachers, and advocates—make of it. Indeed, much has already been *made* of Operation Occupy Wall Street. The so-called movement has enchanted, perplexed and disappointed popular, political and academic commentators alike. Critical inquiry has tended to focus upon the movement's utilization of digital and/or social media[1]; its use of retro and novel protest forms[2]; its diverse and complex politics[3]; and/or its connection to other protest movements across the globe[4]. This essay weaves these diverse threads together, spinning OWS alternately as an important contributor to the contemporary maker movement. We believe that the protest that began at Zuccotti Park became a legitimate contender in the discursive battle against contemporary neo-liberal capitalism not so much because of the *content* of its assertions (e.g. the 99% versus the 1%) but because of the *maker practices* through which these assertions were made possible.

A Movement in the Making: The Contemporary Maker Movement

Humans are makers. We have been making—dwellings, artifacts, systems—for as long as recorded history. This chapter focuses upon a contemporary trend in making as a *political act*. This trend is constituted by a diverse group of formal and informal organizations and platforms (e.g. science cafes, knitting circles, Etsy, artist cooperatives) activities (e.g. DIY home improvement, home brewing, knitting, open source initiatives), and literatures (e.g. makerspaces.com, makezine.com). While activist circles have in the past used making strategically (e.g. "crafting"), this essay focuses upon the transformation of making into a movement in and of itself. This movement is composed of different but related practices and communities—craft, DIY, maker, and hacker—that have been alternatively described as cultures (Anderson, 2012; Bratich & Brush, 2011), spaces (Doorley & Witthoft, 2012), identities (Dougherty, 2013; Mann, 2014) isms (Hatch, 2014; Mann, 2014; Solomon, 2013), ethos (Gauntlett, 2011; Sennet, 2008) and movements (Anderson, 2012; Gauntlett, 2011). According to Dale Dougherty, founder of *Make Magazine*, "all together, makers are seeking an alternative to being regarded as consumers, rejecting the idea that you are defined by what you buy. Instead, makers have a sense of what they can do and what they can learn to do" (2013, p. 8).

Bratich and Brush (2011) are critical of the tendency to refer to this emergent movement as a "revival" offering instead "resurgence" as a more representative articulation of the ways in which maker processes may have been broken or transmutated by industrialization, but have no less "persisted and proliferated in cracks and intertices" (Bratich & Brush, 2011, p. 250). Instead of bringing something back from the dead, as "revival" connotes, the contemporary maker movement represents a revitalization of an enduring maker culture.

The contemporary maker movement has multiple sources, the most obvious of which is the anti-industrial Arts and Craft movement that flourished in Europe and North American in the late eighteenth and early nineteenth centuries (Gauntlett, 2011). The movement's ethos is also rooted in the more recent anti-consumerist, anti-authoritarian countercultural movements of the 1960s, which were spearheaded by educational reformists like John Holt and Ivan Illich. Holt and Illich focused predominantly on the forms of alienation and incompetence inspired twentieth century contemporary institutions. In his book *Makers: The New Industrial Revolution*, Chris Anderson acknowledges the indie/zine music scene and DIY culture of the 1980s, as well as, the digital web as major sources of the movement. About the digital web writes: "the past ten years have been about discovering new ways to create, invent, and work together on the Web. The next ten years will be about applying those lessons to the real world" (p. 17).

Like most "newest social movements" (Milberry, 2014), the contemporary maker movement is both a critique of, and response to, the processes, materials, identities, places, and communities created, altered and/or destroyed by industrialization and (late) capitalism. According to Adamson (2013), the parallel processes of industrialization and capitalism produced a historically unique bifurcation in the world of making, about which he writes:

> What had been an undifferentiated world of making, in which artisans enjoyed relatively high status within a broader continuum of professional trades, was carved into two, with crafts people usually relegated to a position of inferiority. This bifurcation divided the infinitely complex field of human production into a set of linked binaries: craft/industry, freedom/alienation, tacit/explicit, hand/machine, traditional/progressive. (p. 2)

Similarly, Bratich and Brush (2011) suggest that industrialization gave rise to shifts from use-value to exchange-value, artisan to laborer, from guild to factory, and from process-orientation to product-orientation.

The nascent study of maker movements has explored the various ways in which maker movements use the act of making to critique and subvert: oppressive and exploitative forms of labor (Solomon, 2013); environmental destruction and waste (Gauntlett, 2011); consumerism (Orton-Johnson, p. 141); loss of "personal touch" and related connection between producers and consumers (Bratich & Brush, 2011); commercialization of "the commons"; and the declining quality of products. Theorists of maker movements/cultures have also explored the contradictions and tensions at the core of maker movements and practices namely the: romanticization of the handmade (Solomon, 2011); nostalgia for the period prior to the Industrial Revolution (Adamson, 2013); unacknowledged ties to whiteness and the middle-class (Dawkins, 2011); and fetishization of consumer products. Adamson (2013), in particular, reminds us that many features associated with industrialization—speed, separation and specialization, and even exploitative labor—were also practices that belonged to so-called craft (Adamson, 2013).

Though OWS is not a self-described maker movement, this chapter explores the ways in OWS evoked maker practices to critique and subvert two problematic contemporary binaries at the center of capitalist culture: expert/novice and public/private.

Expert/Novice

According to Solomon (2013), "the arrival 20th century introduced the concept of professionalism with the implementation of hierarchies of knowledge and institutions of knowledge/skill affirmation" (p. 12). The concept of the professional implied, and indeed gave rise to, the idea of an amateur, against which professionalism was, and continues to be, defined (Leadbeater

& Miller, 2004). While there are many positive effects of professionalism including highly skilled labor, reliability, and accountability, it has also had negative effects, most notably the "trained incapacities" it instigates in both so-called professionals and amateurs (Burke, 1954).

The emergent maker movement is a reaction to, among other things, professionalization. In reconnecting individuals to the tools, processes and communities by which we are constituted ("in-sourcing" our everyday needs), the act of making reasserts people's power over their own matter/s. Contemporary making projects and communities (most notably the DIY movement) have given rise to a new category that has been dubbed the "Pro-Am" (professional-amateurs) or "advanced amateur" (Leadbeater & Miller, 2004; Solomon, 2013; Stebbins, 1992). According to Solomon (2013), the Pro-Am is a highly skilled or practiced maker who more often than not uses leisure time to produce "work of very similar quality" to professionals. Pro-Ams spend a considerable amount of their leisure time (and often money) learning the skills, techniques and cultures of their craft or activity, often seek and/or create communities in which they can share skills and display work, and in some cases become "disruptive innovators" that change the craft/activity wholesale (Leadbeater & Miller, 2004). In some cases, according to Stebbins (1992), good amateurs often produce better work than mediocre professionals.

According to Solomon, contemporary maker culture, combined with the sharing, displaying and collaborating capacities of the digital web, have "creat[ed] a space for the professional-amateur to thrive" (2013, p. 13). Gauntlett suggests that the rise of the Pro-Am is also evidence of a larger cultural shift from what he calls a "sit back and be told culture" to a "making doing culture" (2011, p. 9). According to the author, "sit back and be told culture" is a product of the institutionalization of schools whereby learning became of a process of transfer. This has been reinforced by many institutions (e.g. media, political, financial), which treat individuals as passive "consumers." "Making and doing culture" on the other hand, is an orientation that "seeks opportunities for creativity, social connections and personal growth" (Gauntlett, 2011, p. 11). This reorientation thus constitutes an ontological shift from passive consumer to active maker. As Gauntlett asserts, the act of making need not be explicitly political—making in one domain naturally "spills over into everyday life," such that the act of making is always already a political act (p. 57). While research on Pro-Ams has focused upon craft and DIY practices and communities, the concept can be very clearly applied to protest movements like OWS where the Pro-Am is very much at play/work.

Public/Private

Recent theorists of place have noted a historically distinct shift in urban scapes from collectivized to market-based wherein "collective activity is largely confined to commercial 'playgrounds' that inevitably exclude some social groups" (Visconti, Sherry, Borghini, & Anderson, 2010, p. 513). According to Gieryn (2000), the loss and/or absence of public places can have devastating implications for individual and collective identity, memory, and history, as well as for psychological and physical wellbeing. It is no surprise then that public places are often both the subject of, and site for, collective action and civil disobedience. The OWS occupation, and eventual encampment, at Zuccotti Park is one such example of the contemporary battle for, and re-occupation of, the "public sphere"—a sphere theorized most famously by Jurgen Habermas as "a social space wherein private citizens gather as a public body with the rights of assembly, association, and expression in order to form public opinion" (DeLuca & Peeples, 2006, p. 246). While an expanding set of literatures turns our attention to the salience of cyberspace in protests and social movements (see for example Langlois & Dubois, 2005; Lievrouw, 2011), physical space remains equally significant in protest rhetoric and tactics. As Gieryn so aptly puts it, "in spite of (and perhaps because of) the jet, the net, and the fast-food outlet, place persists as a constituent element of social life and historical change" (Gieryn, 2000, p. 463).

As "made places," urban sites are both metaphoric and literal embodiments of power relations. According to Hosey (2000), civic architecture has historically provided both material evidence of a society's values and, perhaps more significantly, "a narrative that promotes a desired image of society" (p. 146). In urban sites, like Manhattan's financial district, it is easy to "observe hierarchies where some figure more prominently than others" (Friday, 2011, p. 160). Whereas civic space is typically an ideological shelter confirming assumptions about social privilege, public spaces can and have historically been "redefined and utilized in new ways by groups with the will to challenge deeply entrenched power" (Rosenthal, 2000, p. 39). Indeed, the juxtaposition of the public (horizontal) and private (vertical) spaces occupied by "the 99%" and "the 1%" during the OWS encampment may have provided a powerful symbol (and evidence) of the movement's main grievance: "the division between the 1% and the 99% is profound, unjust and un-American" (Vogelaar, 2015). Nevertheless, the architecture of public places serve more often as contraceptives to, rather than breeding grounds for, civil disobedience and collective action. Indeed, "as public spaces in cities are privatized, stigmatized, avoided or destroyed, the effect is chilling on the possibility of mobilization and public protest" (Gieryn, 2000, p. 484).

The contemporary maker movement is also a reaction to the public/private binary and the increasing "colonization" of public spaces by commercial (private) interests—spatial reconfigurations that have had significant consequences for democratic practices, subjectivities and communities. Using varied practices, the maker movement is a reinvigoration of the idea that: civic participation, community and conviviality are more central to our happiness than capital and consumption (Gauntlett, 2011); "public spaces [communities and subjectivities] are processes under construction" (Bratich & Brush, 2011, p. 247); and the ethos and technologies of web 2.0 might be harnessed to re-imagine and craft belonging in our times.

The following sections examines two specific features of Occupy Wall Street as they embodied the ethos and practices of the maker movement.

Operation Occupy Wall Street and the Politics of Making

Insomuch as it may allow people to reclaim agency in their everyday lives (Chidrey, 2014), inspires joy and accomplishment (Gauntlett, 2011), encourages creativity and connection (Gauntlett, 2011), challenges existing systems of authority (Ratto & Boler, 2014), and invites critical reflection about materials, processes and relationships (Ratto & Boler, 2014), *making is power.* Indeed, we believe that contemporary social movements around the globe are best understood as manifestations of an emergent and pervasive maker ethos. Several recent publications have explored the explicit intersections of making and activism—what Mann (2014) has termed "maktivism" (see Bratich & Brush, 2011; Gauntlett, 2011; Solomon, 2013). Ratto and Boler's (2014) *DIY Citizenship: Critical Making and Social Media* in particular explores the various ways in which "mediated networks and DIY culture are spawning new modalities and expressions of political and civic engagement" (p. 24). According to the authors, "the DIY ethos has seismically reshaped the international political sphere, as can be seen in ongoing global uprisings and the uses of media and communications within a 'logic of connective action'" (2014, p. 1). The OWS movement in particular, "exemplifies DIY citizenship in its refusal of existing hierarchical practices in favor of attempts to embody the utopian ideal of radically different processes of social organization, decision making, and leadership" (2014, p. 25). Our essay builds upon this assertion, spinning OWS as a significant thread in the emerging global maker movement—a movement invested in re/taking (personal and political) agency via varied practices (large and small) of re/making subjectivities, communities, and places. In so doing, the maker movement is an explicit attempt to replace the contemporary ethos of "consumerism" with an ethos of "makerism".

What became Occupy Wall Street began[5] with a laconic call to action issued by the famous, anti-consumerist, culture jamming organization/publication, *Adbusters*. It proclaimed, "On September 17, we want to see 20,000 people flood into lower Manhattan, set up tents, kitchens, peaceful barricades and occupy Wall Street for a few months" (Adbusters, 2011, p. XXX or paragraph). Not long after the initial call to action and subsequent formation of groups who met in New York City that summer, the (in)famous "hacktivist" group Anonymous issued an online call for support, attracting even more attention to OWS (Schneider, 2013). Adbusters is a non-profit organization that produces media and events with the specific goal of critiquing, combating, and transforming contemporary consumer culture and the institutions/individuals/practices that support, maintain and benefit from it. Anonymous is an international, decentralized network of activists and "hacktivists" who use media, technological prowess, and offline events to "hack" and/or attack unjust institutions, individuals and practices. Hacks and attacks are often directed at dominant institutions like large corporations or governmental entities, but they need not be. Both Adbusters and Anonymous have a long tradition of publically criticizing corporate interests and actions, but it was the 2008 "financial collapse" and subsequent "bail out" of the guilty financial institutions that really crystallized, and made public, the profound disparity and injustice at the heart of the present financial regime. Following the events of 2008–2009, it became clear that the present regime has produced an elite class (the so-called 1%) of corporate officers and government officials who use political positions to personal advantage and, alternately, a disenfranchised class of consumers (the so-called 99%) who have forgotten both what democracy and citizenship look like.

It is, of course, no surprise then that Occupy's main grievances were tied to the contemporary (neoliberal, capitalist) financial regime, and more specifically to the perception of Wall Street (and the market as a whole) as being unethical and unjust (Chafkin et al., 2012; Radivojevic & Starosta, 2011; Sommer, 2012; Yardley, 2011). The *Adbusters* call, in particular, described Wall Street as "the greatest corrupter of our democracy" and "the financial Gomorrah of America."[6] Although the movement strategically lacked a unified demand, much of the discourse emanating from movement participants concerned the concentration of power and wealth in the current global financial system.[7]

Like its contemporary predecessors, the Arab Spring and Los Indignados, Occupy Wall Street used public (both offline and online) *demonstrations* to "stage" both the problems faced, *and* possibilities imagined, by movement participants. In retaking and remaking public places like plazas and squares, such demonstrations evoked the battle cry of the global justice movement—"This is What Democracy Looks Like"—displaying the subjectivities (citizens), practices (participatory democracy) and communities (democratic) required to produce a more just society. The remainder of this essay explores two particular "maker" features of the OWS movement—the evocation of the Pro-Am (professional-amateur) as a novel political subjectivity and the manipulation of "public" space—as they demonstrated the problems faced and possibilities imagined by movement participants.

Politics as Pro-Am: Making Democratic Subjectivities

Nowhere are professionalization's discontents more obvious than in the sphere of American politics. To call American politics "professional" is a misnomer in the sense that the sphere does not subscribe to official training/schooling or codes of conduct. That being said, American politics, at the state and national-levels, has transformed largely from a public service model to a paid vocation model. This transformation has had many consequences, one of which is the tendency to keep amateurs (citizens) out of politics. In its most basic sense, Occupy Wall Street was a reassertion of the citizen/amateur in American political culture—a feature celebrated most famously by Alexis de Tocqueville's *Democracy in America* (1988). The mere designation of "the 1%" and "the 99%" drew attention to the fact that American politics has been taken from "the people." More significantly, the movement was run largely on the unpaid, *but often highly skilled,* labor of a small group of a group of motivated participants who used their leisure time (or time of un(der)employment) to produce and spread information, manage logistics, and facilitate interactions and actions.

Like many social movements, Occupy Wall Street was run by Pro-Ams—in this case, citizens with a passion for democratic processes and politics. Indeed, very few of the initial protest coordinators were professional organizers. The first General Assemblies were comprised mainly of artists, teachers, and students. While many members of the initial group had experience in previous protests (both within the US and abroad), there was little (if any) in the way of official political affiliation and/or training (Schneider, 2013). Despite this, the group still maintained lofty political aspirations and ultimately engaged in quite high-level civic discourse. "It was a utopian act, but in the form of realism," writes Schneider about the initial meetings: "With artists mainly in charge, Occupy Wall Street was art before it was anything properly organized, before it was even politics. It was there to change us first and make demands later" (2013, p. 6).

OWS also evoked a unique form of Pro-Am wherein activists matched their "marketable" professional skills (e.g. programming, marketing, PR, writing, managing) directly to the needs of the movement. From the beginning of the occupation, web-designers set up sites to publish group proceedings, as well as wikis to support protestors both local and global (NYCGA, n.d.); graphic designers created posters and signs that became emblematic of the movement as a whole; and social media professionals created and shared memes and messages that spread eventually around the globe (Lang & Lang/ Levitsky, 2012). In each of these cases, traditionally market-based skills were used against the capitalist system that produced them. The skillsets evoked by OWS Pro-Ams were, of course, not always business-oriented. For example, Joe Therrien, a NYC elementary teacher with an MFA in puppetry, worked alongside the Arts and Culture working group to create the Occupy Wall Street Puppet Guild, who in turn crafted a number of very large protest puppets (including a 12-foot Statue of Liberty and a 12-foot Wall Street bull). The Statue of Liberty puppet became a crowd favorite during organized marches (Kim, 2012).

Due to the horizontal and highly democratic structure of OWS, participants had the agency to choose or invent the ways in which they might most meaningfully contribute to the movement. Occupiers most often worked communally in "working groups", applying their own skillsets and learning from others in turn. In this way, the working groups themselves a sort of Pro-Am training ground where participants could apply *and learn* new skills.

Political Platforms 2.0: Putting Politics in Its (Public) Place

Though its tactics were many, Occupy Wall Street was made most visible through its encampment of a Lower Manhattan park—Zuccotti Park. The park was eventually renamed "Liberty Plaza" by the occupiers; a reference to both the park's previous name and its sister movement in Tahir Square. The site became an embodied representation (and in this way a form of prefigurative politics as outlined by Carl Boggs in the 1970s) of its maker ethos (Schneider, 2013). We assert that Zuccotti Park and its sister encampments around the world were large-scale embodiments of what has come to be called "makerspaces"—"publicly-accessible places to design and create"[8] based in the philosophies and theories of radical democracy, space/ place and democratic design, subversive tactics, and the culture of making. Makerspaces are sites that facilitate various practices of making and are guided by a collaborative, anti-consumerism, open source ethos. Many, though not all, makerspaces are explicitly political projects to re-take, re-articulate, and re-design democracy by creating, re/making and/or repurposing (1) spaces (physical, digital, symbolic, philosophical) and (2) objects (art, crafts, food, technologies, clothing) for the purpose of generating identities, communities and ideas and actively participating in world making (large and small). Occupy Wall Street encampments were prototypical makerspaces that served as a physical and symbolic space where rhetorical invention was made possible, where ideas could be generated and circulated, and where Occupiers could create, embody and perform the identities, communities, and political practices they sought—all the while doing the mundane day-to-day work needed to *make* a movement (fulfill the basic needs of participants, talk to the press, negotiate with public officials, rally for support, etc). What is more, we assert that the use of space in the OWS encampment functioned two-fold, both as a means of *deconstructing* the contemporary anti-democratic manipulation and degeneration of public spaces as well as a means of *constructing* new pro-democratic identities, communities, places and ideas.

The difficulties of mobilization and public protest in this new world order were nowhere more visible than in that "privately-owned public space" called Zuccotti Park, in which OWS protesters were daily hassled, and from which they were eventually evicted. While the very public evacuation may have put a damper on the movement's capacity to re-make democracy, it functioned rather brilliantly to reveal the movement's main grievance: America is owned by a ruling few and the engines of civil society (police, local government, mass media) are in their service (Vogelaar, 2015). While no one much thought about the rules of conduct particular to a "privately owned pubic park" like Zuccotti Park (e.g. what rights to free speech and assembly does one have in a private-public park? And, whose job is it to police such spaces?), its eventual evacuation revealed both the obvious oxymoron at the heart of this pseudo-public space, as well as, the degree to which "the construction of these spaces seems to have outpaced our understanding of the relationship of the public to private-public spaces" (Scola, 2011).

The strategic (anarchistic, horizontal) design of the OWS encampment revealed the degree to which movement coordinators understood that "engagement and estrangement can be built-in" (Gieryn, 2000, p. 477). Internal and external communication about the site's architecture indicate that the encampment designers were well aware of the seemingly contradictory design/space theories that suggest on the one hand that well-designed public spaces draw people to them and on the other that the places most conducive to community are not designed at all (Gieryn, 2000). The design also highlighted the conflicting ways in which the formal qualities of a built environment exert a powerful effect on individuals by shaping the possibilities for behavior, all the while individuals shape and infuse their spaces with meaning (Gieryn, 2000). According to Doorley and Witthoft (2012), space design influences the actions and attitudes of a group of people. By creating spaces in which people feel free to play and be creative, makerspaces may bolster makers' confidence and help them develop powerful new competencies.

As a makerspace, the encampment at Zuccotti Park/Liberty Plaza was much more than a place to sleep and congregate; it was a site that both symbolized and made possible the diverse competencies, subjectivities, and communities required to create the democracy OWS participants sought. In providing the fundamental infrastructures and materials at the heart of their grievances of the current system—community, work groups, space, open information, nourishment, healthcare, housing—the encampment embodied and performed what a democratic/horizontal community looks like (all-the-while using these infrastructures and materials to make the movement move).

Initially, the layout of the Liberty Plaza encampment was minimally organized and most occupiers simply gravitated to wherever they could find space. Regarding the distribution of tents and tarps, the majority of the structures of the camp sprang up based on the whims of each individual (NYCGA, 2011). There were, however, clearly defined areas for a few of the working groups. At the center of the park, an area was set aside for distribution of goods. The Supply Group gave out clothes, blankets, and toiletries that had been donated to the movement. Similarly, the People's Kitchen collected donated food and distributed meals to hungry occupiers. Protesters in need of medical care could also visit a specially zoned first aid area, where a group of volunteers provided 24-hour care (NYCGA, n.d.; Saget & Tse, 2011). As the encampment grew, the Town Planning Working Group attempted to create a cohesive and logical scheme with which to organize the camp structures. According to the Occupy Together Field Manual (a wiki published to support and foster new Occupy encampments), the group first examined the ways in which people had begun to use the park space by looking at foot-traffic, empty areas, and assembly locations. From there, they distributed blank maps around the camp, asking for suggestions regarding necessary zones and possible issues. These suggestions were then combined into an organized zoning proposal. Ironically, the main element of this proposed system involved the division between the public sectors and the private sectors. To the northeast, the majority of the working groups housed were those involved with public interactions. This included the Media Working Group, and the General Assembly as a whole, along with an area designated as art space. To the southwest resided the private sector of the park, dominated by the Housing sector, along with the medical and social services. A pathway separated the public and private sectors and allowed for a practical and visual division of the park (Occupy Together Field Manual, n.d.).

In the meetings of the Town Planning Working Group, members discussed not only the setup of tents and zoned areas in the camp, but also the symbolic associations of each of the group's decisions. The group drew a great deal of attention to the idea of (and challenges regarding) personal space versus public space within the camp. They discussed the very nature of tents as "separating devices," both literally (through physical separation) and symbolically (who gets to use these tents if there isn't space for everyone?). Additionally, the group argued over whether the very idea of zoning the park was itself a breach of the horizontal ideology crucial to the occupation. As the working group put it, "the idea of zoning implies that OWS has authority over some people" (NYCGA, 2011). Ultimately, the encampment embodied the fluidity, individuality, and most of all, distinct lack of structural hierarchy characteristic of makerspaces.

While the symbolic and material effect of the basic infrastructures and services for the maintenance of a healthy, equitable democratic community, we believe that the real power of the OWS encampment was in the multitudes of occupier competencies it activated and cultivated. The working group system in particular allowed protestors to self-select into areas to which they were particularly suited to assist in, including but not limited to the Direct Action Group, the Media Group, the Tech Group, the Sanitation Group, the Accounting Group, and the Town Planning Group. Lest work groups fall victim to the problems of specialization, working groups regularly came together under the auspices of The New York City General Assembly. One particularly interesting example of the ways in which the OWS makerspace cultivated competencies that benefited both the movement and individual makers was when tech savvy members of the Tech Ops group set up a variety of informational systems to combat opposition by local and government forces. While members of this "hackerspace" side of OWS were certainly less visible than their Direct Action counterparts, their involvement in the protest was no less important. One key example was the creation of "mesh networks" (another tactic used to great effect in the Arab Spring), battery-powered wifi hotspots that can be used to maintain Internet connections while still being ultra-portable. As many Occupiers feared monitoring and disconnection by local law enforcement, mesh networks allowed for a stable line of communication between the Zuccotti encampment and the outside world. Another technological tool used in the encampment called "the Illuminator" uses optical light systems to broadcast messages to protestors (Drew, 2013). The Illuminator was used to powerful effect in OWS. As artist Mark Read described in an interview with the BBC, the device was used as a "mobile bat-signal," namely a "culturally legible," immediately impactful, and deeply symbolic call to action for protestors and observers alike (Lang & Lang/Levitsky, 2012; Mason, 2012). The images of the device in use are striking—it allows for an artist's design to shape the built environment into a powerful (yet non-destructive) juxtaposed image. These technologies, in addition to extremely savvy use of both Web 2.0 and old-school networking tactics, allowed the OWS group (as well as the Operation Occupy protest groups as a whole) to maintain effective inter-group communications, despite the size and scope of the operation (Drew, 2013).

To an outsider, the Zuccotti Park encampment may have looked like a random, disjointed bricolage of activism, but in reality the protestors afforded a great deal of care and energy to the design and maintenance of the encampment. The (multitude of) messages of the movement, so often rendered invisible by the powerful elite, were brought harshly into view in an intentional and staggering recolonization of public space. Public, open space is an essential element in the staging of political protest precisely because of its powerful rhetorical impact, and thus the resettlement (even temporarily) of such spaces is an important tactic for activism (Hayden, 1995; Lynch & Carr, 1979). The protestors at OWS were responsible for the active *making* of a new space. They repurposed a piece of New York's financial district and remade it in their own image and service—it became both a physical manifestation of their grassroots political message and a source of inspiration. As OWS set up and "took root" within the space, so too did they encourage the growth of new social and political identities.

Conclusions: Making OWS

As we contended at the outset of this essay, much has been made of Occupy Wall Street. It has been variously interpreted, critiqued and celebrated by advocates, politicians, academics and the popular press in many disparate and provocative ways. Our particular contribution to this conversation advocates framing OWS as a maker movement; not only because we believe it is a more instructive way of seeing OWS but also because it is what we (as makers of ideas, knowledge and the next generation) have chosen to make of it. While no movement is without its flaws and contradictions, we believe that the contemporary maker movement and ethos are worthy of our labors. As a reaction to decades of disenfranchisement, displacement, and disconnection on both micro- and macro-scales, making, and the culture that surrounds it (regardless of the medium), is the proactive retaking of the process of creation. Whether it be the act of making crafts or democracy, the maker movement problematizes and corrects several flaws associated with contemporary capitalism namely the "trained incapacities" fostered by cultures of specialization (and elitism), deterioration of communal places wherein citizens can forge identities and communities and the disconnection fostered by automation, mechanization, and outsourcing.

Framing OWS as a maker movement moves us beyond examinations of the movement's message, focusing instead upon its medium—making. And consistent with Marshall McLuhan's argument, the message of OWS was its medium. Consistent with the maker movement, OWS celebrated and promulgated the Pro-Am (skilled but amateur citizen politician), advocated and produced public spaces for connecting and making and most importantly, recast the political process as a process of making. How these identities and practices continue beyond the movement is yet to be seen. This book goes to print in, what has become, the most dispiriting period of the American political process—the Presidential election campaign—a process-turned-event that is the antithesis of democratic. With mass media channels dedicated to the meaningless banter, partisan clichés, theatrical debates, and empty polls, it is a difficult time to assess the legacy of OWS on political maker culture. Once the distraction of an election year has passed, we hope to have a clearer view of what has become of OWS.

Notes

1. See for example Bastos, Mercea, & Charpentier, 2015; Castells, 2012; Costanza-Chock, 2012; DeLuca, Lawson, & Sun, 2012; Gaby & Caren, 2012; Gerbaudo, 2012; Juris, 2012; Milner, 2013; Park, Lim, & Park, 2015; Penney & Dadas, 2013
2. See for example Bennett, 2012; Haugerud, 2013; Juris, Ronayne, Shokooh-Valle, & Wengronowitz, 2012; Liboiron, 2012; Maharawal, 2013; Min, 2015; Razsa & Kurnik, 2012; Vogelaar, 2015
3. See for example Gursozlu, 2015; Lewis and Luce, 2012; Schneider, 2013
4. See for example Bacon, 2011; Castañeda, 2012; Halvorsen, 2012; Kerton, 2012; Sparke, 2013; Uitermark and Nicholls, 2012; Van Gelder, 2011
5. OWS is of course the product of a multitude of people, ideas, events, and organizations including for example the journalist, David DeGraw, whose "99 Percent Movement" is credited with initiating the notion of the "99 percent" (Captain, 2011) and also movements that preceded OW, such as the the Arab Spring in several countries in the Middle East and Africa and Los Indignados in Spain.
6. See http://makezine.com/2013/05/22/the-difference-between-hackerspaces-makerspaces-techshops-and-fablabs/. Also, see makerspaces.com and makezine.com for information about the application of the concept in education and DIY culture.

References

Adamson, G. (2013). *The invention of craft.* London: Bloomsbury.

Adbusters. (2011, July). #OCCUPYWALLSTREET. Retrieved from http://www.adbusters.org/blogs/adbusters-blog/occupywallstreet.html

Anderson, C. (2012). *Makers: The new industrial revolution.* New York, NY: Crown Business.

Bacon, D. (2011). From planton to occupy. *Social Policy, 41*, 42–45.

Bastos, M. T., Mercea, D., & Charpentier, A. (2015). Tents, tweets, and events: The interplay between ongoing protests and social media. *Journal of Communication, 65*, 320–350. DOI:10.1111/jcom.12145.

Bennett, W. L. (2012). The personalization of politics: Political identity, social media, and changing patterns of participation. *The Annals of the American Academy of Political and Social Science, 644*(1), 20–39. DOI:10.1177/0002716212451428.

Bratich, J. Z., & Brush, H. M. (2011). Fabricating activism: Craft-work, popular culture, gender. *Utopian Studies, 22*(2), 233–260.

Burke, K. (1984). Permanence and change: An anatomy of purpose. University of California Press.

Captain, S. (2011, October 7). The inside story of Occupy Wall Street. *Fast Company.* Retrievedfromhttp://www.fastcompany.com/1785918/the-inside-story-of-occupy-wall-street

Castells, M. (2012). *Networks of outrage and hope.* Cambridge, UK: Polity Press.

Casey, R., Goudie, R., & Reeve, K. (2008). Homeless women in public spaces: Strategies of resistance. *Housing Studies, 23*(6), 899–916.

Castañeda, E. (2012). The indignados of Spain: A precedent to Occupy Wall Street. *Social Movement Studies, 11*(3–4), 309–319. DOI:10.1080/14742837.2012.708830

Chafkin, M., Beggs, A., Guiducci, M., Lalinde, J., Nicholas, E., Sacks, R., & Sanders, K. (2012, February). Revolution number 99. *Vanity Fair.* Retrieved from http://www.vanityfair.com/politics/2012/02/occupy-wall-street-201202

Chidgey, R. (2014). Developing Communities of Resistance? Maker Pedagogies, Do-It-Yourself Feminism, and DIY Citizenship, In *DIY citizenship: Critical making and social media* (pp. 101–114). Boston: MIT Press.

Costanza-Chock, S. (2012). Mic check! Media cultures and the occupy movement. *Social Movement Studies, 11*(3–4), 375–385. DOI:10.1080/14742837.2012.710746.

Dawkins, N. (2011). Do it yourself: The precarious work and postfeminist politics of handmaking (in) Detroit. Utopian Studies, 22, 266–277.

DeGraw, D. (2010). The economic elite vs. the people of the United States of America. Originally published on *Ampstatus.com.* Retrieved from https://daviddegraw.org/the-economic-elite-vs-the-people-%EF%BB%BF%EF%BB%BForiginal-99-movement-call-to-action/

Deluca, K. M., & Peeples, J. A. (2002). From Public Sphere to Public Screen: Democracy, Activism and the "Violence" of Seattle. *Critical Studies in Media Communication, 19*(2), 125–151.

DeLuca, K. M., Lawson, S., & Sun, Y. (2012). Occupy Wall Street on the public screens of social media: The many framings of the birth of a protest movement. *Communication, Culture & Critique, 5*(4), 483–509. DOI:10.1111/j.1753-9137.2012.01141.x

DeLuca, K. M., & Peeples, J. A. (2002). From public sphere to public screen: Democracy, activism and the "Violence" of Seattle. *Critical Studies in Media Communication, 19*(2), 125–151.

Doorley, S., & Witthoft, S. (2012). *Make space: How to set the stage for creative collaboration.* Hoboken, NJ: John Wiley & Sons.

Dougherty, D. (2013). The maker mindset. In M. Honey & D. Kanter (Eds.), *Design, make, play: Growing the next generation of STEM innovators.* New York, NY: Routledge. Retrieved from http://unm.eblib.com.libproxy.unm.edu/patron/FullRecord.aspx?p=1154290

Drew, J. (2013). The shape of things to come. In *A social history of contemporary democratic media* (pp. 153–187). New York, NY: Routledge. Retrieved from

Friday, J. (2011). Prague 1968: Spatiality and the tactics of resistance. *Texas Studies in Literature and Language, 53*(2), 159–178.

Gaby, S., & Caren, N. (2012). Occupy online: How cute old men and Malcolm X recruited 400,000 US users to OWS on Facebook. *Social Movement Studies, 11*(3–4), 367–374. DOI:10.1080/14742837.2012.708858

Gauntlett, D. (2011). *Making is connecting: The social meaning of creativity from DIY and knitting to youtube and web 2.0.* Cambridge, UK: Polity Press.

Gerbaudo, P. (2012). "The hashtag which did (not) start a revolution": The laborious adding up to the 99%. In *Tweets and the streets: social media and contemporary activism* (pp. 102–133).

Gieryn, T. F. (2000). A space for place in sociology. *Annual Review of Sociology, 26*, 463–496. Retrieved from http://www.jstor.org/stable/223453

Gursozlu, F. (2015). Democracy and the Square: recognizing the Democratic Value of the Recent Public Sphere Movements. *Essays in Philosophy, 16(1)*, 26–42.

Habermas, J. (1989). *The structural transformation of the public sphere.* (Thomas Burger, Trans. and Ed.). Cambridge, MA: MIT Press.

Halvorsen, S. (2012). Beyond the network? Occupy London and the global movement. *Social Movement Studies, 11*(3–4), 427–433. DOI:10.1080/14742837.2012.708835

Hatch, M. (2014). The maker movement manifesto. In *The maker movement manifesto* (pp. 1–31). New York, NY: McGraw Hill.

Haugerud, A. (2013). After satire? In *No Billionaire Left Behind: Satirical Activism in America* (pp. 187–203). Stanford, CA: Stanford University Press. DOI:10.1111/aman.12090_15.

Hayden, D. (1995). *The power of place: Urban landscapes as public history.* Cambridge, MA: MIT Press.

Hosey, L. (2000). Slumming in Utopia: Protest construction and the iconography of urban America. *Journal of Architectural Education, 53*(3), 146–158.

Juris, J. S. (2012). Reflections on #Occupy everywhere: Social media, public space, and emerging logics of aggregation. *American Ethnologist, 39*(2), 259–279. DOI:10.1111/j.1548-1425.2012.01362.x.

Juris, J. S., Ronayne, M., Shokooh-Valle, F., & Wengronowitz, R. (2012). Negotiating power and difference within the 99%. *Social Movement Studies, 11*(3–4), 434–440. DOI:10.1080/14742837.2012.704358

Kerton, S. (2012). Tahrir, here? The influence of the Arab uprisings on the emergence of occupy. *Social Movement Studies, 11*(3–4), 302–308. DOI:doi.org/10.1080/14742837.2012.704183

Kim, R. (2012). The audacity of Occupy Wall Street. In A. S. Lang & D. Lang/Levitsky (Eds.), *Dreaming in public: The building of the occupy movement* (pp. 27–30). Oxford: New Internationalist.

Lang, A. S., & Lang/Levitsky, D. (2012). *Dreaming in public: The building of the occupy movement.* Oxford: New Internationalist. Retrieved from http://search.ebscohost.com/login.aspx?direct=true&scope=site&db=nlebk&db=nlabk&AN=491061

Langlois, A., & Dubois, F. (Eds.). (2005). *Autonomous media: Activating Resistance and Dissent.* Montréal: Cumulus Press.

Leadbeater, C., & Miller, P. (2004). The pro-am revolution: How enthusiasts are changing our economy and society. Demos, 12–52. Retrieved January 2015 from https://www.demos.co.uk/files/proamrevolutionfinal.pdf?1240939425

Lewis, P. and Luce, S. (2012). Labor and Occupy Wallstreet An Appraisal of the First Six Months. *New Labor Forum, 21(2),* 43–49.

Liboiron, M. (2012). Tactics of waste, dirt and discard in the occupy movement. *Social Movement Studies, 11*(3–4), 393–401. DOI:10.1080/14742837.2012.704178.

Lievrouw, L. A. (2011). *Alternative and activist new media.* Cambridge: Polity Press.

Lofland, J. (1985). *Protest: Studies of collective behavior and social movements.* New Brunswick, NJ: Transaction Pub.

Lynch, K., & Carr, S. (1979). Open space: Freedom and control. In K. Lynch, T. Banerjee, & M. Southworth (Eds.), *City sense and city design: Writings and projects of Kevin Lynch* (pp. 413–417). Cambridge, MA: MIT Press.

Maharawal, M. M. (2013). Occupy Wall Street and a radical politics of inclusion. *Sociological Quarterly, 54*(2), 177–181. DOI:10.1111/tsq.12021.

Mason, P. (2012). Occupy artists take message to streets. *BBC Newsnight.* Retrieved from http://www.bbc.com/news/business-17903474

Milberry, K. (2014). (Re)making the internet: Free software and the social factory hack. In DIY Citizenship: Critical making and social media (pp. 53–64*). Boston: MIT Press.*

Milner, R. M. (2013). Pop polyvocality: Internet memes, public participation, and the Occupy Wall Street movement. *International Journal of Communication, 7*, 2357–2390. Retrieved from http://ijoc.org/index.php/ijoc/article/view/1949

Min, S.-J. (2015). Occupy Wall Street and deliberative decision-making: Translating theory to practice. *Communication, Culture & Critique, 8*(1), 73–89. DOI:10.1111/cccr.12074

New York City General Assembly (NYCGA). (2011). General Assembly Minutes 9/17/11. Retrieved from http://www.nycga.net/2011/09/17/general-assembly-minutes-91711

NYCGA. (n.d.). #OWS Direct Action Info. Retrieved July 29, 2015, from http://da.nycga.net/info/

Park, S. J., Lim, Y. S., & Park, H. W. (2015). Comparing Twitter and YouTube networks in information diffusion: The case of the "Occupy Wall Street" movement. *Technological Forecasting and Social Change, 95*, 208–217. DOI:10.1016/j.techfore.2015.02.003

Peeples, J. A., & DeLuca, K. M. (2006). The truth of the matter: Motherhood, community and environmental justice. *Women's Studies in Communication, 29*(1), 59–87. DOI:10.1080/07491409.2006.10757628

Penney, J. and Dadas, C. (2014). (Re)Tweeting in the Service of Protest: Digital Compostion and Circulation in the Occupy Wallstreet movement. *New Media & Society, 16(1)*, 74–90.

Radivojevic, I., & Starosta, M. (2011). *Nobody can predict the moment of revolution.* New York City, NY: Vimeo. Retrieved from https://vimeo.com/29513113

Ratto, M. and Boler, M. (2014). *DIY citizenship: Critical making and social media.* Boston: MIT Press.

Razsa, M., & Kurnik, A. (2012). The occupy movement in Žižek's hometown: Direct democracy and a politics of becoming. *American Ethnologist, 39*(2), 238–258. DOI:10.1111/j.1548-1425.2012.01361.x

Reed, T. V. (2005). *The art of protest: Culture and activism from the civil rights movement to the streets of Seattle.* Minneapolis, MN: University of Minnesota Press.

Rosenthal, A. B. (2000). Spectacle, fear, and protest: A guide to the history of urban public space in Latin America. *Social Science History, 24*(1), 33–73.

Saget, B., & Tse, A. (2011, October 5). How Occupy Wall Street turned Zuccotti Park into a protest camp. *The New York Times.* New York City. Retrieved from http://www.nytimes.com/interactive/2011/10/05/nyregion/how-occupy-wall-street-turned-zuccotti-park-into-a-protest-camp.html?ref=nyregion

Schneider, N. (2013). *Thank you, anarchy: Notes from the occupy apocalypse.* Retrieved from http://public.eblib.com/choice/publicfullrecord.aspx?p=1251017

Schwartz, M. (2011a). Pre-occupied: The origins and future of Occupy Wall Street. *The New Yorker*, 1–12. Retrieved from http://www.newyorker.com/reporting/2011/11/28/111128fa_fact_schwartz?currentPage=all

Schwartz, M. (2011b, November 28). How Occupy Wall Street chose Zuccotti Park. *The New Yorker.* New York City. Retrieved from http://www.newyorker.com/online/blogs/newsdesk/2011/11/occupy-wall-street-map.html

Scola, N. (2011, October 2). For the anti-corporate Occupy Wall Street demonstrators, the semi-corporate status of Zuccotti Park may be a boon. *POLITICO* New York. Retrieved form http://www.capitalnewyork.com/article/politics/2011/10/3583314/anti-corporate-occupy-wall-street-demonstrators-semi-corporate-stat

Sennett, R. (2009). The craftsman. *Penguin press.*

Sommer, J. (2012, December 22). The war against too much of everything. *The New York Times*. New York City. Retrieved from http://www.nytimes.com/2012/12/23/business/adbusters-war-against-too-much-of-everything.html

Solomon, R. (2013). Homemade and Hellraising through Craft, Activism and Do-It-Yourself Culture. *PsychNology, 11(1)*, 11–20.

Sparke, M. (2013). From global disposession to local reposession: Towards a worldly cultural geography of occupy activism. In *The Wiley-Blackwell companion to cultural geography*, 387–408. John Wiley & Sons.

Stebbins, R. A. (1992). *Professionals. Amateurs, professionals, and serious leisure*. Montreal: McGill-Queen's University Press

Tocqueville, Alexis de. (1988). *Democracy in America*. New York, NY: Harper Perrennial.

Town Planning [Wiki]. (n.d.). Retrieved July 29, 2015, from https://web.archive.org/web/20120928115340/http://occupytogether.wikispot.org/Town_Planning

Uitermark, J., & Nicholls, W. (2012). How local networks shape a global movement: Comparing occupy in Amsterdam and Los Angeles. *Social Movement Studies, 11*(3–4), 295–301. DOI:10.1080/14742837.2012.704181

Van Gelder, S. (2011). *This changes everything: Occupy Wall Street and the 99% movement*. San Francisco, CA: Berrett-Koehler Publishers. Retrieved from http://public.eblib.com/choice/publicfullrecord.aspx?p=799363

Visconti, L. M., Sherry Jr., J. F., Borghini, S., & Anderson, L. (2010). Street art, sweet art? Reclaiming the "Public" in public place. *Journal of Consumer Research, 37*(3), 511–529. DOI:10.1086/652731

Vogelaar, A. (2015). Staging revolution: The OWS encampment at Zuccotti Park. *Media Fields Journal: Critical Explorations in Media and Space, 9*. Retrieved January 15, from http://mediafieldsjournal.squarespace.com/staging-revolution/#13

White, A. F. (2011). Starving where people can see: The 1939 Bootheel share croppers' demonstration. *TDR/The Drama Review, 55*(4), 14–32.

Yardley, W. (2011, November 27). The branding of the occupy movement. *The New York Times*. New York City. Retrieved from http://www.nytimes.com/2011/11/28/business/media/the-branding-of-the-occupy-movement.html

12. *The Détente Model of Managing Divergent Values in the Maker-Sphere*

ANN LIGHT
University of Sussex

This chapter describes two studies undertaken to explore the behaviour of groups that are knowingly political in intent, resistant in attitude and determined to be productive in output. I describe design practices that avoid overt negotiation of conflicting social values, specifically because of the political content in these making activities. I use Spivak's concept of *strategic essentialism* and de Beauvoir's idea of *the Other* to offer an analysis of how self-organising groups manage divergence of values in designing and making as part of social action. In doing so, I begin to show how making itself is part of that negotiation.

Introducing the Maker-Sphere

In this chapter, I argue that radical self-organising makers regularly negotiate the complex issue of agreement on goals and outcomes in collaborative design work for social change; however, they manage much of this without direct appeal to values. To make this argument, I draw on research undertaken as part of a long-term study of activists who design to support social change in Britain. And I frame these insights using post-colonial and feminist scholarship.

In setting out, I should first distinguish this work from that considering makers who do not see their mission as principally political. It is quite possible that other makers use similar tactics to resolve choices and decision points, but I did not focus on more general practices of teaching skills, making products or passing the time. My interest has been to understand the management of tension in contexts where a social change agenda was paramount. One side effect of this is that some of the material I am quoting comes from interviews for which I promised deep anonymity. Therefore, in the first study, I allude to neither individual and group identity, nor what is being made. Referring to either in such an interwoven social space would make interviewees recognisable. This ethical compromise, to protect people sharing their views, is a trade-off for frank discussion.

Second, I understand the related ideas of 'making' and 'hacking' in an open way, seeing these practices as part of movements with a long socio-technical history and liking the definition found in Raymond's New Hackers Dictionary (1996) that "hacking might be characterised as 'an appropriate application of ingenuity'" (p. 506). The contexts I studied involved people using ingenuity, both in making, and in managing the processes round this core activity.

Last, I include myself within the definition of social activist, so, while I may not have been hacking the system in writing this chapter, I do not lay claim to an outsider perspective, but rather draw on empathy and experience to make suggestions about what I have seen and heard.

Theoretical Background

Much has been written on the post-colonial concepts of strategic essentialism (Spivak, 1987) and 'the Other' (de Beauvoir, 1949). Here I intend to use these constructions as a means to explore the activity of social activists involved in making. Consequently, I will give an overview of each position and the reason for interest in it, rather than consider the latest scholarship in either area.

"Strategic essentialism" is a term introduced by the postcolonial scholar Gayatri Spivak to describe the way that groups with limited but powerful common interests can group together to advance these interests, appealing to a common identity for strategic reasons. At a more profound level, it is also an acknowledgment that there are no essential characteristics to identity, but, rather, social norms and discourses position different parts of society and different societies in relation to each other. Spivak describes it as "strategic use of positivist essentialism in a scrupulously visible political interest" (1987). The groups that Spivak is referring to have a perception of themselves as part of non-dominant groupings; in other words, they may be groups with reference to which dominant positions are taken and defined, but they do not have straightforward access to power.

Beyond my own adoption of this framing, in the technology literature I can also look to Dourish's use of the term when he discusses the alliances that make up the environmental lobby. He refers to ways in which "marginalized social groups may temporarily put aside local differences in order to forge a sense of collective identity through which they band together in political movements", noting that even problematic and unstable groupings—that erase significant differences and distinctions—can support important political ends (Dourish, 2008). In doing so, they aspire to gain the mass, stature and/or discernibility needed to challenge more dominant positions.

In addition to this concept, I invoke Simone de Beauvoir's (1949) concept of "the Other", now such a popular cultural term that it has spawned the verb 'to other'. de Beauvoir was speaking about gender relations, seeking a way to talk about and address woman's subjugation: 'She is defined and differentiated with reference to man and not he with reference to her; she is the incidental, the inessential as opposed to the essential. He is the Subject, he is the Absolute—she is the Other. ... no group ever sets itself up as the One without at once setting up the Other over against itself.' Dominant norms are identified and policed by casting those outside those norms as Other.

Whereas strategic essentialism is a knowing and deliberate co-option of a set of status and identity choices for collective ends, 'othering' can happen to anybody who finds themselves outside the dominant norms at work in their environment. Nonetheless, the status of 'the Other' too can become one that is perceived and exploited. This more ironic approach can, for instance, be seen at work in movements, such as the campaign for gay marriage, which challenge the legal marginalisation of those considered different.

If we take these insights and combine them for our purposes here, it is possible to describe the behaviour of some groups as the pragmatic alignment with others regarded as the Other, taking strategic essentialist means to achieve particular ends in counter-cultural or radical processes. This would seem particularly pertinent in looking at the behaviour of groups that are knowingly political in intent, resistant in attitude and determined to be productive in output. How do disparate individuals with differing shades of opinion align to produce new stuff together?

Values and Making

Change-makers tend to be committed to higher-than-self goals (Crompton, 2010), passionate about their beliefs, and vocal in expressing them. Gauntlett (2011) looks at the new trends in making as a powerful way of connecting with others, mediated by activity and materials, with more or less political intent. Nafus (2012) and Toubin (2014) consider the gendered nature of 'openness' in the context of making FLOSS software and creating feminist hackerspaces, respectively, revealing how implicit cultural values work to shape allegiances and opportunities.

Within the new literature on 'hackerspaces', some commentators look specifically at values and the broader political context. A recent study of a small town hackerspace illustrates the multi-layered aspects of making as a social activity:

> At the beginning of our ethnography, the hackerspace seemed to us to be very tool-focused, with a constant emphasis to visitors about what tools are available in the space, what workshops they can take that will teach them to use certain tools, and which members are experts on which tools. However, having spent over a year with them, a different social understanding of the space emerges: the space has more to do with providing a social atmosphere for its members, operating as a third space, one that is neither home nor the office, for members to relax and visit with each other, where members can work on projects that fail without fear of judgment, where members learn to engage with their materials and tools on a deep level. (Toombs, Bardzell, & Bardzell, 2014)

This contrasts the overt practices and values of making (described also as pragmatism and ad-hocism and here symbolized by the emphasis on tools), with the more subtle culture of friendship, care and support which the researcher gradually intuits in his experience of becoming part of the membership (developed in Toombs, Bardzell, & Bardzell, 2015). Elsewhere, Milne, Riecke, and Antle found that, despite valuing support, their Canadian makers chose to work alone a lot of the time because that was more productive (2014). Tanenbaum, Williams, Desjardins, and Tanenbaum describe the balance of personal enjoyment and political motivations in the maker movement, noting that '[m]akers in the developed world represent a growing group within the middle class that is re-negotiating the social contracts around the production and consumption of technology. This is a form of resistance that manages to hide in plain sight' (2013), while Lindtner and Li (2012) look at sites of making in the context of China's wider politics as sites of resistance.

The importance of discussing values in designing technology has its own literature, including Flanagan and Nissenbaum (2007), who urge 'designers and producers to include values as the set of criteria by which the quality of a given technology is judged, to strive actively for a world whose technologies are not only effective, efficient, safe, attractive, easy to use, and so forth, but that promote the values to which the surrounding societies and cultures subscribe'. They suggest these values might include 'liberty, justice, inclusion, equality, privacy, security, creativity, trust, and personal autonomy' (ibid). Friedman's pursuit of Value-Sensitive Design (e.g. 1996) suggests there are global values that should be accommodated in system design and articulated as part of the design process. Not only do both research groups argue for articulating high-level values, but, in positioning values in this way, neither commentary allows for explicitly counter-cultural work.

Within the main discourses of designing technology, work that comes closer is that of participatory design (e.g. Simonsen & Robertson, 2013), which, as a movement, assumes the participation of a diverse range of people in the design process and often has the democratisation of the design process as a political goal. However, situations described in the literature involve designers coming from outside to a community, workplace or other formation and drawing representatives from that context to help in the design of some product and/or process. By contrast, the research here looks at the values work undertaken by groups that self-organise to conduct design. Here the wall between roles of commissioner, beneficiary, user and producer crumbles, though there must still be some arbitration as to what is created and how this relates to the beliefs and social/political values of the participants.

The Studies

This paper presents two studies, drawn from 10 years' research into ad-hoc and self-organising design activities for social action (e.g. Light & Akama, 2014; Light & Miskelly, 2008). It looks at the values work that the groups in two studies perform in seeking to be inclusive and tolerant of heterogeneity, yet productive, as they tackle social issues through making.

The first study is drawn from many years of participant observation. It concerns the articulation of divergent beliefs during agreement of design processes and outcomes in groups with a political agenda. It draws on several interview accounts to show how members of these (anonymised) groups self-censor their contributions and use a focus on *making* to resolve differences of priority and concern. This strategy was common across groups, presented each time as a means to ensure productivity.

The second was a co-research study, with collaborative roles for academics and partners from social action organisms. The purpose was to learn together more about motivations and management in making social change. At outset, chosen ground-rules included that no participant's social action activities would be 'studied'. The chapter gives the rationale for this and other conditions, the ensuing research design and what was learnt as they were adhered to, thereby going deeper into understanding the generation of working practices.

Study 1: Discourses of Production

In a long-term study of activists who design or configure digital technology to support social change in Britain, several interviews about reconciling values were collected. The initial interest in how these groups orientated towards each other was stimulated by observations in a workshop group for a two-year research project called *Practical Design for Social Action* (PRADSA), where individual practitioners came together to study social activist making (see Light & Miskelly, 2008). The group formed in response to the question: "Are you shaping the tools or techniques that help other people shape their world?" But, in this workshop, frictions occurred whenever values were touched on, whether about the nature of software (open source vs off-the-shelf) or about the nature of how the world should be (even more polarised and diverse), although all participants believed that change was needed to improve the world.

In the words of a report to research group members about the culture of the workshop: 'This group is not united by one specific, previously articulated set of values and argues ferociously about any objective for PRaDSA below the level of the rather bland statements such as improving quality of life … we operate on détente. We work together to do things and we can share contacts and ideas, but we do not have space in the structure of the wider group to develop some aspects beyond the agree-to-disagree line and deeply held views would make it impossible, and potentially destructive, to arrive at highly specific statements of general objectives. There are many areas of conflicting beliefs that might never be resolvable—such as FLOSS [free/libre/open-source software] vs off-the-shelf approaches to technology. This group seems used to operating in these détente situations to get things done. It seems to be a tenet of many ad hoc activist practices.' (internal memo, Light, A. & Miskelly, C., 2008).

From this observation, a series of research questions was formed about how self-organising design groups around social activist themes manage themselves, when talking potentially leads to arguing. While it is counterintuitive that things can be achieved without agreement, it appeared that there is a highly pragmatic and opportunistic emergent design practice that works on a need-to-decide basis and leaves certain outcomes open.

Three Examples of Managing Production

Over the next couple of years, further activist-developers were interviewed about their practices in particular instances of co-creation. The following excerpts are from interviews drawn in this way.

> Interview 1
>
> … we often work most effectively together when we don't spend too long discussing the precise boundaries of who we think 'belongs' in our camp, and who we want to work with. So when we get into questioning the boundaries or what forces we regard as 'progressive' (and whether progressive is a useful category for selecting our allies, etc.) then we not only spend a lot of time on abstract discussion, but we also start to highlight our differences and making us all less effective. On the other hand, when we identify concrete actions and each evaluate them from our own particular ethical or political perspectives, we can get further faster.

In this first excerpt, we can see that the speaker has thought about the way that the work goes ahead without explicit discussion of values, which are too 'abstract', and how 'concrete actions' allow ethical and political positions to be held. We can see that whereas the actions are collective, the positions held on it are personal and to some extent private. This, in his words, allows for getting 'further faster', so another—implicit—value is productivity. Last, we observe an acknowledgment of boundaries as to who is acceptable to work with and the idea that there is a camp to belong to. These concepts are raised to be dismissed, but it is clear from this that group identity holds a core, if tacit, role in the work.

> Interview 2
>
> Discussions are facilitated. Rules about taking turns, etc, are stringently enforced. This insistence on non-hierarchical 'fairness' tends to produce a fair degree of harmony and a good sense of genuine consensus.
>
> How do we know we want to work together if we don't discuss values? … A major shared value. Obviously, we do discuss politics with individuals we're closer to—it's public discussion that people shy away from. But, for example, although it pisses me—and the other dykes—off that the women-only thing is 'excused' by women's alleged fragility, we know that the space is fragile and could be damaged by a ruck about it, so we hold our peace. The energy overall is extremely positive and the space is genuinely buzzy and feels supportive. It makes us feel protective.
>
> We also avoid reference to the fact that some women are straight and some are dykes. … the women involved come from diverse cultures and backgrounds and that [causes tension] in some areas—notably women-only, sexual orientation and all that stuff.

In our second excerpt, the speaker identifies 'a major shared value' as the means by which a group that does not discuss values manages its commitments. This is the basis of strategic essentialism and we can see it being achieved through the work of collaborating carefully and allowing each person there own priorities. We hear the benefits of working in this way—that the space is buzzy and positive—and the value this has for the speaker who holds her peace because she feels protective. As in the first example, the management of the collaboration is to 'avoid' the topics identified as most contentious and all formal political views are kept for private discussion.

Interview 3

I can either say 'that's an interesting model'… or say 'that's an interesting, but impossible, model'. And because I was quite grumpy I went for the second one. I was glad I pushed that, but quite often I wouldn't do it. I don't want to hijack the conversation and in the end I stepped back and said 'let's carry on the conversation' because we weren't going to get anywhere with it. His question to me eventually was 'are you saying I shouldn't do this because it's not inclusive?' And I said 'I'm not saying you shouldn't do it but I am saying you should try and find ways that are more inclusive'. But you could see that I'd killed the conversation in a way, because half the people it hadn't occurred to and they were just sitting there. So that's the kind of thing … normally I wouldn't have [raised the topic] but there I felt strongly enough about it.

In the first two examples, the speaker is describing what the group does not talk about in public session, how it is managed and why. In the last excerpt, the speaker recounts an occasion when she breached her understanding of the rules of conduct and raised a concern around (the value of) 'inclusiveness', how normally she would not make this breach, and how it killed the conversation. In this way, she demonstrates the working of this mechanism and her awareness of it. She shows the work of collaboration by holding out for an ideal at the expense of easy social process.

In all three examples, it is apparent that a very deliberate tactic is being employed so that people with strong and divergent political and social opinions (or values) can be productive together, how fragile the peace is and the extensive work that is done to protect it.

Understanding Détente

As has been noted, the PraDSA project was riven with argument. This was true of the research team too, since it brought together many researchers from different institutions, with strong feelings and opinions and different disciplinary commitments. Referring to the same internal memo written for colleagues: 'The PRaDSA team is learning to operate similarly [to the groups it is studying] by taking out certain value laden terms from our discourse and instead recognising that a 'field' for a value is open and contested, awaiting filling by the individual member (to use a database analogy)'. (internal memo, Light, 2008).

In this way, the team involved in the research project navigated using the collective learning that working with the practitioners had yielded. We found that making together became more than the ambition of a productive team, but also a tool that promoted understanding and trust, allowing us the space to negotiate our differences harmoniously. This gave us useful insight into how these radical self-organising groups work to leave certain kinds of discussion out, while being wholly values-oriented in their political making work. Indeed, it seems to be the very strength of their desire for change in society that dictates that discussion of high-level values is not profitable, but that begs the question of how a group brings its values into its making. And given the concerns that parts of the design research community has raised about incorporating values in design (e.g. Borning & Muller, 2012), we can ask if silence on broader values matters.

To answer, we must contrast broad-based social values and incorporation of thought-through design features in specific contexts of making. In other words, 'liberty' is harder to operationalise at the design stage than 'ease of use', but some design choices are more in line with concepts of liberty than others and these 'feature' discussions are part of the emergent design.[1] To give an example, the making of a FLOSS screen reader for those who find it difficult to read might be an act of radical inclusion. Another product might not have *inclusive* values incorporated because its focus might be a different form of emancipation—something evident in Nafus' (2012) and Toubin's (2014) discussions of an 'openness' that excludes women, see above. We see a partial answer in the intervention in our third example, where a person speaks up to ensure that something 'alternative' on another dimension is shaped with inclusiveness in mind.

But there is a marked difference between discussion of a feature, which is a how/what discussion in the moment of making, and loftier conversation about *why* something needs to be made. Negotiation with regard to a design's detail reflects the broader thinking of members of the group. But this negotiation is attached to instances of design work, not a precursor to it. This distinction underpins the practices of members whose value systems are so strongly conceived and held that they could not collaborate together on other terms.

Study 2: Effectiveness in Action

The second context examined here was a coming-together of social activists to explore aspects of their practice in a purpose-made research project, facilitated by the authors. Participants came from multiple small fluid organisations and represented the informal end of the change-making spectrum, employing 'liquid organising and choreographic leadership' to stay light-footed (Gerbaudo, 2012).

In reporting this study, I focus on researcher-researcher relations, not least because this was a co-research experiment as well as an opportunity to explore activist maker values. The authors' research practice is anyway to attempt to create conditions for reciprocation, with learning and practical outcomes for all (see Light, Egglestone, Wakeford, & Rogers, 2011), but the co-research/co-production ambitions as well as the social activism theme threw a focus on the group dynamics.

I have already explained that deep anonymity was a condition of publishing interviews with the activists spoken to in the first study. The agreed position on talking about the work in this project requires me to say here that this section and the analysis that follows are my thoughts and do not reflect on those of others involved in this project. That is the ethical position taken in this project that impacts upon how we report it.

As mentioned above, ground-rules included that no participant's social action activities would be observed or 'studied'. The arguments for not reviewing existing practices too closely were manifold. It was written in the final report that:

- Some practice is too indeterminate, evolving or new to define meaningfully;
- Tactics get exploited in an escalation of confrontation when shared (as with the adoption of 'fluffy' tactics by security and police in road protests);
- Analysis and comparison are not useful to developing practice, whereas group reflection is;
- Definition might stultify growth and exploration, eg reducing 'play' to a 'concept of play';
- Toolkits abound and participants did not want their practices reduced to a formula.

All these are plausible reasons why a straightforward study of practice might yield little on one hand and compromise the groups' activities on the other. But another interpretation might be that, by watching others, we editorialise their work with our own values and language. It is a form of colonisation. With reluctance to analyse practice came reluctance to synthesise findings or speak as a single voice on diverse, highly contextual and idiosyncratic issues. We developed a policy of juxtaposing plural ideas to convey meaning and giving all work disaggregated authorship. This then evolved into a strategy for organising outputs.

Valuing pluralism and discretion over 'normal' research practices suggested a specific research approach. Unlike the participant observation work in the first study, this meant that the project could not rely on activity that any one collaborator group might have been considering anyway. Instead, the core team staged events at which people could come together, using these collaborations to articulate aspects of activist work—both in the staging and the content. To act as a focus for, and summary of, these articulations, the group agreed to make something together as a support for other activists and use this vehicle to frame our understandings so that we could communicate learnings about activism in a context that it might resonate.

In the end, the main output was a book. The material was taken from the words of the change-makers—people active in civil society across the UK—who had been brought together in the three co-produced events. It was edited collaboratively by the core project participants in a long slow process of consensus finding, which began in workshops and continued asynchronously and remotely over emails and cloud storage editing facilities. The ideas in it are intended to circulate, so postcards, to tear out as mementoes or pass to others, were included at the back of the physical copy of the book. The pdf of the book can be found here: https://db.tt/lgeP0Uoc.

Again we see much the same process in action. The book becomes multiple private spaces in which we hold our values intact and it is the juxtaposition of these positions that becomes the learning about social action, not a synthesis, which would represent the negotiation—and perhaps diminishment—of personal views. The group comes together to make this collective statement as a source of nourishment to other activists, but no statement within it is collective. And the book has less than normal comprehensibility outside the frame of the work and its intended audience.

Hence, the book is the only group statement of the project and it is a series of occasionally fragmented, mixed-up pieces, each with their own author credit. Far from anonymous, every author stands alone and named; any composite meaning comes from juxtapositions and from reader interpretation, rather than the aggregation of findings more usual in research.

These choices made in editing the book reflect the 'major shared' values of the group as they came together, emergent in the negotiation of group norms. Again, the making made the negotiations possible between groups using very different forms of action (from direct action to art installations), even while the growing culture made possible the book. Thus, the designing of the book became a microcosm of the learning from the process elements of the project. The choices are striking in tone, colour palette, aggregation of graphic design styles and emphasis on aesthetics, as well as content.

The core team reviewed 48 hours of spoken material as well as emails, blog accounts and other reflective texts. Days were spent agreeing a policy for this in such a way that credit was given, voices were heard, no authorial or managerial perspective dominated and, though an elaborate organising system was created, the organisation was not linear or terribly obvious to readers. Agreeing (perhaps reluctantly) that it must, nonetheless, be readable by the social activists who were its audience, some concessions were made, such as explanatory passages dotted throughout.

Last, since, at the very least, the academics in the group were keen to be able to discuss what was achieved, it was agreed that anyone was free to give their perspective on the work as long as they did not attempt to speak for others. This largely outlawed the use of the term 'we' and later led to the presentation of a co-design session on 'There is no We' by members of the core team.

The ground rules we had negotiated served us relatively well to keep conflict about the designing to a minimum across the group, while allowing for considerable decision-making. Personal political differences did manifest and issues of power, control and relations with the establishment, in the form of a university and the state funding apparatus behind it, did come up. These inevitably created a challenge to a common purpose, but focus on an agreed output with an agreed design and method repeatedly kept the activities alive.

Bottom-up Making

This creation process in this study is different from those reported in the first study in being the artefact of the research activity rather than the observation and report of engagement with DIY political making. It is different in that we could see the ethos of the group form, whereas the accounts from makers, above, are all snapshots from more established groups and we do not see the original negotiations that led to their informal policies.

However, one of the things that the latter study has in common with the former is the use of making as a way of productively negotiating different positions. Although collectively we were a research project, born of a common curiosity about our practices as well as a political ambition to create something new, we used a similar approach—of bringing people together round a task—which we agreed as part of deciding a way of working across groups with different sympathies and political affiliations. This was explicit. We did not want to eliminate conflict, but we were keen as a group (with some complementary understanding of how schisms form) to manage our disagreements without entrenchment. This happened best when we had a focus outside ourselves, coupled with a common purpose. This approach was recognisable to the makers and the campaigners in the group—expressing different parts of the social action scene—as how they managed their social process in their main groups of allegiance. What differed in the group we were forming is that it had no collective history to draw upon. Unlike the practices described in the first study, we were making our practices as we went.

Discussion: How Values Are Negotiated

In this piece, the hypothesis that there exists a highly pragmatic and opportunistic emergent design practice working on a need-to-decide basis and leaving certain outcomes open has been supported by interviews with radical self-organising design groups and the creation of a making-based research project with social activists. In the studies, no one spoke of contrary strategies, but many described characteristics of this 'détente model' of operating, and our attempts at conducting work together in both research projects showed the same tendency towards using activity to provide unity. Coherence is achieved through making, not discussing. Alignment comes through engagement in actions taken together over time and through harmonious interactions leading to respect and trust. This is epitomised in our editing by consensus. Other ways to find agreement worked less well. We see this when PraDSA spent time trying to examine the values that had brought people together, in the way recommended by design researchers, and things ground to a halt, the atmosphere grew tense, productivity fell and the whole project hung in the balance.

What kept us together if we could not agree on the higher-order targets that inform the purposes of design activities? Seemingly, it is an underlying belief that the world needs to be changed and a 'major shared value' as to how. Identification as radical brought together everyone who gathered. In other words, certain conditions—such as a perception of other group members as part of the non-dominant Other—would seem a precondition of any engagement together at all. Initial definition of membership in each group took place through reference to a dominant state that required challenging and a sense of functioning outside the dominant (or of oneself as Other).

Once someone was perceived to belong to this class of outsider—and it could be as modest a position as awareness of the potential to be Other or as committed as identifying as 'dyke'—then they could be accepted. The resultant grouping might be considered a form of strategic essentialism, made valuable to those associating as a unit of productivity and production as well as, or instead of, a political lobby. It is interesting to try and gauge how far the groups described here were not defined by their common beliefs, but by their common antipathy to people who believe otherwise. We are reminded of the saying that *my foe's foe is my friend.*

So, we see the social elements described in Toombs et al. (2014) and others, the membership over time that leads to bonds based on group identity and the trust that comes from working together, but the interview material also shows the care that runs behind the decision not to raise certain areas of known conflict or challenge ways of working. People have an understanding that the work they are doing is important and is to be protected by careful navigation. And part of the recognition of importance and protectiveness relates to a common opposition. It is interesting to speculate too whether the emphasis that Toombs saw on tools and making in his initial visits comes in part because the tools, as symbols of making, are also the means to achieve connection across diverse cultures and communities. Access to good tools and people who will help you use them may be reason to join a hackerspace, but perhaps there is something deeper in members' investment in showing them off, which relates to the tools' function as safe bridges.

In all these situations, despite a research presence, the groups engaged in making social change were self-organising and purposeful in their activities (even if they altered their usual practice in the second study so as not to be watched). They were not generally using a shared space to do different projects, as happens in many maker- and hackerspaces, and they were also functioning outside straightforward commercial design or traditional community development. This contrasts with many contexts, where the design or facilitation element comes in from outside (see also Light & Akama, 2012, 2014). Such, more traditional, exogenous approaches might learn from these radical processes of engagement in addition to the designer voices that urge consideration of values. In any context, ways of working have been established and the incomer has to earn a position in the social and political structures around them. Being mindful of the level of sophistication of groups' functioning and of the negotiated détente that may exist where a group has constituted itself in relation to other concerns is a good supplementary consideration alongside techniques for values elicitation, especially in community or professional situations where emotions are charged (Light & Akama, 2012). Aspiring to design first and using this as a means of negotiating values may be a good alternative to working towards more specified social virtues, especially if these might be highly disruptive to the group or laborious to elicit. Instead, aspects of the design process can be facilitated to encourage a collaboration based on détente understandings, where making precedes other defining activities.

The tactics here bring to mind another strand of research that emanates from the earliest activist studies. Light (2011) talks of a politically-motivated design meta-value. The meta-value is of leaving space in the way that technology is designed so that alternative social structures can be born. To do this, the technology must keep a degree of play in the specification of social, cultural and political elements. I argue for 'the oblique route that puts the flexibility in the system for diversity to flourish for ... whatever a future generation, allowed the freedom to conceive of them, considers the key virtues to be.' (Light, 2011).

As we noted earlier, the work of making without specifying our more abstract political values (as to how we would like the world to be) can be likened to leaving a 'field' for values open, waiting to be filled by the individual member. This is, perhaps, a strategy related to the one found in the description of meta-values: choices are not closed down; the potential for change is promoted without trying to dictate the outcomes. Design, as a material political intervention, makes social change that cannot be known at the time of enactment. This is acknowledged in the détente model. New products, processes and systems are released into the world, intended to be on the side of the angels, intended to tip the balance modestly towards whatever *each of us believes* is a better existence. The uncertainty of the impact is matched by the play designed in the system for each member of the group to appropriate the *purpose* of the thing that is made in their own chosen way.

Conclusion

This paper reviews a practice and presents a potential explanation of how it is sustainable. I have not examined that practice in terms of outcomes. One upshot of this enquiry is to suggest that more work is needed on how design emanates from groups of social activists intent on preserving a productive harmony, and whether what is designed is as open and generous as the means that produced it. The ideas presented here are offered—in keeping with the spirit of the piece—to throw up questions and keep the debate alive.

What we can conclude is that groups of change-makers tend to be committed to higher-than-self goals, passionate about their beliefs and vocal in expressing them, yet able to navigate many related opinions about the nature of the change to be made so as to be able to collaborate. One navigation strategy that emerges is to collaborate not so much with those of like mind as those with a common opposition to other positions in society (seen here as strategic essentialism) and another is to avoid discussion of detailed beliefs—and so bypass the potential to mire action in argument—by focussing on what action to take and not why. In this, I suggest, we are seeing the workings of a highly critical and sophisticated reading of society and a tendency to self-Other. This positioning as Other complements those values that lead to the desire for change and those that put a stress on the productive force of making the tools for change.

Note

1. Although concepts of liberty are also very diverse.

References

Borning, A., & Muller, M. (2012). Next steps for value sensitive design. *Proc CHI'12.*

Crompton, T. (2010). *Common cause: The case for working with our cultural values.* London: WWF-UK.

de Beauvoir, S. (1972/1949). The second sex (H. M. Parshley, Trans.). London: Penguin.

Dourish, P. (2008). Points of persuasion: Strategic essentialism and environmental sustainability. *Proc. Pervasive'08 Workshop on Pervasive Persuasive Technology and Environmental Sustainability* (pp. 63–66). Retrieved from eprints.qut.edu.au/18247/2/18247.pdf

Flanagan, M., & Nissenbaum, H. (2007). A game design methodology to incorporate social activist themes. *Proc CHI 2007.*

Friedman, B. (1996). Value-sensitive design. *Interactions, 3*(6), 16–23.

Gauntlett, D. (2011). Making is connecting: The social meaning of creativity, from DIY and knitting to YouTube and Web 2.0. Cambridge: Polity Press.

Gerbaudo, P. (2012). Tweets and the streets: Social media and contemporary activism. London: Pluto.

Light, A. (2011). HCI as heterodoxy: Technologies of identity and the queering of interaction with computer. *Interacting with Computers, 23*(5).

Light, A., & Akama, Y. (2012). The human touch: From method to participatory practice in facilitating design with communities. *Proc. PDC 2012.*

Light, A., & Akama, Y. (2014). Structuring future social relations: The politics of care in participatory practice. *Proc. ACM PDC 2014* (pp. 151–160).

Light, A., Egglestone, P., Wakeford, T., & Rogers, J. (2011). Participant-making: Bridging the gulf between community knowledge and academic research. *Journal of Community Informatics, 7*(3).

Light, A., & Miskelly, C. (2008). Brokering between heads and hearts: An analysis of designing for social change. *Proc. DRS 2008*, Sheffield, July 2008.

Lindtner, S., & Li, D. (2012). Created in China: The makings of China's hackerspace community. *Interactions, 19*(6), 18–22.

Milne, A., Riecke, B., & Antle, A. (2014). Exploring maker practice: Common attitudes, habits and skills from Vancouver's maker community. *Proc. FabLearn 2014.*

Nafus, D. (2012). 'Patches don't have gender': What is not open in open source software. *New Media & Society, 14*, 669–683.

Raymond, E. S. (Ed.). (1996). *New hackers dictionary* (3rd edn.). Cambridge, MA: MIT Press.

Simonsen, J., & Robertson, T. (Eds.). (2013). *Routledge international handbook of participatory design*. London and New York, NY: Routledge.

Spivak, G. (1987). *In other worlds: Essays in cultural politics.* London: Taylor and Francis.

Tanenbaum, J. G., Williams, A. M., Desjardins, A., & Tanenbaum, K. (2013). Democratizing technology: Pleasure, utility and expressiveness in DIY and maker practice. *Proc. CHI'13* (pp. 2603–2612).

Toombs, A. L., Bardzell, S., & Bardzell, J. (2014). Becoming makers: Hackerspace member habits, values, and identities. *Journal of Peer Production, 5.*

Toombs, A. L., Bardzell, S., & Bardzell, J. (2015). The proper care and feeding of hackerspaces: Care ethics and cultures of making. *Proc. CHI'15* (pp. 629–638).

Toubin, S. (2014). Feminist hackerspaces: The synthesis of feminist and hacker cultures. *Journal of Peer Production, 5.*

Section IV. Case Studies Introduction

Jeremy Hunsinger
Wilfrid Laurier University

The chapters in this section are case studies of makerspaces, hackerspace, and hacklabs in western and non-western contexts. They examine specific countries and places where cultural specificities undermine easy generalizations about hacking and making. That is, they allow us to examine, understand, and even challenge how people act collectively in particular situations. Our inclination to include case studies is that specific trajectories and transversals in the world of hacking and making cannot be seen from broader overviews and global/hegemonic perspectives. Therefore, the chapters in this section should be read in relation to the rest of the chapters in this book, as they fill in niches and clarify possibilities raised elsewhere.

Pip Shea, in her study of Farset Labs and its use of the Raspberry Pi, introduces us to questions of social innovation in relation to appropriate technologies. Her exploration of the community of practice in the lab describes the critical interplay of material and cultural discourses that surround social innovation and technology. To the outsider, that system of relations might seem simple, but this case study emphasizes the complexities internal to the operation of the hackerspace in its context. Shea's exposition demonstrates the how hackerspaces become gendered by drawing on feminist theories of community and knowledge. Tensions derived from her feminist analysis drive her demand for more pragmatic interventions in the relations between social innovation and hackerspaces.

Nicholas Balaisis engages questions of Cuban hacker and maker practices, particularly the the work of Ernesto Oroza, using a political economic and cultural studies framework. He draws on the rhetorics of creativity in Cuban culture to consider tactics as a mode of analysis. His exposition demonstrates a discursive link between labor practices and rhetorics of creativity in Cuba.

Specifically, Balaisis critiques the "creative class" concept developed by of Richard Florida, and places it in relation to earlier political economic frameworks such as Fordism and entrepreneurialism. In doing so, he lays bare the discursive constructions of creativity in the rhetorics of hacking and making as they occur in Cuba.

Following Balaisis, Xin Gu engages us with a case study of the maker movement in China. She shows that some of the potentials that are widely assumed to exist in the hacker and maker movements are not realized in China because making was uncritically received and has produced only constructions of a new industrial model. These ideological constructions derive, in part, from the recent normalization of middle class lifestyles of many makers. In other words, Xin recognizes the maker movement is challenged by the context of creativity and production with their stabilizations and traditions, deflating easy claims of a democratic equalization of productive power.

The final chapter in this section is by Karen Louise Smith, whose case study examines Hive Toronto. Hive is a Mozilla stewarded network that provides local services including educational opportunities. Smith looks at Hive Toronto from 2012 to 2015 in an attempt to analyze the infrastructures in place that enable hacking through events such as HackJams. She uses education and equity to argue that Hive participants were transformed in these events from users of the web to producers. One of the infrastructural examples that she uses is the network of Remixable Open Educational Resources affiliated with Hives. Hives can use these resources to create new educational experiences that are unique to their own Hive. Emphasizing the role of Mozilla as infrastructural provider, she argues that transforming from silos and walled gardens to a more open system would broadly benefit hacking and making.

These case studies substantiate a plurality of approaches to hacking and making that manifest in specific countries and locations. They provide moments of critique, but also points of recognition of difference that might empower alternatives. These points of connection across cases center around issues of identity, equity, creativity, architecture, and infrastructure. These concepts cut across the worlds of hacking and making, as exemplified by discourses and practices such as re-imagined notions of hobbies, innovation, entrepreneurialism in their specific regional/national contexts that exist outside of some of the imagined political economic hegemonies of the worlds of hacking and making. Ultimately, these crosscutting assemblages of discourse and practices inform the other chapters in this volume and provide evidence toward the growth and broad institutionalization of hacking and making worldwide.

13. *Hacker Agency and the Raspberry Pi: Informal Education and Social Innovation in a Belfast Makerspace*

PIP SHEA
Monash University

Introduction

Hackers circumvent to invent. They see the potential in *things* (their affordances) as well as *themselves* (their abilities and capabilities) to create anew with existing materials. Hackers model new structures, devise alternative infrastructures, or exploit systems. This process of material reimagining can help people gain a greater understanding of things and systems, but what forms of agency are required to hack in the first instance? The following chapter responds to this question, supported by an investigation of hacking activities in Farset Labs, a makerspace in Belfast, Northern Ireland. It argues that informal education activities performed at Farset Labs are helping individuals and groups to understand how to support hacking activities. This is framed as nurturing *hacker agency*: enacting the conditions to support the inversion, subversion, or reconfiguration of things. The inquiry unfolds around Farset's Raspberry Jam program—a monthly workshop exploring Raspberry Pi single-board computers—situating it as an appropriate socio-technical system to nurture hacker agency. Empirical data was collected over a period of one year at Farset Labs. My observations were gleaned while participating in the co-working space, attending public events, participating in member meetings, and volunteering at three Raspberry Jam events.

This investigation of hacker agency in the makerspace context in Belfast attempts to provide new sightlines for "critical hacktivism" as proposed by McQuillan (2012). Critical hacktivism is an approach that values "messing

Farset Lab's Raspberry Jam was chosen to investigate the conditions of hacker agency because the Raspberry Pi was designed as both a computer and an educational tool; and, because Farset's Raspberry Jam offers empirical evidence where the conditions of hacking are supported. While acknowledging the spectrum of definitions and practices associated with hacking, this chapter focuses on the systems, tools, initiatives, and environments that help people build their capacity to reveal the affordances of sociotechnical systems; or, *hacker agency*.

Hacker Agency: Enacting the Conditions of Hacking

Hacker agency is an asset when designing and developing appropriate technologies for social contexts. This is relevant to makerspaces in Northern Ireland as they are situated in relation to a sectarian conflict that began in the late 1960s. Colloquially known as *The Troubles*, this period was shaped by paramilitary violence from groups contesting the political sovereignty of Northern Ireland, and how this was affecting the rights and responsibilities of the citizenry. The two groups in opposition were those who fought for the reunification of Northern Ireland with the Republic of Ireland, against those loyal to Britain who fought to remain part of the United Kingdom. These two communities are often identified as either Catholic or Protestant. Efforts to build peace culminated in 1998 with the signing of the Good Friday Agreement by Northern Ireland's political parties, and the Irish and British governments. Although violence has subsided, reconciliation is ongoing, and Northern Ireland's makerspaces are actors in this peace process. They run various programs that actively engage with the politics of "good relations" between Catholic and Protestant communities, and these activities are increasingly framed as *Digital Social Innovation*. This chapter promotes hacker agency in these social contexts as materially remodeling technology can lead to more appropriate solutions.

I understand *hacker agency* to be the enactment of desire, personal capacity, time, tools, and space that leads to material remodelling. Having the desire to perform hacking activities is directly impacted by a person's capacity to experiment, explore, or problem solve as well as their capacity to configure, or reconfigure the material in question. Having access to tools and the knowledge to work them is also crucial, as is having physical space where messiness and disorder is permissible, or at least tolerated for periods of time. Having, or making the time to hack also figures prominently, as experimental or playful work can sometimes be difficult to justify within a busy modern schedule. The enactment of these conditions can support material engagement to help reveal the affordances of technology and things. A previous mention of the term "hacker agency" describes a scene where "high-tech workers—no matter how inexpert—can interrupt, upset, and redirect the smooth flow of structured communications" (Ross, 1991, pp. 92–93, cited in Lin, 2004). The conditions of hacker agency are changeable. Optimal arrangements for hacking—or material remodelling—are not achievable at all times. This framing of *agency* draws on Barad's (2007) conceptualization of "agential realism": that phenomena and possibilities are constantly being reconfigured to reveal different conditions.

Different groups might emphasize different rationales for nurturing conditions for hacking: large technology firms, governments, and NGOs might reference free labor and the promise of innovation; small businesses might be driven by pragmatism born from resource limitations; while grassroots organisations such as makerspaces might emphasize the desire for alternatives. Individual motivations associated with hacking also vary: people might have the need or desire to repair something, or to develop a greater understanding of the potential for materials to be used in novel ways. These actors all share an interest in the social—and economic—benefits of material reimagining.

Material reimagining can lead to new understandings of the politics of artifacts and their entanglements with people and systems. In 1980, Langdon Winner famously argued that explicit attention should be paid to the idea that artifacts have politics (1980). In one example, he exposes the design of bridges in Long Island—structures deliberately designed too low for public buses to drive under—to sharpen his point that designed systems can embody specific forms of power and authority. Winner makes the point that we are so accustomed to infrastructure like roads and bridges that "we see the details of form as innocuous, and seldom give them a second thought" (1980, p. 123). Winner's later focus on the unintended consequences of designed things (1986) is most relevant to this study as it supports the argument that engagement with the materiality of technology can cast a critical lens over artifacts.

Researchers are increasingly adopting methods that get people to engage more critically with the stuff of technology, as evidenced by what has been described as *material turns* across scholarly disciplines (Fuller, 2006; Gillespie, Boczkowski, & Foot, 2014; Marres, 2012; Ratto, 2011). Artists critically reimagining digital networks include Julian Oliver, who does so to develop a more "rigorous personal relationship" with software and digital technologies (Bucher, 2011). Oliver is part of the Critical Engineers working group with Gordan Savičić, and Danja Vasiliev. Oliver and Vasiliev have collaborated on numerous projects, including *Newstweek*, a "network intervention" and "reality distorting device" (Oliver & Vasiliev, 2011). This project disrupts "public" wireless hotspots typically deployed in coffee chains such as Starbucks. They do this by installing an innocuous wall plug device in the shop that wirelessly interferes with the display of major news websites. The artists provide a separate website where people in the coffee shops can add to or edit fake news stories.

Having the skills to materially reimagine and remodel are becoming an asset in various areas of life, such as contributing to participatory government initiatives like civic hackathons. These events are often public forums that encourage the remixing and reinterpretation of government datasets. They are legitimized by the premise that citizens are participating in the systems that will shape their lives, but are limited in scope as they involve only those who *know* how to hack. Irani (2015) proposes civic hackathons celebrate a Silicon Valley-esque "entrepreneurial citizenship," that favors "quick and forceful action with socially similar collaborators" over approaches that account for the politics of difference. Civic hacking practices have also been situated as "friction-rich endeavours" (Perng & Kitchin, 2015) because contention lingers in the design and production of technologies. This contention surrounding emergent modes of civic participation provides evidence that becoming familiar with the conditions that optimize hacking might be a useful personal development trajectory.

Processes of material reimagining can help people gain a greater understanding of things and systems, but what forms of agency are required to hack in the first instance? And, how can we better understand the conditions that contribute to material engagement and reimagining? These questions have inspired the following investigation in to Farset Lab's Raspberry Jam. It offers evidence of the conditions for hacking being nurtured, resulting in practices that encourage the circumvention of constraints through bricolage, modification, or adaptation. These events are offered for free, and made available to those who are willing and able to spend one Saturday afternoon a month in Belfast's city center. The Raspberry Pi is the technical focus of the workshops, but the activities they perform, build people's capacities to hack. Through material engagement, they create anew with existing materials to build new systems and exploit incumbent ones.

Farset Labs' Raspberry Jam

Farset Labs is a non-profit company and registered charity based in Belfast, Northern Ireland. It began as a hackerspace in 2012 and is now also referred to as both a hackerspace and a makerspace. It is funded primarily by membership fees, is volunteer-run, and has a co-working space, events room, workshop, kitchen, and lounge area. There are communal computers, dedicated high-speed internet, a 3D printer, a vinyl cutter, a milling machine, drills, soldering irons, cabinets of cables, and boxes of "redundant" technology on offer for repurposing. Farset Labs is a "social workshop" (Hunsinger, 2010) providing physical space, communications infrastructure, and access to a community of people who are enthusiastic about hacking and making. Beyond these standard offerings, Farset runs public events that build the capacities of participants: CoderDojo events invite young people to the space to learn software programming skills, and hackathons provide incubators for social learning in an atmosphere of experimentation, invention, sleep deprivation, and pizza.

The Raspberry Pi can be situated as part of a broad technology activist project, but is aimed at schools and young people. This has led to comparisons being made to the *BBC Micro*,[3] a computer estimated to have been in 80% of UK schools during the 1980s. Launched in 1981 by the British Broadcasting Corporation as part of a computer literacy project, the BBC Micro helped expose UK school children to computer programming. The Raspberry Pi project is supported by organizations that are developing interoperable software and hardware for the device, and informal education bridges to extend the reach of the device, such as Raspberry Jams.

The Farset Labs makerspace plays host to a free, monthly Raspberry Jam. It began in 2012, led by Andrew Bolster, one of the three Farset Labs' founding directors. Nineteen-year-old Andrew Mulholland—a computer science student at Queen's University Belfast—now runs the program. The event is "aimed at anyone from 10–110 years old" and places an emphasis on making things, and having fun while doing so (Bolster & Mulholland, 2013). The sessions take place on the first Saturday of the month and last for 4 hours. The organizers distinguish the Farset Raspberry Jam from other similar events because it provides all the necessary equipment for attendees (Bolster & Mulholland, 2013). Much of the labor required to successfully run a Raspberry Jam involves promoting the event, managing volunteers, managing expectations of participants, plugging in equipment, moving desks, buying doughnuts, and looking for funding to sustain all of the above. Code must also be tested so that activity sheets can be written, photocopied, and distributed. These tasks require a number of people with a range of skills. Currently, there are *just* enough volunteers to sustain the program.

Participants at the Farset Raspberry Jam are given a choice of tutorials that focus on such things as music making in *Sonic Pi*, and model making in *Minecraft Pi Edition*. They also use *Scratch*, a program that uses a graphical user interface to teach programming. One activity also encourages participants to program in Python, in turn exposing them to digital networking operations. Collective problem solving is designed in to some of the tutorials, as is the goal of hacking other participants' computers. Farset's Raspberry Jam also connects participants with the physical computing potential of the Raspberry Pi. The operating system used is the Linux-based *Raspbian*, the most common OS for Raspberry Pi hardware.

Farset Labs' Raspberry Jam aims to "inspire and enthuse the next generation of technologists, through an experiment-focused programme of supported exploration" (Bolster & Mulholland, 2013). Through the enactment of software and hardware exploits using Raspberry Pis and various digital and analogue peripherals—the Raspberry Jam is building participants' capacities to hack. Through social activities that frame hacking as fun, participants are revealing the affordances of electronics and digital technologies. This process of nurturing hacker agency is made possible by an appropriate mix of people, place, time, and space.

Revealing Affordances Through Pi Play

The Raspberry Pi was chosen as a focus for this study as it has the potential to contribute to a multitude of projects and systems across digital networks and in the physical world. Its hidden affordances abound. It can be thought of as part of "hardware's long tail" (Buechley & Hill, 2010), as it facilitates niche projects that are reshaping physical computing. The Raspberry Pi's original project—to revolutionize computer science education—emphasizes learning computing through play. Programs such as *Minecraft Pi Edition* have been specifically adapted to support this aim. Farset Labs' Raspberry Jam extends this educational paradigm through social making and learning.

As a contemporary form of informal education, Farset's Raspberry Jam finds support in Douglas Thomas and John Seely Brown's (2011) framing of the "new culture of learning" as the cultivation of play, questioning, and imagination. They assert that a combination of self-directed online learning must be balanced out with "bounded and structured environment(s)" that enable "unlimited agency to build and experiment" (2011, p. 19). The Raspberry Jam, along with Farset Labs, are nodes in a system that has been described in humanities research as *Connected Learning*, the foci of which are social learning and the making of things that happen across different sites and locations, online and offline. The major hypothesis of a 2013 report published by the Connected Learning Network (Ito et al., 2013) argued that young people require "caring adults, supportive peers, shared cultural references, and authentic ways of contributing to shared practices in order to mobilize their skills and knowledge." Farset Labs' Raspberry Jam achieves this by creating the conditions that nurture hacking practices: activities that reveal the affordances of technologies and things.

Minecraft is used to reveal the affordances of software both on the Raspberry Pi platform generally, and at the Farset Labs' Raspberry Jam. The objective of Minecraft is to use blocks to make models and create environments, and it doesn't take long to teach someone how to play. It is an entry point for hacking activities, as it is the biggest selling PC game of all time (Campbell, 2014, cited in Apperly, 2014) and familiar to many people. One Minecraft tutorial uses a join-the-dots game with conductive paint to trigger a hack in the game. Conductive paint is another entry-level activity, used to engage participants with the material stuff of electronics. Hacking Minecraft gives participants a taste of controlling software through programming, while also exposing them to the practice of computer game modification.

Another Raspberry Jam exercise, that demystifies the process of hacking analogue objects, uses a decommissioned traffic light affectionately known as *wizard signal*. A member of Farset Labs who has an interest in road infrastructure saw the traffic light being dismantled by the roadside. She asked if she could take it as it was being replaced with a newer model. She took the traffic light to Farset Labs to play around with it. The facilitator of the Raspberry Jam took an interest in the light, and worked with her to hook it up to a Raspberry Pi. The wizard signal is now used to teach Raspberry Jam participants about IP addresses and linking digital and analogue devices across a local area network (LAN). Through simple programming activities participants are taught to turn the traffic light on and off. This becomes a race to see who can turn the different colors on first. This example exemplifies how the Raspberry Jam is revealing affordances through play, as it encourages participants to become the directors of network traffic through the metaphor of road traffic signals.

Learning how to hack in informal education contexts can be thought of as an important developmental step to reaching the critical mindsets outlined in Santo's (2011) "hacker literacies" literature. Increasingly, having the skills to contest the design of technologies by questioning the intentions of its creators is being framed as an important critical digital literacy. Learning about computing and hacking at workshops in community makerspaces also challenges normative understandings about the purpose of computer science, modes of play, methods of making, and the contexts in which hacking might take place. Raspberry Jams can be considered a node in the connected learning ecosystem that helps people learning to see the potential in themselves through hacking things. And, the more proficient a hacker becomes the better positioned they are to create affordances as well as reveal them.[4]

The hardware and software ecology of the Raspberry Pi, especially its price and size, makes it appropriate for many projects across different contexts, proof it has the potential to radically increase the scale and reach of physical computing. But this hardware long tail will be shaped by who uses the devices and for what ends. Raspberry Jam informal education programs are helping to diversify the Raspberry Pi user base, which in a small but significant way, contributes to a more distributed technological ecosystem. This, combined with a focus on hacking for fun, increases the potential for the exploitation of the visible, perceived, and hidden possibilities of objects or systems.

Farset Labs and Digital Social Innovation

The promise of *Digital Social Innovation* (DSI) policy and practice includes increased participation in civic life, building literacy capacities, and new modes of production, communication, and evaluation. This chapter promotes critical material engagement—or "critical hacktivism" (McQuillan, 2012)—as a future trajectory for DSI initiatives, as it helps to untangle preconceived articulations and assemblages to make way for more robust and inclusive processes of remodeling or reassembly. McQuillan's conceptualization of critical hacktivism builds from Von Busch and Palmas' (2006) "abstract hacktivism" thesis—where they position hacking as a new conceptual model through which we can understand and approach the world—and the critical pedagogy of Paolo Freire (1972). Previous studies of hacktivism include Jordan and Taylor's (2004) account of an emergent counter culture focusing on hacking as a political activity, and Alleyne's (2011) reminder that hacktivism is a descendant of other counter-cultural activities, such as pirate radio.

Hacking activities are an entanglement of the socio-technical. When they take place, relations enacted from people, places, and things are enfolded into the hack. When designing for social contexts, the details of these entanglements can lead to technologies being appropriate or inappropriate for the communities engaging with them. This is certainly the case in the Northern Irish context, where "good relations" projects engage people whose experience of daily life varies greatly. This provides further rational for using critical hacktivism philosophies to encourage appropriate DSI projects.

Farset Labs is already engaging in critical hacktivism. In late 2014, a group of Farset members were involved in a *Social Innovation Camp (Sicamp)*.[5] Sicamp matches social development projects with technology practitioners in a peer-to-peer learning and making environment. The Sicamp model overtly champions the enactment of social innovation through material engagement, and is directly linked to the philosophy of "critical hacktivism" as it was co-founded by Dan McQuillan. The model was developed to extend the scope of social technology applications and to encourage social start-ups outside of institutional constraints. McQuillan has characterized the process as "organising the moment of self-organisation."

Farset Labs also hosts events that aim to critically reimagine social contexts. It hosted the first "policy hack" organized by the Northern Ireland Council for Voluntary Action (NICVA). NICVA is a membership and representative umbrella body for the voluntary and community sector in Northern Ireland. Through its policy hack series, this agency is borrowing "the techniques and ethos of the tech sector, (to) apply them to a number of social policy topics and come up with solutions to key social, economic and environmental problems."[6] The call to action from this third sector agency to "hack our way to better outcomes" provides further evidence that processes of inverting or subverting incumbent systems is gaining traction in Northern Ireland, and that Farset Labs is an actor in this social innovation movement.

Discussion: Challenges and Opportunities

The changeable conditions of hacker agency—desire, capacity, time, tools, and space—are thoroughly entangled in the politics of place, things, and systems. This creates multiple challenges and opportunities for the ideas and initiatives discussed in this study. Despite the positive rhetoric surrounding the Raspberry Pi, some have queried whether it will encourage people to code and make, or whether it will only be used by the usual suspects: those already familiar with computing. Informal education programs like the Raspberry Pi succeed in exposing this as a false choice, as *nerds* and *noobs* come together to learn, socialize, and have fun. But the existence—and success—of the Raspberry Jam in Northern Ireland also points to the importance of bridging initiatives that bring specialists in contact with enthusiasts. This position is supported by Quinlan's (2015) NESTA report that revealed only 12 per cent of UK parents felt informed enough to signpost digital making activities for their children.

In the case of Farset Labs, challenges to enacting hacker agency may also be seen as opportunities, and vice versa. For example, some may view the physical premises of Farset Labs as a barrier, as its location next to a Loyalist[7] residential area might be intimidating to a person outside of this culture. However, Farset Labs' proximity to Belfast city's major train station might be parsed as a positive factor for others. Some might consider the membership fees low-cost, while others may find them difficult to justify. Farset Labs is also supported by its landlord through flexible rent arrangements; but the more sustainable the organization becomes the more that is expected of them financially. And, this could impact on the organisation's capacity to deliver informal education programs.

At Farset Labs, certain cultures are developed, supported, and celebrated. As a result, these spaces are not for everyone. The ambiguity surrounding the operations, purpose, and social protocols of Farset Labs, has created some issues for the organizers of the Raspberry Jam. On occasion, parents have brought their children to the Raspberry Jam thinking it is government funded. This has created expectations of the program, around financial resourcing, the type of service it is, how it is run, who is allowed to attend, and the longevity of the program. Such dynamics reveal the limited capacities of this current informal education model—one that relies on a handful of altruistic personalities with limited financial resources—to contribute to the nurturing of making or hacking practices.

Due to their involvement in the Farset lab's Raspberry Jam, Andrew Mulholland and Andrew Bolster were approached to develop and deliver a Raspberry Pi training program specifically for teachers in the Greater Belfast Area in 2013. The local agency driving the program were responding to an increased need for continuing professional development activities focusing on computer programming in Northern Irish schools (Carson, 2013). The Raspberry Pi was chosen because it had been developed specifically to target school students. The program's evaluation highlighted that more training of this kind is needed, offering evidence that an opportunity exists for Farset Labs to establish a "train the trainer" initiative. Such a program could also be used to address gaps in the abilities of current and future Raspberry Jam volunteers. This could relieve pressure from the main trainers, while building the capacities of others who may be more appropriately positioned to train specific cultural groups.

Farset Labs exists alongside two other Northern Irish makerspaces that focus on digital technologies and making practices: FabLab Belfast and FabLab Nerve Centre in Derry. They form part of the MIT-supported FabLab network, and funding is linked to the running of "peace-building" programs.[8] Both FabLabs have full time staff, which enables them to host school and community groups on a regular basis. They teach practices that focus on digital fabrication, providing an entry point for people to consider the materialities of digital technologies. By encouraging material engagement through turning *bits in to atoms* (Gershenfeld, 2005), these organizations are nurturing literacies and competencies that are aligned with hacking practices. Farset Labs have already had involvement with these organizations, but further opportunities exist for knowledge exchange, particularly in the realm of critical making and design processes.

A broader view of the hacking imperative is Gregg and DiSalvo's description of hacking as "today's preferred economic identity" (2013). This brings to attention how values form around hacking processes, the types of things that are hacked, and the partnerships formed in order to hack and make things. It also raises questions about who is exposed to hacking practices, and in which cultures, or sub-cultures, hacking is condoned. *Messiness* is one such condition of hacking that might seem more acceptable for some and not others. This is a particularly gendered paradigm, in that women have been traditionally charged with the responsibility of keeping things clean. Scholars have highlighted other gendered dynamics, such as the suggestion that maker culture often elevates the status of traditionally male activities over stereotypically female ones (Powell, 2012). To counter this, Powell recommends a reframing of maker culture as a research community to account for women's histories and innovations of cultures past. Toupin (2014) adds to the debate proposing that the increasing visibility of feminist hackers will "open up the hacker ecology to further diversity and nuance" (2014).

This chapter has built a case to support further investigations in to the nexus between hacking and social innovation in Northern Ireland. As the purposes and practices of hacking become more visible in the third sector context, opportunities exist to use material engagement to help bring political questions in to the light (Von Busch & Palmas, 2006). Hacking paradigms could increasingly be used to navigate and negotiate overtly political contexts in Northern Ireland. New forms of material participation could offer levels of engagement not conducive to other public forums. This type of activity might even lead to "adversarial design" solutions (DiSalvo, 2012), which focus on the productive frictions of political conflict.

Opportunities also exist for makerspaces in Northern Ireland to harness momentum surrounding the Digital Social Innovation (DSI) policy moment. A more visible culture of critical hacktivism within the lab would further establish it as a site for DSI. Promoting the idea that the ethics of technologies should be disentangled in socially focused projects is one approach. This serves to highlight the strength of the makerspace: that those who are materially engaged with technologies are more likely to understand their politics. If the third sector starts to associate Farset Labs with socially engaged technology practices, they may be more inclined to seek input from its socially engaged practitioners.

In sum, the continuation of Raspberry Pi bridging initiatives will help diversify access to its technology. Organizational forms like the Farset Labs and its Raspberry Jam initiative require constant gardening if they are to continue on their critical paths. Maker cultures and the physical spaces that support them must remain open to new inputs without forgetting important historical markers. While theory, policy, and practices surrounding "social innovation" must be in constant dialogue to encourage more appropriate technologies.

Conclusion

The combination of desire, capacity, time, tools, and space helps reveal the affordances and the entanglements of socio-technical systems. This study proposed that Farset Labs makerspace and its Raspberry Jam program are enacting these conditions. Farset Labs facilitates hacking activities in an environment that questions assumptions about the political entanglements of people, technology, and things. It is a community of practice that gathers around shared infrastructure to engage in critical material and discursive practices, and is a dedicated and sanctioned space for hacking. The Raspberry Jam offers an informal, quarantined, discreet allocation of time that guides making, modeling, and hacking activities using Raspberry Pi computers. This mix of time, tools, and space encourages participants' desire to hack, in turn building their capacities to hack.

The interplay between Raspberry Pi technologies, makerspaces like Farset Labs, informal education programs such as the Raspberry Jam, practices of "social innovation," and the aspirations of Northern Ireland's peace processes requires further research. Ongoing scholarly engagement with these dynamics will support processes of disentanglement to provide scaffolding for better programs, practices, and policies that emphasize material processes. Critical questions emerging form this study provides sightlines for future scoping. The focus on external conditions for nurturing hacker agency highlights the changeable conditions surrounding physical sites, and how this can affect people's capacities to identify and reveal affordances. So, how might we prepare ourselves for these changeable conditions? And, how might we fortify participants against fears associated with working on unstable ground?

Better methods for designing *play* in to social innovation processes are also welcome. Resources that help non-designers playfully engage in material processes could help critical hacktivism philosophies become integrated in to everyday social innovation. Beyond this, philosophies of "critical hacktivism," "hacker agency", and "social innovation" require deeper probing and more pragmatic interventions, lest they become branded as marginally useful theoretical perspectives.

Notes

1. '3D printers used for prosthetic limbs in Syria conflict,' BBC News, 10 March 2015 http://www.bbc.co.uk/news/uk-northern-ireland-foyle-west-31812040
2. 'Review: Raspberry Pi. Can a $35 computer persuade kids to put down their smartphones and try their hands at programming?' MIT Technology Review, 4 September, 2012 http://www.technologyreview.com/news/429048/review-raspberry-pi/
3. 'Raspberry Pi 2 vs BBC Micro Bit: How do the DIY computers compare?' Trusted Reviews, 25 July 2015 http://www.trustedreviews.com/opinions/raspberry-pi-2-vs-bbc-micro-bit
4. The notion of the hacker as creator of affordances was developed in conversation with Andrew Schrock.
5. Sicamp was hosted by Northern Ireland's Building Change Trust, an organization established in 2008 to build capacities within the voluntary and community sector in Northern Ireland.
6. 'Hacking our way to better outcomes,' 29 January, 2015, NICVA http://www.nicva.org/article/hacking-our-way-better-outcomes
7. Ulster Loyalism is a political ideology that emerged from sectarian conflicts in Northern Ireland. It is aligned with Protestantism, and supports the preservation of Northern Ireland as a part of the United Kingdom as opposed to part of Ireland. It emphasizes loyalty to the British monarchy and is often associated with paramilitary activities.
8. FabLab NI is partially funded by the EU Programme for Peace and Reconciliation in Northern Ireland and the Border Region of Ireland http://www.seupb.eu/programmes2007-2013/peaceiiiprogramme/overview.aspx

References

Alleyne, B. (2011). We are all hackers now: Critical sociological reflections on the hacking phenomenon. *Under Review*, 1–32.

Apperly, T. (2014). Glitch sorting: Minecraft, curation and the post-digital. In D. M. Berry & M. Dieter (Eds.), *Post-digital aesthetics: Art, computation, and design* (pp. 232–244). London: Palgrave Macmillan.

Barad, K. (2007). *Meeting the universe halfway: Quantum physics and the entanglement of matter and meaning*. Durham, NC: Duke University Press.

Bolster, A., & Mulholland, A. (2013). Getting kids excited about computer science in Northern Ireland. Retrieved April 5, 2015 from https://www.scribd.com/doc/250366471/Getting-kids-excited-about-Computer-Science-in-Northern-Ireland

Bucher, T. (2011). Network as material: An interview with Julian Oliver. Retrieved June 24, 2015 from http://www.furtherfield.org/features/interviews/network-material-interview-julian-oliver

Buechley, L., & Hill, B. M. (2010). LilyPad in the wild: How hardware's long tail is supporting new engineering and design communities. *Proceedings of the ACM Conference on Designing Interactive Systems* (pp. 199–207).

Campbell, E. (2014). Minecraft sales surpass 15 million copies on PC. Retrieved July 9, 2015 from *IGN Australia* http://au.ign.com/articles/2014/04/29/minecraft-sales-surpass-15-million-copies-on-pc

Carson, M. (2013). DCAL Raspberry Pi pilot project 2013 evaluation report. Retrieved from http://www.scribd.com/doc/250367086/DCAL-Evaluation-Report-Raspberry-Pi-Project

DiSalvo, C. (2012). *Adversarial design*. Cambridge, MA: MIT Press.

Dunne, A., & Raby, F. (2001). *Design noir: The secret life of electronic objects*. Basel: Birkhauser.

Freire, P. (1972). Pedagogy of the oppressed. Harmondsworth, Middlesex: Penguin Books.

Fuller, M. (Ed.). (2006). *Software studies*. Cambridge, MA: MIT Press.

Gaver, W. W. (1991). Technology affordances. In *SIGCHI Conference*, New Orleans.

Gershenfeld, N. (2005). *Fab: The coming revolution on your desktop: From personal computers to personal fabrication*. New York, NY: Basic Books.

Gibson, J.J. (1977). The theory of affordances. In R. Shaw & J. Bransford (Eds.), *Perceiving, acting and knowing* (pp. 67–82). New York, NY: Wiley.

Gillespie, T., Boczkowski, J. P., & Foot, K. A. (Eds.). (2014). *Media technologies: Essays on communication, materiality, and society*. Cambridge, MA: MIT Press.

Gregg, M., & DiSalvo, C. (2013). The trouble with white hats. Retrieved February 19, 2015, from *The New Inquiry* http://thenewinquiry.com/essays/the-trouble-with-white-hats/

Hostick-Boakye, S. (2014). *Turning up the dial: Digital social innovation in Northern Ireland*. The Young Foundation, report commissioned by Building Change Trust. Retrieved from http://youngfoundation.org/wp-content/uploads/2014/04/Turning-up-the-Dial-report-FINAL.pdf

Hunsinger, J. (2010). The social workshop as PLE: Lessons from hacklabs. In *Proceedings of The PLE Conference 2011, 10th – 12th July 2011*, Southampton, UK.

Irani, L. (2015 June). Hackathons and the making of entrepreneurial citizenship. *Science, Technology, and Human Values*, 1–26. DOI:10.1177/0162243915578486.

Ito, M., Gutiérrez, K., Livingstone, S., Penuel, B., Rhodes, J., Salen, K., … Watkins, S. C. (2013). *Connected learning: An agenda for research and design*. Irvine: Digital Media and Learning Research Hub.

Jordan, T., & Taylor, P. A. (2004). *Hacktivism and cyberwars: Rebels with a cause?* London: Routledge.

Levy, S. (1984). *Hackers: Heroes of the computer revolution*. Garden City, NY: Doubleday.

Lin, Y. (2004). *Hacking practices and software development: A social worlds analysis of ICT innovation and the role of free/libre open source software* (PhD Thesis). University of York.

Marres, N. (2012). *Material participation: Technology, the environment and everyday publics*. London: Palgrave Macmillan.

Maxigas, & Troxler, P. (2014). Editorial. *Journal of Peer Production, 5*. Retrieved from http://peerproduction.net/issues/issue-5-shared-machine-shops/

McQuillan, D. (2012). Critical hacktivism. Retrieved November 25, 2014 from http://www.internetartizans.co.uk/critical_hacktivism

NESTA. (2015). Top findings from the open dataset of UK makerspaces. Retrieved May 5, 2015 from *NESTA* http://www.nesta.org.uk/blog/top-findings-open-dataset-uk-makerspaces

NESTA et al. (2015). Growing a digital social innovation ecosystem for Europe. A study prepared for the European Commission DG Communications Networks, Content & Technology. Retrieved from http://www.nesta.org.uk/sites/default/files/dsireport.pdf

Oliver, J., & Vasiliev, D. (2011). *Newstweek*. Berlin. Retrieved from http://newstweek.com/

Perng, S., & Kitchin, R. (2015). Solutions, strategies and frictions in civic hacking. The Programmable City working paper 10. Paper presented at MediaCity 5. Plymouth: Plymouth University.

Powell, A. (2012). Cultures of the "Maker" Movement. Retrieved February 19, 2015 from http://www.alisonpowell.ca/?p=522

Quinlan, O. (2015). *Young digital makers: Surveying attitudes and opportunities for digital creativity across the UK*. London: NESTA.

Ratto, M. (2011). Critical making: Conceptual and material studies in technology and social life. *The Information Society: An International Journal, 27*(4), 252–260.

Ross, A. (1991). *Strange weather: Culture, science and technology in the age of limits*. London: Verso.

Santo, R. (2011). Hacker literacies: Synthesizing critical and participatory media literacy frameworks. *International Journal of Learning and Media, 3*(3), 1–5.

Sawyer, M. (2012). Everyone wants a slice of Raspberry Pi. Retrieved May 12, 2015, from *The Guardian newspaper*. Retrieved from http://www.theguardian.com/technology/2012/nov/04/raspberry-pi-programming-jam-cern?CMP=twt_gu

Schrock, A. (2014). "Education in disguise": Culture of a hacker and maker space. *InterActions, 10*(1).

Söderberg, J., & Delfanti, A. (2015). Repurposing the hacker. Three temporalities of recuperation. Retrieved from SSRN: http://ssrn.com/abstract=2622106

Thomas, D., & Seely Brown, J. (2011). A new culture of learning: Cultivating the imagination for a world of constant change. CreateSpace. Retrieved from http://www.newcultureoflearning.com/

Toupin, S. (2014). Feminist hackerspaces: The synthesis of feminist and hacker cultures. *Journal of Peer Production*, 5. Retrieved from http://peerproduction.net/issues/issue-5-shared-machine-shops/peer-reviewed-articles/feminist-hackerspaces-the-synthesis-of-feminist-and-hacker-cultures/

van Dijck, J. (2013). *The culture of connectivity: A critical history of social media*. Oxford: Oxford University Press.

Von Busch, O., & Palmas, K. (2008). Abstract Hactivism: The Making of a Hacker Culture (published online, 2006).

Wark, M. (2004). The Hacker Manifesto. Cambridge: Harvard University Press
Winner, L. (1980). Do artifacts have politics? *Daedelus, 109*(1), 121–136.
Winner, L. (1986). *The whale and the reactor: A search for limits in an age of high technology*. Chicago, IL: The University of Chicago Press.

14. *Hacking as a Way of Life: "Makers" at the Margins of Global Digital Culture*

NICHOLAS BALAISIS
University of Waterloo

"I am a knowledge worker ... I live on my wits" (Charles Leadbeater)

"Hay que inventar!" ("One must invent!")
Colloquial expression in Cuba.

The emergence of hacking and making into popular vernacular must be understood within the broader economic context of late modernity and the shift from a Fordist industrial economy to a post-Fordist, digital economy. As a number of scholars have outlined, post-Fordism is constituted by (among other things), more flexible or precarious employment, shorter and more varied career cycles, less division of labour hierarchies and union support, and more personal stake in financial security (Bell, 1973; Kasvio, 2001; Masuda, 1990). In short, there are fewer "jobs for life" or "cradle to grave" employment with workers changing careers more frequently and often cobbling together multiple contracts to make do (Reich, 1992). The conditions for success within this new economic climate are often linked to characteristic such as creativity, flexibility and an entrepreneurial spirit: a kind of "career hacking." This is the overarching theme of Charles Leadbeater's book on work in the information society, *Living on Thin Air: the New Economy*. Leadbeater argues that the worker in post-Fordism will succeed if they are able to be creative, innovative and entrepreneurial. Creativity and "self-making" (and re-making) are thus central to the new worker ethos: "I live by my wits" (1999, p. 1) While this new model is riskier and less secure than the Fordist economy, it offers the potential payoff of a more fulfilling career geared by personal interest: "I get paid for my own interests." Creativity—the capacity to constantly innovate and re-invent yourself—is the central pillar of the new economy worker.

New media and technology are closely wedded to this discourse of post-Fordist work. For many proponents of post-Fordism, successful workers and entrepreneurs will be creative early adopters and innovators in the fields of new media production. For Leadbeater, this kind of innovation is at the level of knowledge capitalism: to best be able to mobilize and market the knowledge capital and knowledge data that you produce. This link between creativity and new media is stressed in Richard Florida's influential book on urban planning and the new economy, *Rise of the Creative Class* (2002). For Florida, cities must attract creative workers and social innovators. These workers include conventional "creatives" such as artists but for the most part refers to workers in the new information economy, workers who produce "wondrous new hardware and software" (Szeman, 2010, p. 31).

This paper asks what it means to think about creativity and innovation at the "back end" of technology adoption cycle: the late adopters of media and technology. As new media workers are celebrated in the post-Fordist economy, are those at the "back end" of the technology spectrum—the so-called *laggards*—rightly excluded from the discourse of innovation and creativity? In other words, does the culture's current focus on innovation and creativity at the vanguard of new media and technology ignore a wider scope of innovation practices at the digital margins, those tinkering and toiling with obsolescent or residual tools and devices? My contention here is that the current discourses around creativity and innovation are limited by their narrow and often fetishistic focus on the "new." These discourses are also neoliberal in their focus on innovation or creativity as exclusively entrepreneurial activities. This paper seeks to recover a broader sense of innovation and creativity—to disentangle them from links with start-up or entrepreneurial culture—through a discussion of "maker" or "hacker" practices from Cuba, a country that, at least technologically, is most certainly a *laggard*. Specifically, I look at the writing of Cuban industrial designer Ernesto Oroza, who argues that everyday practices—material and technological—offer examples of hacking or creative ingenuity as everyday necessity. By looking at Cuban "maker" or "hacker" practices, I aim to disassociate creativity from its increasing link with neoliberal economics, particularly in an urban and civic discourse increasingly framed by the ideas of Richard Florida (Mellonder, Florida, Asheim, & Gertler, 2014). Creativity is something that has been increasingly sought and encouraged by local governments in order to bolster civic development (Balaisis, 2015), and is one of the key components of DIY maker and hacker discourse (Gauntlett, 2011). Secondly, I wish to recover an expanded sense of the term "hacking." I hope to highlight in particular an aspect of hacking that to disassembling and managing or making do, as opposed to its dominant association with subversive or illegal computer activities. From this angle, hacking can be seen as a practice of subsistence and survival, or as creative solutions and innovations to material and technological scarcity: hacking as a way of life.

My discussion of Cuban maker or hacker practices, terms that I use inter-changeably in the essay, are informed in part by Michel de Certeau's notion of tactics. Tactics describe the ways in which everyday strategies of people working within defined and often constrained "nets" of discipline, can re-order and re-structure these broader schemes of disciplines through ordinary practices. He gives the example of *la perruque* (the wig) to describe the mechanisms of this process. *La perruque* describes the way in which a worker performs his own work under the guise of work for his boss; nothing is stolen except for the boss's time. "It differs from absenteeism in that the worker is officially on the job. *La perruque* may be as simple a matter as a secretary's writing a love letter on 'company time' or as complex as a cabinetmaker's 'borrowing' a lathe to make a piece of furniture for his living room" (pp. xix–xv). De Certeau's description here is useful for an account of everyday design or hacking practices in Cuba in the late socialist period because it offers the prospects of agency within delimited circumstances such as those of politics or economics. The Cuban practices that Oroza documents reflect a similar clandestine "theft" or intervention into the political and economic macrocosm, an intervention that I would extend beyond Cuba, to read as a critique of material and techno-culture more broadly. What is useful in de Certeau's notion of creativity, and what distinguishes it from David Gauntlett's broader definition of creativity as a "process" and a "feeling" (p. 17), is the idea of creativity as a form of agency within conditions of restraint. This is an important framework for Cuba given, on the one hand, the political climate of restraint, but more specifically that of technological restraint—that of limited resources, and the absence of new tools and component. This concept provides an alternative framework for thinking about cultural and media practices, and specifically, creativity, outside of a neoliberal and entrepreneurial context.

"Tools for Makers:" New Media and the Rhetoric of Creativity

In contemporary discourses, new media are often associated with creative work and creative people. This is most obvious in the advertising for Apple and their products aimed at creative makers. Much of the advertising for

Apple emphasizes the fact that their products can be used for creative practices either by individuals or by workers. As Lisa Nakamura argues, the rhetoric of creativity in Apple marketing works ideologically, constructing consumers as creative "makers" and thus encourages the frequent updating of Apple products. This rhetoric of creativity, she adds, fetishizes a certain kind of work over another: it values the work that people do with their iPhones or Apple computers and ignores the often dangerous and precarious work that goes into the production of the iPhone itself.

> For the iPhone has, from its inception, been set apart from other cellphones; it is both marketed and understood as a tool for "making" as well as a communicative device [...] To call oneself a "maker" is to claim an exalted cultural status, and the term "creative" has achieved the status of a noun to describe workers such as designers and artists who are "makers." The iPhone is part of this fetish of self-sufficiency and creativity, and thus exempt from critiques of conspicuous consumption that have arisen as a logical response to a worldwide recession. (2011, p. 1)

Nakamura rightly asks about the other labour that underwrites the creative "maker" culture solicited by Apple: those who who make "the tools for the makers" (Nakamura, p. 2). The branding strategy of Apple links "makers" with "creativity" and posits creativity at the "front end" of the technology spectrum. Here creativity is linked with a certain kind of production in a very narrow area of expertise—new media technology. Creatives are those who do things with the tools of digital culture, and in particular, with the most recent and updated versions of those tools, whereas those who actually make the tools themselves are excluded from the very discourse of creativity and making.

This discursive link between creativity and certain kinds of labour practices is also evident in the discourse of the creative city or the creative class, popularized by business professor Richard Florida. For Florida, the creative class represents a group of workers committed to create "meaningful new forms" (6). These workers represent the vanguard of innovation and as such, exemplify the term creativity. Though they may range in their fields of expertise, from scientists to artists, engineers to academics, their "economic function is to create new ideas, new technology and/or creative content" (8). The creative class is thus marked by its link to the forefront of innovation, whether that is in technology or in management or labour. For instance, independence and autonomy are seen as key markers of this new economic and social force, who are not committed to the old ways of but often invent new job titles, roles, wardrobes and work hours for themselves. The key point for Florida is that "creatives" are good for economic growth and innovation. Cities and companies should thus seek out these workers and design cities that are attractive to them in order to be successful. New York and San Francisco have done well to attract the creative class, whereas Buffalo and New Orleans have not. These ideas have influenced and informed urban policy (for example in New Orleans in the wake of Katrina) in order to shape city planning in order to attract these kinds of workers.

This discussion of the creative class echoes other discussions of the new economy, or information society, where success in the post-industrial economy is linked with creativity and flexibility. Charles Leadbeater describes the existential challenges faced by the post-industrial economy where worker roles and jobs have been re-defined and former career paths liquidated. While he acknowledges that the new economy produces challenges for identity and workers used to formal timetables and consistent roles, the potential rewards of the new economy are vast, and promises new kinds of self-fulfillment and opportunity to adept workers. For Leadbeater, the key elements for workers in the new economy are creativity and flexibility: being able to adapt and change at a moment's notice. His book traces an optimistic vision of the future in the new economy, one where early adopters of these changes will have more rewarding career paths than before. Creativity—the capacity to generate something outside an institutional Fordist framework—is central to Leadbeater's new economy.

> We are living in an age of unprecedented productivity and creativity, in which science and innovation are bestowing upon us families of new products with exciting possibilities: genetic treatments for disease, powerful computers the size of a television remote control, robots smaller than the size of a coin [...] Our children will not have to toil in dark factories, descend into pits, or suffocate in mills, to hew raw materials and turn them into manufactured products. *They will make their livings through their creativity, ingenuity, and imagination.* (p. ix)

Similar to Florida, creativity becomes a key figure around which the new economy subject is articulated. It is a closed loop, where creativity begets more creativity. There is a link here between creativity and self-making, where creativity is connected to the production of one's own life as well as one's own schedules and way in life. It is an updated version of the self-made man rhetoric of Horatio Alger (Nackenoff, 1997). As Leadbeater describes proudly, he is very much a "maker" in his life and career, generating income through the "thin air" of the new economy. He does not produce in the old-fashioned and barbaric manner he links with the industrial age of brick and mortar companies. Rather, he produces information and services for the new economy, and in so doing, he produces himself. In describing his work, there is a link between his ability to conjure work from "thin air" and a new (and privileged) knowledge and information production: "When people ask me, 'What do you do?' I find it hard to come up with a clear, concise answer. [...] I am one of Charles Handy's portfolio workers, armed with a laptop, a modem, and some contacts. Peter Drucker, the management expert, anointed people like me 'knowledge workers.' *Put it another way: I live by my wits*" (1). Again, the stress is on the link between creativity—wits—and self-production. I do not produce things, but I do something more meaningful: I produce *myself*. The "maker" in the knowledge economy is thus privileged not because they make things but because they make themselves. What seems to privilege the post-industrial "maker" is the fact that they do not make things, but they conjure material from thin air: money, ideas, and most importantly, themselves. This emergent "maker," or "selfpretreneur," is mapped at the "front end" of the technology spectrum. They are innovators, and early adopters, engaged in "new" practices and thus, as Nakamura notes, in need of the newest tools and technologies in order to do their work.

"The Maker Manifesto," or, the Neoliberal Discourse of "Making"

The neoliberal link between making and entrepreneurialism or self-making can be seen explicitly in some of the more dominant cultural expressions of maker culture, currently undergoing resurgence in North America. In the intentionally analog compilation *Critical Making* (2014) a number of researchers in the field offer critiques of the dominant expressions of maker culture as fetishistic and apolitical. Many of the scholars in the compilation are skeptical of the popularization of "making" in both popular and university expressions. For Alexander Galloway, the development of maker spaces in universities is a disguise for the "neoliberal-fication" of the university, where seminars are transformed into laboratories for entrepreneurship under the guise of being "maker spaces" (2015). In another critique of dominant attention to "making" Matt Ratto argues that magazines like *Make* focus too much on the "technical knowledge" of making and are motivated by a kind of "gee whiz" fascination with tools and new technology (2014, p. 7). These kinds of sites invite amateurs to become as close as possible to engineers at the expense of a more critical making practice that is attentive to "the conditions under which things are made" (Ratto, 2014, p. 7). Finally, McKenzie Wark equates the maker movement with a broader cultural fascination and attention to aspects of working class culture, of which Brooklyn is the epicenter of what he describes as "gratuitous making." For him, a certain aspect of maker culture "makes a fetish of the artisanal quality of labor as another way of avoiding the question of labor" (2014,

p. 5). Critics of maker culture remain committed to certain aspects of maker culture, while skeptical about its dominant expressions. Matt Ratto maintains that making, in the context of a critical paradigm, can be a productive way of fostering a more material engagement with the world around you. Critical making should "help people see our environment as a made environment, made in particular interests and serving particular interests" (2014, p. 8).

This neoliberal tendency is expressed in the recent book by maker advocate Mark Hatch in his book *The Maker Manifesto*. As CEO of Tech Shop, the largest commercial maker shop in North America, Hatch has a vested financial interest in promoting "making" as both a creative leisure activity as well as pillar a new creative entrepreneurialism. Techshop operates as a larger-scale version of some smaller maker spaces in North America. It charges a monthly fee to members who then gain access to the use of modern tools such as 3-D printers and laser cutters in an environment conducive to the DIY spirit: self-motivation, creativity and community-formation. Hatch makes a number of bold claims about the role of maker spaces in the post-industrial, post-Fordist economy in the United States. These claims are on par with those of Florida and Leadbeater, who see an opportunity for the link between entrepreneurial success and creative self-actualization. Like Florida, creativity is the anchor of Hatch's argument about maker culture and the ethos of making. Hatch's manifesto on making stresses that making is "fundamental" to what it means to be human, and this making involves high degrees of creativity and play which allows humans to become better, more fulfilled and whole versions of themselves. "Embrace the change that will naturally occur as you go through your maker journey. Since making is fundamental to what it means to be human, you will become a more complete version of you as you make" (Hatch, 2014, p. 2). The creativity of working in a workshop with 3-D printers and laser cutters is not just a means in itself, however, but also a crucial part of what Hatch envisions as a post-Fordist culture in North America. While maker spaces may indeed provide people with productive leisure and sociality like other kinds of hobbies, sports and leisure activities, what distinguishes maker spaces are their ability to foster economic growth for individuals and communities at large. Hatch sees maker spaces as potential anchor tenants that would attract so-called creatives to certain parts of the city, spurring parallel and spin off economic activity:

> What I've described in this chapter is the emergence of a new kind of community that grows up from a community that has a density of creatives in it. But to grow this community, it needs the infrastructure, design, purpose, support, and even building codes to flourish. Those cities that get this right will develop a vibrant creative cluster, and that cluster will produce culture, music, art, start-ups—and jobs. (p. 67)

There is a parallel here to the new vision for cities and economies promoted by Richard Florida in his notion of the creative class, which amounts similarly to making cities amenable to artists and creative people and the local economy will move outward from there. One of the critiques of Florida or Hatch is that they adhere too strongly to a neoliberal paradigm, where society is seen to function best when individuals are best served when their entrepreneurial freedoms are liberated within a system of market rules and logic and free trade and free enterprise (Harvey, 2007). In *The Maker Manifesto*, individuals are best served both creatively and economically when they have access to key "tools of innovation" such as 3-D printers and prototyping. In this model, the individual is freed up from institutions and able to rent high-end technology in order to prototype objects and products that can begin the life force of a new company.

The critiques that can be levied at this discourse are numerous. From the perspective of critical making discourse, it sees making exclusively as an entrepreneurial activity, as a way of developing a product to be sold for market etc. Making in this instance is not a means by which to engage in a proactive practice to understand how things work or for engaging with the world in a material way (though Hatch would see this as a fringe benefit). Second, it idealizes a type of maker space worker, one who has free time, energy and resources in order to pursue—in their leisure time—creative and entrepreneurial activity. Finally, it doesn't take into account the various world of labor that would ultimately underscore the kinds of design initiatives proposed by a lay-designer. This model privileges the leisured labor of the designer at Techshop, but does not acknowledge the offshore labor that underwrites this design work. In this way, it is congruent with the discourse of labor and creativity marketed by the likes of Apple, which privileged the labor of design in Cupertino California over the labor of production in China (Nakamura, 2011). While there is an intriguing attention to materiality, creativity and new urbanism in *The Maker Manifesto*, its attention remains solely on the labor and creativity in the so-called developed world and does not take into account the labor that underwrites this design work in the global economy.

Creativity at the Back End the Technology Spectrum

Cuban industrial designer Ernesto Oroza offers an important intervention into this discourse on maker and hacker culture in his charting of Cuban everyday design "hacks" in the post-Soviet period. Oroza is attuned to the emergent incorporation of hacking and making into the global vernacular, and wishes to problematize and expand this discourse in his writing on Cuban cultural and material practices from the 1990s, practices rooted in the economic and political crises in Cuba after the collapse of the Soviet Union. Oroza's work is important in that it helps to expand the discourse around hacking and making to include creative practices at what I am calling the "back-end" of the technology spectrum, that is, with media, objects and technology that constitute old or residual forms of media and material culture. My use of residual media comes from Charles Acland and his critique of what he calls the "myth of new media." He argues that the reigning myth of media is "that technological change necessarily involves the 'new' and consists solely of rupture from the past" (Acland, 2007, p. xix). This view necessarily ignores continuities with the past, and the persistence of older forms, media or otherwise. I argue that Cuban material practices described by Oroza offer examples of novelty and "new" objects precisely through the re-imagining and dismantling of old media and matter. By drawing these practices hacks or maker projects, I hope to widen the discursive scope around popular, and neoliberal, strains of maker discourse, and draw contiguities with critical maker discourse. I also wish to de-couple creativity from associations with new media and technological innovation and entrepreneurship.

Ernesto Oroza has worked to document and theorize Cuban media and design practices within Cuba's severely limited technological and economic climate. This climate is marked particularly by one of the world's lowest levels of new media penetration and use. While Cuba has a higher economic standard of living than many of the poorer nations in the world, its levels of internet use are some of world's lowest. For example, only roughly 26% of Cubans accessed the internet in 2012, up only just under 4% from a decade earlier. In addition, there were only 100 internet cafes on the island in 2013 serving less than 1% of the population. By comparison, Angola, which has a similar population, 44% of the population are served by internet cafes. Cuba is thus what some accounts of digital culture has claimed as an "internet black hole," (Al Jazeera) a place with limited or severely restricted internet access and connectivity. In this way, Cuba is situated alongside places like North Korea or Myanmar, on the far end of the "digital divide." The reasons for this are largely attributed to the combined effects of a longstanding economic blockade with the US, which has prevented the export of telecommunications companies and technologies, and the political resistance on the part of the Cuban government, which has regarded access to modern technologies of communication such as the web as a threat to state hegemony (Vigíl, 2008).

Beyond access to advanced digital technologies, Cubans have suffered from general material shortages since the collapse of the Soviet Union and the loss of their primary economic partner since the 1960s. The 1990s ushered in a period known officially as the Special Period in times of Peace, where rations were expansive and Cubans were encouraged to make-do and find new uses for a limited stock of goods and materials at hand. Oroza describes, for instance, a state manual that speaks of how to simulate a steak (very rare in Cuba at the time) with the rinds of a grapefruit (over abundant in Cuba). Cuban scholar Ariana Hernandez-Reguant describes the infamous strategies for making do in this period, that in Cuba, now constitute part of the national mythology: "Raising pigs in bathtubs, making omelets without eggs and pizzas with melted condoms, getting married for the state-allocated free case of beer, and other epic takes of survival, seldom void of black humor, form the lore of the time" (Hernandez-Reguant, 2009, pp. 1–2).

In his curatorial and written work, Ernesto Oroza has worked to document and archive the strategies of survival in this period, as well as to theorize these efforts as a form of media or technology theory. To this end, Oroza has chronicled two principle kinds of innovations made by Cubans: housing and everyday objects. This corresponds to distinct material conditions in Cuba. One of them being the fact that house purchasing (until very recently) was illegal in Cuba and thus Cubans were often forced to expand and make do with their own house to accommodate new people rather than purchase or seek larger or more appropriate accommodation. The other relates to the fact that there are very few imports in Cuba (as a result largely of the US embargo) and the legacy of the Special Period, which stopped subsidies from the Soviet bloc from reaching Cuba. Oroza has built a theory of technology and architecture around the kinds of material practices that he has traced in Cuba: what has termed "architecture of necessity" and "technological disobedience."

Oroza maps these processes in a host of other innovations and "hacks" that exceed bare necessity and extend to include objects of aesthetic beauty and lifestyle. His website is filled with numerous examples of everyday "hacks" ranging from large scale housing and auto repairs and re-inventions to smaller scale electrical home appliances. For example, one project shows a small electrical toy car affixed to a small home radio. This works as a vernacular antenna that is used to "steal" the radio signal for state programming. Oroza notes that these vernacular radio antennas are found all over Cuba and are commonly hidden in small plastic boxes or more commonly, electrical toys like cars. In other small household examples, a razor blade is fashioned to a pencil to create a shaving tool, and discarded bottle caps are fastened as wheels to a glue bottle to make a children's toy car. Other vernacular projects include an improvised stove and outdoor grill. In the first case, an iron is used on its head and modified to work as a stove for brewing everyday items like coffee, while in the latter, a metal chair is used as a base for grilling meat, with a fire added to the bottom of the chair.

In some cases, there are relatively elaborate technical revisions and modifications needed to fashion the new object, and in other cases, the user has simply envisioned a new use for the object using its existing form (the chair grill). In some cases, the objects respond to very pressing household needs, like cooking and shaving, while in other cases, the objects represent less immediate needs and fill the function of a object of pleasure or luxury (such as a toy, or better radio reception). These practices, for Oroza, express what he calls "technological disobedience." This refers to a relationship to an object or technology not as an engineer, designer, or even a user, but as an *artisan*. Cuban vernacular use of objects does not respect the original purpose or intent of the design but invents uses that reflect the needs and imagination of the user. This attitude towards objects and technology is expressive of the particular material and ideological conditions in socialist Cuba, and specifically, the period after the collapse of the Soviet Union and the period of great economic struggle and material scarcity. Within this context, Cubans were both encouraged by government and their own survival to invent creative solutions to material scarcity.

> While reinventing their lives, an unconscious mentality emerged. As a surgeon becomes desensitized to wounds, Cubans became desensitized to designed objects. They stopped seeing the original purpose of the object; instead it became a sample of parts. This is the first Cuban expression of disobedience in their relationship with objects—a growing disrespect for an object's identity and for the truth and authority it embodies. (Oroza, 2012)

In Oroza's reading of Cuban vernacular "making" objects, even recognizable and identifiable brands such as Sony or Swatch do not contain authority or aura, and thus may be repatriated to serve different purposes than they initially were meant to. The object or brand does not exert the force of authority that dictates how it may be used, repaired or replaced. If it can be repaired creatively, it will. If a Sony part can be used to service a stove or a washing machine, it will. Another example of an object's authority relates directly to the Cuban experience, with an anecdote he describes of Fidel Castro speaking on Cuban television and selling Cubans a Chinese pot to boil water with. An example he discusses as exemplary is the "fan-phone."

> An improvised repairman remembered, when his fan's base broke, that he had kept somewhere a broken phone from Communist Germany. He recalled it because the Orbit fan base somewhat resembled the prismatic pyramidal shape of the phone; the inspired creator was interested not in associations or meanings but in the formal analogy based on size and structure. The repaired, rebuilt, and repurposed fan was, at the same time, an outline of the cunning abilities of the individual, a diagram of the accumulation in his house, and an image of his disobedience. (Oroza, 2012)

In the example of the fan, the "hacker" looked at the commodity or object in a different way than we may look at objects in a consumer commodity framework. Rather than see the immediate use-value of the object—seeing the object through the lens of what it is sold and marketed as doing—the maker in this case sees it for its "formal" properties. In other words, Oroza's argument is that in the economic landscape of post-Soviet Cuba, Cubans were compelled to see objects differently, through the lens of use and necessity rather than as commodities with brand auras. The repairman recognized similarity in shape and size, rather than through the object's intended use and function.

In recent years, Oroza has worked to export these practices through design workshops that he has given in Europe to students immersed and excited by the hacker and maker ethos. Students in his workshops produced a series of videos posted online documenting their Cuba-inspired hacks. The videos reveal a hybrid formation where they evidence both the enthusiasm of playing with new media technologies such as 3-D printers, and the exoticism of Cuba as a framework for these hacks. For example, in one video produced at a workshop in Milan, students created a karaoke machine out of a showerhead and an old tape recorder. The project in one way evokes some of the projects exhibited by Oroza in Cuba, such as hybrid media examples of an electric toy car turned into a radio. The video also evidences the technological disobedience articulation by Oroza, as they enthusiastically tear apart a tape machine and weld an mp3 player into it. In this way, their object represents a transplanted incarnation of Cuban practices from the special period, where users are invited to see machines and technologies not as auratic, holistic entities to be used and replaced, but as a "sample of parts" (Oroza, 2012).

Oroza's work highlights "making" and "hacking" as an economic and social necessity rather than as a hobby or entrepreneurial activity, and thus offers a comparative lens onto emergent maker discourses in North America. In this way, the practices and methods described by Oroza are contiguous in part with the emergent discourse in *critical* making, a discourse that has recently taken root in a number of university research labs motivated by "making" as a complement to humanistic inquiry (Hertz, 2014). Though in the Cuban cases, these practices emerge, according to Oroza, from a confluence of political and economic forces specific to the time and place of 1990s Cuba.

They are not critical by design but by necessity. By bringing these examples into a discussion of emergent maker culture in North America, however, they offer an important contrast to the increasingly neoliberal tone of the discourse offered by entrepreneurs such as Mark Hatch. They are not objects with patent or entrepreneurial potential but they do reflect the creative innovations of domestic survival. In this way they trouble the link between creativity and neoliberal articulations of selfhood, popularized most explicitly by Richard Florida and his idea of the "creative class." They reflect an ethos rooted not in liberal fantasies about self-fulfillment and entrepreneurial success but in the material necessities of everyday life.

These Cuban practices also alert us to the broader etymological definition of hacking. As I have mentioned, hacking has three predominant meanings in English: as a verb, an activity, and as a practice. The first is the chop or cut: to hack off the limb of a tree, for example. This draws on the old English noun origins of the word, referring to chopping or cutting into pieces: *haccian*. The second, and likely the most popular meaning in terms of media culture, is of gaining unauthorized access to something, i.e. a database or computer hardware. This unauthorized access cuts two ways in popular discourse, referring both to counter-cultural or counter-hegemonic practices (such as wikileaks, Snowden etc.) as well as to practices of state power and state hegemony (i.e. the NSA "hacking" into everyone's email account without their knowledge). The final definition refers informally to managing or coping, usually in the negative: "you can't hack it in this business," for example. Hacking here refers to hanging on, managing, or coping in what is likely a difficult or stressful situation. All three of these meanings of the terms hacking are evident in the Cuban cases described by Oroza, but specifically, the first and third: to cut into pieces and to manage or cope. This again constitutes a kind of critique of discussions of DIY "life hacking" that is often seen as a means of progressive, counter-consumer home activism (Ratto & Boler, 2014). Or at the very least, they expose the inherently bourgeois notions of home and leisure evident in domestic making or hacking practices promoted on websites such as Pinterest.

Finally, the cultural practices mapped by Oroza can be seen as a critique of modernization and globalization theory more generally (Canclini, 2014; Gaonkar, 2001), particularly as it concerns media and new media theory. Much of the rhetoric around new media and technological adoption reinforce the "catch-up mode" of modernization—that there is a technological curve which governs the flow and evolution of modernity. For example, the logic of the digital divide presumes that knowledge-capital in the information age (i.e. to be successful) is reliant upon access to up-to-date digital tools (computers etc.) and reliable internet connectivity. This binary reinforces conventional binaries between the developed and developing world, and privileges certain kinds of creativity over others. From this view of modernity and modernization, Cuba stands at the back end of this curve, as late adopters or technological "laggards." Baskar Sarkar, however, argues that if we think about modernization in these terms, as the global circulation of technologies and commodities—from west to the rest—then within this "itinerary the parts of the globe that are variously known as the 'third world' or the 'postcolonial world' are forever playing 'catch up.' By the time they seem to be drawing level, the stakes have become higher; so they are always falling behind" (Sarkar, 2008, p. 48). Sarkar proposes alternative ways for thinking about this cycle of globalization and modernization, by thinking about the detours and delays in the cycle of goods, commodities and technologies as a "second logic" that resists and recasts "the rational teleology of modernization" (48). Oroza's documented practices reveal such a "second logic" in that they offer a vision of innovation and creativity decoupled from a neoliberal paradigm of entrepreneurial success and globalized labour markets. In particular, they decouple the link between creativity and innovation and "new" objects and technologies, challenging a popular discourse that privileges tools at the vanguard of the commodity life cycle, rather than at the residual obsolescent end.

References

Acland, C. R. (Ed.). (2007). *Residual media.* U of Minnesota Press.

Al Jazeera, 'Cuba Calling.' Retrieved July 16, 2015 from http://www.aljazeera.com/programmes/viewfinder/latinamerica2014/2014/02/cuba-calling-201421163356349778.html

Balaisis, N. (2015 Winter). Factory nostalgia: Industrial aesthetics in the digital city. *Architecture_MPS.* Retrieved July 31, 2015 from http://architecturemps.com/los-angeles/

Bell, D. (1973). *The coming of post-industrial society.* Harmondsworth: Penguin.

Canclini, N. (2014). *Imagined globalization.* (Yúdice, G., Trans.). Durham, NC: Duke University Press.

de Certeau, M. (1997). *The practice of everyday life.* Berkeley: University of California Press.

Florida, R. (2002). *The rise of the creative class: And how it's transforming work, leisure, community and everyday life.* New York, NY: Basic Books.

Galloway, A. (2015). *Critique and making.* In Hertz, G. Conversations in Critical Making. Vancouver, Canada: Ctheory Books.

Gaonkar, D. P. (2001). On alternative modernities. In D. Gaonkar (Ed.), *Alter/native modernities.* Durham, NC: Duke University Press.

Gauntlett, D. (2011). *Making in connecting: The social meaning of creativity from DIY and knitting to YouTube and Web 2.0.* Cambridge: Polity.

Harvey, D. (2007). *A brief history of neoliberalism.* Oxford: Oxford Press.

Hatch, M. (2014). *The maker movement manifesto: Rules for innovation in the new world of crafters, hackers and tinkerers.* New York, NY: McGraw-Hill.

Hernandez-Reguant, A. (2009). Writing the special period: An introduction. In A. Hernadez-Reguant (Ed.), *Cuba in the special period: Culture and ideology in the 1990s* (pp. 1–20). New York, NY: Palgrave.

Hertz, G. (2015). Conversations in Critical Making. Vancouver, Canada: Ctheory Books.

Kasvio, A. (2001). *People, cities and the new information economy.* Helsinki: Palmenia.

Leadbeater, C. (1999). *Living on thin air: The new economy.* London: Hodder and Stoughton.

Masuda, Y. (1990). *Managing in the information society: Releasing synergy Japanese style.* Oxford: Blackwell.

Mellonder, C., Florida, R., Asheim, B. T., & Gertler, M. (Eds.). (2014). *The creative class goes global.* Oxford: Routledge.

Nackenoff, C. (1997). The Horatio Alger myth. In P. Gerster & N. Cords (Eds.), *Myth America: A historical anthology*, Vol. II. St. James, NY: Brandywine Press.

Nakamura, L. (2011). Economies of digital production in East Asia: iPhone girls and transnational circuits of cool. *Mediafields, 2.* Retrieved from http://www.mediafields-journal.org/economies-of-digital/

Oroza, E. (2012). Technological disobedience. *Makeshift Magazine, 3.* Retrieved July 31, 2015 from http://mkshft.org/2012/07/technological-disobedience/

Ratto, M., & Boler, M. (Eds.). (2014). *DIY citizenship: Critical making and social media.* MIT Press

Reich, R. (1992). *The work of nations: Preparing ourselves for 21st century capitalism.* New York, NY: Vintage.

Sarkar, B. (2008). The melodramas of globalization. *Cultural Dynamics, 20*(1), 31–51.

Szeman, I. (2010). Neoliberals dressed in black; or, the traffic in creativity. *ESC: English studies in Canada, 36*(1), 15–36.

Vigíl, M. L. (2008). The Cuban media. In P. Brenner, M. R. Jiménez, J. M. Kirk, & W. M. LeoGrande (Eds.), *A contemporary Cuba reader: Reinventing the revolution.* Plymouth: Rowman and Littlefield.

Wark, M. (2012). Making New York. In G. Hertz (Ed.), *Critical making.* Hollywood: Telharmonium Press.

Wikipedia. Technology adoption lifecycle. Retrieved July 30, 2015 from https://en.wikipedia.org/wiki/Technology_adoption_lifecycle

15. *The Paradox of Maker Movement in China*

XIN GU
Monash University

A 3D printer became the focus of the 2015 Cultural Expo in Shenzhen—but more interesting was the story behind it. A few young makers set up a street stall near the *Huaqiangbei Market*, China's biggest electronic market, trying to sell the idea of 3D printers to the public. One of the onlookers was a woman whose job was as an "ear cleaner". She asked whether 3D printers could help making mini torch to put on her finger so that she could get a better look inside her client's ears. A few days later, she got her very own mini torches and she was overjoyed, so were the makers who were eager to join the league of creative entrepreneurs.[1] Stories like these provide convincing narratives of the democratising power of the Maker movement in China—a convergence of young university graduates turned creative entrepreneurs and peasant workers using high-tech.

China's first hacker space in Shanghai, *Xinchejian* ("new factory unit") does not immediately strike one as home-grown Chinese. David Li (Taiwanese educated in the US), Min Lin Hsieh and Ricky Ng-Adam (foreign expats) founded it. It was setup in 2010 in a co-working space *Xindanwei* ("new work space") a co-working space run by entrepreneurs Liu Yan and Chen Xu. Both were educated abroad—Liu in Holland and Chen in UK. Chen Xu has also worked for John Howkins who was the self-proclaimed "father of the creative industries"—creative industries is another a "fast policy" meme transposed to Chinese cities (O'Connor & Gu, 2015)—and was a key figure in introducing a UK version of the "creative industries" to Shanghai (Keane, 2006; O'Connor & Gu, 2006, 2012, 2015). David Li often described *Xinchejian* as a place for locals to practice English for free, for there is a flow of foreign engineers, scientists, IT experts and hobbyists passing through. *Xinchejian* like many other maker space in Chinese cities has quickly become connected with an affluent urban middle class looking for new "cool" and fancy ideas from abroad. A media scholar based in Shanghai, Lindtner (2014) observed, China's maker culture advocators are mostly middle and upper-middle class people. This is perhaps true of almost all maker culture across the world (cf. Tanenbaum, Williams, Desjardins, & Tanenbaum, 2013), but *Xinchejian* was the attempt of the Chinese makers to contest the cultural logic of the western Maker Movement. The founders were also trying to re-root the Chinese makers within a lower socio-economic class of manual workers and a localized claim to cultural value that sets the story of Chinese maker culture apart.

On the surface, the Chinese maker industry resembles other emerging innovation industries that the country has prioritized for development.[2] They are mostly located in the many "Science and Technology Parks" run by local governments. These businesses receive tax breaks and the opportunities for cross sectoral collaboration by being in the clusters. Foreign direct investment (FDI) is prioritized in investing in these businesses, which are seen to bring new knowledge to the process of industry re-structuration (Pei, 2006).

Maker industry clearly fits the innovation bill. The industries spun out from the "Maker Movement" can largely be categorized as "foreign facing" businesses. They are often joint partnerships between Chinese and foreign entrepreneurs. To learn from those who possess the knowledge is very much the operational logic here. By the early 2000s, the market for makers was clearly driven by the spillover of creativity outside of China looking for cheap manufacturing solutions to actualize creativity (Lindtner & Li, 2012). Maker and creative knowledge is an exception to the operational logic. Their knowledge is tacit, hard to translate into formulas and impossible to predict the outcome.[3] Most importantly, maker culture is rooted in localized creative eco-systems attached to a variety of culture values (Luckman, 2015). Its fundamental situatedness poses challenges to Chinese Maker Movements because the imported Maker movement is based on a western cultural ethos. The concept of "everyone can be creative", arrived in China as a byproduct of the globalization of the Maker movement rather than emerging from within.[4] And this is partially what the Chinese makers try to contest—the creative power controlled by this western cultural maker ethos. In addition, the resources for developing alternative creative ecosystems have been controlled tightly by the authoritarian government. The need for opening up and sharing limited resources is key to the success of Chinese Maker Movement. By contesting powers in both systems, a unique third space is opening up for constructing new meanings for the Chinese Maker movement—a space between the neo-liberal marketization of ideas and the authoritarian controlled sphere in

China. I will discuss some of the challenges faced by Chinese Maker movement caught in these two power systems.

The Arrival of Maker Movement in China

Established in 2008 in the city of Shenzhen, *Seeed Studio* is one of those occupying a unique position between the manufacturing capacity of China and the global creative industries. Its main business model was to supply open source hardware to overseas makers.[5] Most of *Seeed*'s clients are outside of China. Its CEO Eric Pan understood that tapping into Silicon Valley's distribution channel would be the only way for his company to grow. "Shenzhen is the Hollywood of hardware for makers", as Pan exclaimed at an International Maker Industry Assembly in Shenzhen.[6] Pan is clearly not satisfied with being at the bottom of the supply chain. *Seeed* became part of a global maker industry by building a strong presence on the crowd funding website Kickstart and is also developing local creative talent through its Maker Space—Chaihuo.[7] The mayor of Shenzhen declared official support for Shenzhen to become a global hub of makers in 2012.[8] Many maker spaces have since appeared in Shenzhen and tend to cluster around the *Huaqiangbei* electronics market.

Aside from the hardware centered vertical integration model, maker industries have developed horizontal links to other cultural industries, in particular visual art. Its representative is the *Beijing Maker Space* run by Wang Shengling based on the idea of the "convergence" of art and technology. Wang's maker space has had collaborations with the *Ullens Center for Contemporary Art*[9]—one of the most internationally renowned foreign art galleries specializing in Chinese contemporary art. Such cross boundary collaboration has driven wider public participation in the contemporary art scene. Maker spaces like these are a new extension of the meaning of culture making. They are not linked to China's manufacturing capability but to an artistic re-imagining of the "handmade" that has prospered in some developed countries including Britain, Australia and Norway. Luckman's (2015) book "Craft and the Creative Economy" captured the ethos of "handmade" in the post-modern age. She argued that the renaissance of handmade items in developed economies around the world is backed by substantial social and economic reasons, including its importance for the process of post-industrial urban regeneration.

The question as to how both of these tendencies (technical and cultural) within "maker culture" have been transposed into China and their popularization by a wide array of industry actors and government departments requires closer analysis. The normalization of these foreign cultural values, "do it yourself" or "craftsmanship", in post-industrial Chinese cities, is a particularly interesting process. The adoption of these values is clearly constrained by economic capital (only those who have the luxury of time and space to undertake it as a hobby), as such it is centered within an elite group of urban consumers. But it also makes claim to local grassroots cultural values for constructing new meanings and resisting any global conflicts in the late modern age. As Dirlik (1996) suggested,

> the local that is at issue here is not the "local" in any conventional or traditional sense, but a very contemporary "local" that serves as a site for the working out of the most fundamental contradictions of the age. (p. 23)

Framing the local in this way is appropriate here in understanding the paradox of the Maker movement in China. The Maker movement represents the possibility of democratizing "making" and the knowledge associated with it, in a context where industry is dominated by state owned or controlled industries. At the same time those pushing the Maker movement are reluctant to get too involved in truly vernacular making cultures—such as *shanzhai*—because it focusses to narrowly on the local.

The "shanzhai" culture, which I will discuss in detail below, is the Chinese answer to the global Maker movement. Despite makers' eagerness to rebranding "shanzhai" as the authentic Chinese maker culture, they are more likely to join the global maker culture revolution than to disrupt the global dominating powers that the shanzhai companies have attempted. The makers' careful selection of a set of terminologies that is agreeable both to the global maker industry and to the powerful state makes this clear.

When David Li labeled his space "the first hacker space" in China, it caused obvious controversy, for the term evokes a negative meaning of "disruption" in the Chinese language. To convince the government that makers are not "trouble makers", Li and his maker friends decided to change the term to *Chuangke*—that is, "creatives". Li was careful also not to equalize "making" to mass produced, low quality and derivative manufacturing that many local shanzhai companies have engaged in. *Chuangke* clearly helps to articulate alternative values to those of mass manufacturing, such as innovation and entrepreneurship, which are key values in the creative industries[10] which is another area that was popular amongst policy groups.

The creation of a new category "Maker Industry" as one of the key creative industries in recent government documents (State Council Report, 2016) has opened up new funding opportunities. Hacker spaces are now essential to the cultural infrastructures of large metropolitan cities such as Shanghai and Beijing. Many of them are not located on the fringes of these cities where the rents are cheaper but in the cultural zones that are attractive to foreign expats and new middle class consumers. They are not the "shadowy" informal economy of the city but an essential ingredient in the mix of "cool" urban creative industries. They are exactly what the Chinese government want the world to see—a free spirited, creative, entrepreneurial cool new breed of Chinese millennials.

Maker Culture—The Answer to the Future of Manufacturing

The rise of Chinese maker culture is both ideological and economic. Its high profile is the result of the combined forces of China's ambitions to global "soft power" and the need for industrial restructuring (SCA, 2016). China's close association with mass manufacturing has allowed it to become the world's largest exporter, the second largest economy and a constant source of superlatives as to the speed and scale of its social and economic transformation. At the same time however, this mass manufacture is the mark of a developing rather than developed economy—dirty, low-skilled, technologically derivative and generally unsuited to the profile of a global hegemon (SCA, 2016). Such a downbeat view of the manufacturing industry was further compounded by a global shift towards niche manufacturing set against all values of "made-in-china":

> A convergence of computer hackers and traditional artisans … Makers tap into an American admiration for self-reliance and combine that with open source learning, contemporary design and powerful personal technology like 3D printers. The creations, born in cluttered local workshops and bedroom offices, stir the imaginations of consumers numbed by generic mass-produced, made-in-China merchandise. (Voight, 2014)

The Maker movement captured a deep-seated longing within Western developed economies for a return to "manufacture", a kind of selective re-industrialisation. Chris Anderson (2014) has suggested that this new wave of maker culture represents a "third industrial revolution" by which America and other post-industrial countries will re-take the power of making from the hands of Chinese industrialists. Mark Hatch substantiated this argument in his "maker manifesto" laying out the toolkit for developing niche manufacturing industry for developed economies (Hatch, 2013).

The Maker movement captured much recent media attention too (cf Bajarin, 2014; Tierney, 2015; Voight, 2014), for it seemed to address two fundamental challenges involved in the re-structucturing of post-industrial economies since the 1970s. The first involved the employment downside of de-industrialisation. The post-industrial imaginary was constructed around a binaries of old and new, utilitarian and symbolic, low and high value; dirty and clean. Despite the fact that the developed economies continued to hold significant "competitive advantage" over developing economies in terms of profits derived from controlling logistical and financial flows, and from the "new", "weightless" and "creative" economies protected by intellectual property rights (increasingly enforced since the establishment of the World Trade Organisation), it seemed that the material, "weighty" side of industrial production provided jobs, underpinned relatively equalizing welfare state regimes, and contributed to coherent social identities and communities. The Maker movement can be seen as a rediscovery or reframing of manufacture in ways that try to recoup at least some of the aspirations and identities associated with the now receded "golden age" of Fordism.

Secondly, it addresses some of the barriers to innovation brought about by the rising cost of R&D across all industries. More than a simple appeal to "open innovation" to off-set some of these costs (Ettlinger, 2014) the Maker movement foregrounds the complex relationship between craft, manufacture and creative input. It recognizes the socio-cultural context within which small and micro businesses innovate, involving networked ecologies of craft skills, material supply chains and manufacturing know-how intertwined with the high-level of situated symbolic competence evoked in the post-Fordist literature (Lash & Urry, 1994; Piore & Sabel, 1984).

This latter proposition has led many to embrace the "techno-democratisation" model: "everyone can do it" is the default slogan framing the Maker movement (Powell, 2012). But many are critical of the potential for easy commercialisation of the Maker movement—grassroots creativity being absorbed into the big corporations (Ettlinger, 2014). Andrew Taylor, an innovation consultant quoted by Voight (2014) observed an increased intensity of large corporations attempting to shift R&D cost to smaller players in the maker industries—a practice long established in the cultural or creative industries (Hesmondhalgh, 2007). Taylor warned that SMEs face formidable challenges in retaining their niche identity in the face of corporate predators. Mostly preferring to ignore these warnings, the Maker movement in the West has launched itself into a re-definition of the value of making and a celebration of the context within which material knowledge and craft skills are shared between different actors in a local (though globally connected) ecology. In many respects it is an attempt to recreate the "industrial district" for post-industrial, post-digital times.[11]

In China, the Maker movement is also related to a re-valuation of craft skills, but these are more closely linked to the challenges faced by manufacturing industries in recent years—a struggle to keep up speed of innovation, a shift of the consumer base from global to domestic markets, and an aspiration to upscale its value chain from "Made in China" to "Created in China" (O'Connor & Gu, 2006). Since the early 2000s, the *shanzhai* industry (small scale production units manufacturing copies of branded goods with a Chinese twist) emerged in mass manufacturing industrial cities like Shenzhen as a response to all these changes. It was linked to grassroots entrepreneurialism and exhibited a strong antipathy to control by global corporations in the digital electronics market.

The subsequent emergence of *maker spaces* in key cities drew aspiration from the grassroots *shanzhai* culture. It is therefore not surprising that makers in China view themselves as the offspring of the *shanzhai* industry or *shanzhai 2.0*. The maker industries have increasingly become a tool to lift productivity and inject much needed creativity into the existing manufacturing industries. But the question for China's Maker movement is whether its increasing institutionalization by the powerful state has rendered it unlikely to connect with local grassroots creativity.

Shanzhai—The Politics of Chinese Maker Movement

In 2008, a techno-cultural phenomenon emerged in China gaining wide popularity amongst locals—*shanzhai,* a way of re-appropriating product ideas for local uses, in particular mobile phones. According to scholar Andrew Chubb (2015), *shanzhai*'s wide appeal as an emergent form of Maker movement in China evokes "ingenious Chineseness, marginality and independence, and playfulness and critique" (p. 272). The word "*shanzhai*" literally means "mountain fortress" and it designates to a group of outlaw entrepreneurs who operate outside of the authoritarian control of economic and cultural production.

Prior to this, most Chinese manufacturers of mobile phones were engaged in build-to-order businesses, low value added, labour intensive and production-line style manufacturing controlled by global mobile phone giants including Nokia, Ericsson and Sumsung (Yu & Yan, 2015). With the invention of MediaTek's circuit board, the design and manufacture of new kinds of mobile phones suddenly became viable.[12] In addition, in 2007 the Chinese government relaxed rules around licensing in the mobile phone manufacturing industry. Making mobile phones suddenly became very easy (Liao & Chen, 2008). It involved buying the MediaTek circuit boards, some generic parts and assembling. Benefiting from the clustering of hardware companies, many small companies emerged in Shenzhen operating on this DIY model of making *shanzhai* mobile phones. Instead of employing hundreds of workers, they might have ten or less people, and finish a short order of less than 5,000 units in less than 24 hours. Design and functionality improvement can be added quite quickly in response to local demands. This new way of production reduced the cost of mobile phone considerably in China, thus aiming primarily at the low cost end of mobile phone consumer market and highly driven by the needs of local consumers (Zhu & Shi, 2010). By 2008, *Shanzhai* mobile phone sales captured 30 percent of the country's entire mobile phone market, and it became a major employer in the city of Shenzhen and doubled the number of mobile phone users in China.

It might be that *shanzhai* was the only force in China capable of disrupting the market dominance by global mobile phone giants at the turn of millennium. Anna Greenspan (2014) argued that *shanzhai* fits into a unique market place in China "where the large, 'best' companies are simply not looking." (2014, p. 200) *Shanzhai* clearly speaks to the need of mass Chinese consumerism which was largely ignored by the global mobile phone industry. Qiu (2009)'s research on the Chinese working class' use of communication technology, in particular mobile phones, supports the theory of a bottom up innovation driven by urban working class consumers. This was echoed by researches on technomobility amongst migrant workers who were excluded by these technological developments in the country (Wallis, 2013; Yang, 2011). In this sense, *shanzhai* can be viewed as form of local resistance to the globalization domination of media powers.

Analyzing the origin of the global maker culture, it seems that *shanzhai* shares unique social and political footprints with some of the early global cybercultural movement. *Shanzhai* evokes the politics of anti-global corporatisation which has deep roots within the counterculture movement in the US in the 1960s (cf. Turner, 2006). The emergence of *shanzhai* can also be set against the background of both the Asian Financial Crisis of 1997–1998 and the later 2008 Global Financial Crisis when Chinese mass manufacturing industries were forced to scale down their production capacity and adopt more specialized production processes (Lin, 2011). *Shanzhai* companies set an example for other Chinese manufacturing industries on how to restructure in the face of more challenging market situations. As Chubb observed, *the* resistance to power was expressed through the "simultaneous worship of and jeering at authority" (Chubb, 2015, p. 263). Unlike other imitations, *shanzhai* doesn't pretend to be the real deal. Its mimicking relies as much on its "look alike" as on its "not quite the same".

The disruptive spirit of the *shanzhai* industry spread quickly to other social and economic areas of life. By the end of 2010, Greenspan observed that "the word *shanzhai* can be used to describe anything that is non-official, underground and inexpensive with acceptable quality. *Shanzhaiism* has become a philosophical term denoting a Chinese style of innovation with a peasant mind-set. Shanzhai's cool DIY spirit has a nationalistic pride but it is rooted not in the strength of the state but in the flexible, creative culture of the street" (204). Zhang and Fung (2013)'s research on the Shanzhai Spring Festival Gala on the other hand pointed to the possibility and limitations of such homegrown digital culture when faced with a powerful state.

Despite the tight Chinese media control and censorship, *shanzhai* managed to impose a new strategy bypassing the rule of the law. Yu Hua saw the word *shanzhai* as the most significant cultural phenomenon in contemporary China, exhibiting a unique anarchist spirit within Chinese culture that no other Chinese word is able to convey (Yu, 2012). Applying Dirlik (2001)'s global modernity theory to the emergence of *shanzhai* in China, Chubb argued that *shanzhai* culture was a fine example of the different kinds of local resistance in the era of global capitalism—in this case, by not confronting the global corporations head on, but through a careful rebalance of the state, local businesses and global corporations.

Perhaps because of its "anarchic" anti-authoritarian ethos, *shanzhai* culture and its industries never achieved the political status that the urban Maker movement was to do, despite its economic success and wide cultural significance. From as early as 2009, there were growing policy debates around the legitimacy of *shanzhai* industry and *shanzhai* culture (Wei, 2008). It is in this climate that Shenzhen, where *shanzhai* industry is most concentrated, began to embrace the Maker Movement.

Maker Movement: Shanzhai 2.0

Shenzhen was seen as the "holy grail" by Chinese makers. *Chaihuo Space* established by Pan Hao openly endorsed the *shanzhai* spirit in nurturing Chinese creativity—"the spirit of *shanzhai* is close to that of the 'maker culture' in that it's an open source movement—it believes that innovation is a bottom up process." Pan made this statement proudly during his talk at the Shenzhen Assembly in 2016.[13]

But *shanzhai* and the Maker movement differed in more than just their labels. *Shanzhai* has all along been negatively viewed as simply flouting intellectual property rights whilst the maker culture makes a principled challenge to the legitimacy of "intellectual property" in the name of the "open source movement". Whilst *shanzhai* companies are still haunted by the legality of the industry they operate in, makers have managed to transcend this regime (at least theoretically) by affiliating with the global techno-cultural movement. “Intellectual property rights should not be exclusive rights, nor should they be political tools to by which one country penalizes another. I believe IP should be used to promote more innovation not to indulge individual rights.” (interview in 2014) David Li has succinctly summarised the difference emphasis between the two terms. Li wanted to retain the anti-authoritarian and grassroots spirit of *shanzhai* but he is not naïve about the controversy of such an approach in the context of state control in China. He thus proposes a model of a “new shanzhai”, of retaining the bottom up innovation spirit, along with the fast paced, efficient and networked nature of *shanzhai* industry, whilst abandoning the contentious ethical issues surrounding the original *shanzhai*.[14]

The Maker Movement’s detachment from the original *shanzhai* ethos, has played a key role in its rise in China—it shows that makers, unlike those in the *shanzhai* industry, are not necessarily troublemakers. In fact, the Maker movement has increasingly been supported by a government that focuses great energy on the potential links between innovation and entrepreneurialism in local creative industries. Innovation plus entrepreneurialism has subsequently become the dominant ethos of the Chinese Maker Movement.

This, however, raises questions about the substance of the Maker Movement’s aspiration to give “power to all”, though this is by no means a uniquely Chinese problem. Media critics like Evgeny Morozov warned about the underlying problem with supporters of the Maker movement in the West,

> For Anderson, such innovation is the prelude to a great business: when hobbyists cluster together to work on obscure technologies, someone eventually gets rich. But it’s misleading to view the Homebrew Computer Club solely through the prism of innovation and entrepreneurship. It also had, at least at first, a political vision.

The point is well made, and is especially apt in regard to many accounts of the transfer of Maker movement from the West to China. Its significance relies on its grassroots and countercultural ethos, where corporate control of media production and distribution channels was viewed as largely restricting civic participation and creativity. But such local resistance to one power source can be read as a celebration of a value system espoused by another power source. In the case of China, anti-global corporate control can be read as nationalism and countering the authoritarian control can be read as empowering global consumerism. In the same way, the role of large corporations is not simply one of exploitation and control as in many instances they offer crucial tools in social empowerment in China, where state control of the media industry has been a notorious problem (cf. Liu, 2011; Rofel, 2007). *Shanzhai,* as a unique form of Chinese maker culture, denotes grassroots creativity which has resulted in the empowerment of working class technology businesses. But such digital democratization also puts itself in the rather uncomfortable situation of challenging the powerful Chinese state.

Thus to adopt a position of countering corporate control is not so straightforward. For a start, the Maker movement is linked to the entrepreneurial spirit of the grassroots creative sector. But this entrepreneurial spirit is derived from a highly westernized interpretation of "cognitive capitalism" (Peters & Bulut, 2011). It does not have a natural link with China's homebrewed entrepreneurialism as exemplified by the *shanzhai* industries (Link & Qiang, 2013). Secondly, in China it has a strong techno-nationalism aspect similar to the development of other "creative industries" in the country. It welcomes originality, entrepreneurial energy and the DIY ethos, but it is all to be framed within the "soft power" of China (Nye, 2005). As such the Chinese Maker Movement's promise of individual creative freedom may be configured very differently, even—in the eyes of their western colleagues—compromised.

The top down version of creativity and innovation can be observed in the many large-scale urban regeneration projects in major cities like Shanghai (Gu, 2014). Culture and innovation are the key ingredients for its development, where no idea is too grand for Shanghai. Since as early as 2005, the city has developed over one hundred creative industries clusters in order to boost creativity and innovation—seen as future "replacement" industries. The ambition to transform "Made in China" into "Created in China" was quickly translated into urban policies, attempting to replace a manufacturing industry by high value services and other forms of the "knowledge economy". But there is an ambiguity in China's approach to "creativity" and "innovation", in that the two terms are interchangeable, at times making it hard to distinguish a cultural policy from a science and technology one. As has frequently been the case in modern Chinese history, "modernisation" has meant learning not just technical and administrative solutions but Western cultural forms and values, involving persistent questions as to "what can be mobilized from one's own tradition, what and how much to borrow, and what social, cultural and political changes might follow in their wake" (O'Connor & Gu, 2006).

The creative industries discourse in the West emphasizes individual, often unorthodox creativity, taken largely from avant-garde artistic practices emphasizing systemic rule-breaking and transgression (O'Connor, 2010). In the 1980s "thinking outside of the box" or seeking "the shock of the new" were increasingly linked to the newly promoted values of entrepreneurialism, in turn linked to the move away from mass consumption and Fordist production to niche markets and small scale, flexible production (O'Connor, 2010). The Maker movement certainly represents a "material" turn away from the "weightless" new economy, but it is very much rooted in the image of the creative entrepreneurs and start-up milieus mobilized by the creative industries (Leadbeater & Oakley, 1999). And this is certainly the way it is positioned in China. This poses real challenges for that version of Chinese Maker Movement; how can its versions of free, individualized, artisanal innovation emerging from the wreckage of mass manufacture be translated into China's own discourse of state-led innovation strategies. In many ways the Maker movement invokes a new set of social structures and spaces which, as with the "creative milieu" of the creative industries, are unlikely to appear in anything like they are envisaged for in the context of China.

The Maker movement illustrates well the compromise between "digital democratization" and "authoritarian control". The de-politicalisation of the Maker movement in China has ultimately led to a cultural consumption model about middle class consumers and an urban culture economy as I will show below. Its institutionalisation within the framework of highly economic-instrumental "creative industries" in China further blunts any potential for democratic or grass-roots empowerment it might have held out.

Practices in the Construction of Maker Industry in China

The Chinese government has always been keen to promote innovation within its domestic industries. One of the key strategies it employs is to institutionalize new innovation by investing in research and development, value chain development, marketing and trading. This has been effective in many science and technology based industries but has been less effective in promoting creative industries as the latter was made up mostly of SMEs and operate within a different learning ecology (O'Connor & Gu, 2012). The promotion of the Maker movement has thus been accompanied by softer approaches to learning dissemination, offering local cultural adaptations through workshops aimed at school kids, urban farming workshops, and maker-led creative entrepreneurship development. Most of these initiatives were designed and spearheaded by the new breed of western educated Chinese creative entrepreneurs who are familiar with the "open source movement", "hacker spaces" and "co-working" in the West, and are keen to transpose these ideas to China. These programs play an important role in raising the profile of the Maker movement within policy circles.

Entrepreneurialism—A Non-business Business Model

Everyone can be creative: mass entrepreneurialism plus mass innovation has become the new politics of the Maker Movement. Shenzhen—"the city of makers"—has used the opportunity to propel itself to a new political importance. Premier Li Ke Qiang visited Pan Hao's hacker space in Shenzhen after the country's most important policy document listed "maker" as a key word in 2015. In an interview with the national news media, the mayor of Shenzhen proudly announced that Shenzhen was to become the centre of an international maker culture.

The government's emphasis on entrepreneurialism—"turning ideas into real business opportunities"—is crucial to understanding the Chinese Maker Movement. Interviewed by the media on the relationship between cultural creative industries and the Maker Movement, Peng who advised on many cultural industries park designs in China suggested that the Chinese creative industries bottleneck—lots of investment, little real success—stems from the failure to actualize the many good ideas into profitable business opportunities. (China Culture). He referred to the maker industry as a potential example of how creative industries can be made more profitable—of extracting value from bottom-up creativity.

Entrepreneurship development is thus built into the practice of maker culture in China and conflated with ideas of DIY, coolness and individuality. Aside from teaching members how to use new technologies in product design, the maker space or *Fab Lab* as they like to call them, provides essential training for members on how to develop startup businesses based on these new ideas. A series of workshops run by the Sino-Finnish Research Centre at Tongji University for the maker culture community in Shanghai, were doing the same. Many other maker spaces also offer networking events bridging local industries with makers. Despite little evidence as to the viability of a maker business model, expectations are running high within China that the maker industry will be one of the pillar industries of the future—with high profit margins and high value added (the white paper of Chinese maker industry). The government is hoping to sponsor over one hundred fab labs across China by the end of 2015 as part of this objective.

However, for many in the Maker Movement, the most obvious business model was that of a state-driven real estate model, in which hacker or maker spaces are developed as part of science and technology parks or creative industries clusters. Following the announcement of Intel's partnership with the state in funding maker industries over the next five years, Chinese makers are likely to enjoy sufficient state funding and tax rebates allowing them to occupy space in these parks or clusters, but not necessarily driving innovative manufacturing industries. As David Li admitted, *Xindanwei* is still a non-for-profit business model for communities. It is a label that allows him to do other things that are not strictly within the maker industry but draws on his high profile in the maker industry (Interview 2014).

Maker Faire and Mobile Makers

Familiar with the link between *Maker Faire*[15] and innovation and creativity, *Xinchejian* organized China's first Maker Carnival in 2013. Located in containers within one of the largest science and technology parks in Shanghai, the Maker Carnival was branded as a very different approach to innovation, not top down but bottom up. As David Li said, mobile maker labs demonstrated that "the Maker Carnival is an open platform for everyone interested in maker culture to share their ideas. You don't have to have hardware or software knowledge just ideas."

Xinchejian's cross-disciplinary peer-to-peer learning and sharing spirit is the exact opposite of the operating principles of China's vast numbers of innovation parks. They are largely enclosed, with management companies working as gatekeeper on behalf of the government/or private property developers. Only those financially successful or politically high profile companies are let in. Maker and creative SMEs often found these innovation parks inaccessible for many reasons (O'Connor & Gu, 2012).

However, whether the public will take this state sponsored Maker movement seriously or treat it as yet another fancy PR campaign or some latest imported fad, is an open one. The container in the creative park did not attract enough attention from the public and was moved to a commercial district in the hope to bring more people in—most of whom turned out to be kids. Similar problems were found in other mobile maker projects in the country, "mobile makers" seemingly unable to generate wide participation. More research is needed to understand this but at least its lack of popularity proves that technologically enabled creative democratization only comes if such knowledge is widely embedded in a socialized public domain. This is perhaps why *shanzhai* technology (copying and mimicking) was more likely to be understood, shared and celebrated amongst Chinese audiences. This can be further evidenced by the fact that most of those having an enduring interest in maker culture are people with engineering and IT backgrounds, despite its initial claim that "everyone can do it".

Education Reform

In the US, where the Maker movement took off, promoting the value of maker culture amongst the younger generation through educational initiatives such as the *Maker Faire* has been one of the corner stones in defining maker Culture. Now supported by national bodies such as the National Science Foundation (NYSCI), the Maker movement is expected to deliver learning outcomes in the fields of Science, Technology, Engineering and Mathematics (STEM) where academic performance among American school kids has continuously declined over the past decades. *Maker Faire* was part of the Obama government's attempt to enhance the future of innovation in America: "I want us all to think about new and creative ways to engage young people in science and engineering, whether it's science festivals, robotics competitions, fairs that encourage young people to create and build and invent—to be makers of things, not just consumers of things" (President Barack Obama at the National Academies of Science) (April 27, 2009).

In similar fashion maker culture is expected to alter the Chinese education system in significant ways—to move the emphasis on rote learning more towards nurturing individual creativity. *Xinchejian* has hosted regular weekend kids' maker workshops since its opening. David Li likes to use the case study of Teacher Huang when discussing the educational value of maker culture. Teacher Huang is a local high school physics teacher who became interested in maker culture's philosophy and adapted it to teaching physics to local kids—not by memorizing things from books but by applying knowledge when making things. Huang now runs regular LED lighting workshops on the website of "Make for Kids", a platform to facilitate the Makers' cross-disciplinary peer to peer learning and to promote the "Makers' sharing spirit, open source software and hardware and open innovation to the education".[16] The educational value of the Maker movement was soon recognized across the nation linking lead industry bodies with educational bodies for innovation spin-offs with the participation of scientists, politicians and industry experts.

Nevertheless, such experiments offered by the Maker movement have had little impact on the Chinese educational system other than at a superficial level. In the controversial documentary "Prometheus"[17] (dir. Deng 2013) on the Chinese education system, educators despaired at the possibility of rewarding creativity and innovation within the Chinese system, as such experiments were highly unlikely to be incorporated into the official system, though they might become extra curricula activities paid for by the wealthy middle class.

Urban Farming

We might mention here a related movement around "urban farming", also closely linked to class and education. In the case of the Shanghai Maker Culture, the romantic idea of growing your own vegetable/fruits not only requires a living condition that the majority of Chinese families are unable to afford, it also relies on accepting "farming" as a cool and fun thing to do, instead of a term affiliated with uncultured "peasants".[18] Writing about the urban farming phenomenon in the US, Banet-Weiser (2012) asserted that "the contemporary urban farmer is a privileged subject position and not an 'ordinary gardener' and 'the spaces of the contemporary farmers' market, community agricultural efforts, and urban farming are often bounded by discourses of whiteness, manifest in color-blind ideologies and assumptions of universalism." (pp. 163) The prejudice of class, social and economic status are visible in the Chinese practice of urban farming. The individual DIY entrepreneurs in the maker culture are now pioneers of such "lifestyle choice" when the majority of the local residents are living with a housing and food crisis.

Conclusion

Although these maker fairs and hacker spaces generate an image of "creativity" and "democratic participation", their lack of any critical stance positions the Maker movement as more of an emerging "new industry practice" than a cultural social "movement". Shying away from critical messages that could be positioned as "anti-establishment" or "anti-authoritarian", the Maker movement remains a marginal form of urban cultural consumption with an exclusive audience from educated and privileged middle class background.

The closest the makers get to a radical cultural movement is to declare that they are promoting authentic creative talent as a way to counter China's "creativity deficit", and that somehow it is driven by grass roots, intuitive creativity. But in fact, these are highly westernized concepts requiring audiences with the ability to learn and decode foreign messages in order to participate. It is a position highly mediated through the work of foreign educated "intermediaries" able to translate global cultural memes into a local context. We might keep open the possibility of a grassroots maker creativity, but we need to be clear as to the dominant the role the state plays in promoting both industry and cultural development—combined in the creative industries.

Investigating maker culture in China is like looking through the "Magic Mirror". In it we can see the emergence of "Cool" DIY makers and fab labs in its cosmopolitan cities. But we are also told that the real makers lie somewhere else—on the street, in the not so fancy factories in the countryside, in the *shanzhai* markets. The separation of Maker movement and *shanzhai* culture, as the government promotes the former as another phenomenon of foreign-inspired modernization, and tries to suppress the latter as disruptive, dispersed and relatively uncoupled from direct state supervision, will have a significant impact on China's creative future. Without real a connection with everyday life and culture, the Maker movement has limited capacity to drive real creative participation, and equally, the shadowy *shanzhai* culture has a dwindling future if its illicit status remains. A dialectical middle ground could be possible—combining a creative nucleus (cultivated through "maker culture" movement) with the mass production capacity of *shanzhai* industry. The question for the state would be how to come up with a creative industries strategy more geared towards facilitating locally rooted initiatives rather than heavy handed top down promotion of the business interests of state owned enterprises.

The Maker movement in China shows the real challenges of China's creative environment, a government controlled arena in which state directed enterprises attempt to institutionalize innovation for its own purposes. There is certainly a concentrated urban grassroots movement searching for an alternative platform for self-expression and self-actualisation, just as there is a wide-ranging ambition to set up small businesses able to operate outside the state-directed sector. These ambitions go back to the town and village enterprises of the 1970s and 1980s, as well as the rise of small businesses emerging in the wake of the massive de-industrialisation of China underway since the 1980s (Huang, 2008). It is however questionable how far the Maker movement can be the driver of China's small enterprise innovation whilst holding to the narrow techno-democratization principle that most Maker movement in the West were based upon. This principle is, on its own, incapable of tackling the issues at hand in China, where the big corporations and the controlling state both have an interest in promoting creativity and innovation within their models and control.

Notes

1. This story was published during the Chinese Cultural Expo (*Wen Bo Hui*) in 2015. The Expo is the first national cultural event themed around *Maker Culture*. See news article 'Cultural Expo has entered the era of Makers', *Time Weekly*, May 19th 2015. Accessed 1st May 2016, http://www.time-weekly.com/html/20150519/29721_1.html
2. These policy driven development strategies are outlined in the country's topdown planning scheme which takes place every five years. The most recent round of

planning—the 13th Five Year Plan has, for the first time, included maker industry as key to build China's manufacturing future—'made in China 2025 (zhongguo zhizao 2025)', *State Council Announcement No. 28.* Accessed 1st May 2016. http://www.gov.cn/zhengce/content/2015-05/19/content_9784.htm

3. The characteristics of cultural goods are understood to be significantly different to other goods. These unique features have given policy makers the reason to list them as a new industry sector 'the creative industries'. For example, the British governments have produced a sequence of policy documents outlining agendas around 'creative industries'.
4. The philosophical understanding of Chinese creativity is beyond the scope of this paper and there are a wealth of publications on this subject, for example, Niu, W., & Sternberg, R. J. (2006). The philosophical roots of Western and Eastern conceptions of creativity. *Journal of Theoretical and Philosophical Psychology, 26*(1), 18–38. DOI:10.1037/h0091265.
5. Seeed Studio, https://www.seeedstudio.com/
6. Eric Pan's talk at the Shenzhen Assembly 2016 can be viewed at https://youtu.be/bEVtgAN-pWk
7. Premier Li Ke Qiang visited Chaihuo in 2015. This sends a signal that the government is right behind the development of maker industries in China. http://www.chinadaily.com.cn/china/2015twosession/2015-03/09/content_19760292.htm
8. Official coverage of Shenzhen Maker faire which is a launch event for Shenzhen's claim to be the 'City of Makers'. http://news.xinhuanet.com/tech/2015-06/23/c_127938960.htm
9. Ullens Center for Contemporary Art: http://ucca.org.cn/en/about/index/
10. For a detailed account of the development of creative industries policy, please refer to O'Connor and Gu (2006).
11. For a detailed account of the 'industrial district', see Alfred Marshall's *Principles of Economics.* Published in 1920 by Macmillan and Co.
12. This technical breakthrough was well documented by a series of publications on innovation and entrepreneurship in the sector. Cf. 'Handbook of East Asian Entrepreneurship' (2015) by Tony Fu-Lai Yu and Ho-Don Yan. Routledge.
13. This speech can be viewed on youtube: https://youtu.be/bEVtgAN-pWk
14. Li has contributed to numerous journal articles since 2014 to lament the term 'the new shanzhai'. Cf. 'The new shanzhai: democratizing innovation in China', Paris Tech Review, December 24, 2014. http://www.paristechreview.com/2014/12/24/shanzhai-innovation-china/
15. Maker Faire is a global event celebrating the Maker Movement. It was established in New York City in 2006. http://makerfaire.com/#
16. http://makeforkids.com,
17. Prometheus (盗火者) is a 2013 ten-part documentary offering a critique of Chinese educational system. First broadcasted on Phoenix TV.
18. Makerplus.org (non-for-profit organization sponsored by Swissnex—a Swiss-China business and trade organization) is a representative of urban farming amongst the Maker community in Shanghai. David Li's hacker space also took an active role in promoting urban farming.

References

Anderson, C. (2014). *Makers: The new industrial revolution.* New York City: Crown Business.

Bajarin, T. (2014). Why the maker movement is important to America's future. *Time.* Retrieved May 19, 2014 from http://time.com/104210/maker-faire-maker-movement/

Banet-Weiser, S. (2012). *Authentic: The politics of ambivalence in a brand culture (critical cultural communication).* New York, NY: NYU Press.

China Culture, 'Culture maker industry cluster: how to achieve win-win for makers and creative clusters?' *China Economy,* 13/06/2015, http://www.ce.cn/culture/gd/201506/13/t20150613_5633861.shtml Accessed 01/07/2016

Chubb, A. (2015). China's Shanzhai culture: Grabism and the politics of hybridity. *Journal of Contemporary China, 24*(92), 260–279.

Dirlik, A. (1996). The global in the local. In R. Wilson & W. Dissanayake (Eds.), *Global/local: Cultural production and the transnational imaginary* (pp. 21–45). Durham, NC: Duke University Press.

Dirlik, A. (2001). *Global modernity.* Boulder, Colorado: Paradigm Publishers.

Ettlinger, N. (2014). The openness paradigm. *New Left Review, 89,* 89–100.

Greenspan, A. (2014). *Shanghai future: Modernity remade.* London: Hurst & Company.

Gu, X. (2014). Creative industries, creative clusters and cultural policy in Shanghai. In L. Lim & L. Hye-Kyung (Eds.), *Cultural policies in East Asia: Dynamics between the state, arts and creative industries.* London: Palgrave Macmillan.

Hatch, M. (2013). *The maker movement manifesto: Rules for innovation in the new world of crafters, hackers and tinkerers.* New York, NY: McGraw-Hill Education.

Hesmondhalgh, D. (2007). *The cultural industries* (2nd ed.). London: Sage.

Huang, Y. (2008). *Capitalism with Chinese characteristics: Entrepreneurship and the state.* Cambridge University Press.

Keane, M. (2006). From make in China to created in China. *International Journal of Cultural Studies, 9*(3), 285–296.

Lash, S., & Urry, J. (1994). *Economies of signs and space (theory, culture & society).* London: Sage.

Leadbeater, C., & Oakley, K. (1999). *The independents.* London, U.K.: Demos.

Liao, Z., & Chen, X. (2008). *Why the entry regulation of the China mobile phone manufacturing industry collapsed: The impact of technological innovation on institutional transformation.* Paper presented at the 2008 Beijing Workshop of Institutional Analysis organized by Ronald Coase Institute.

Lin, Y.-C. (2011). *Fake stuff: China and the rise of counterfeit goods.* London; New York, NY: Routledge.

Lindtner, S. (2014). Hackerspaces and the internet of things in China: How makers are reinventing industrial production, innovation, and the self. *China Information, 28*(2), 145–167.

Lindtner, S., & Li, D. (2012). Created in china: The makings of China's hackerspace community. *Interactions, 19*(6), 18–22.

Link, P., & Qiang, X. (2013). From grass-mud equestrians to rights-conscious citizens: language and thought on the Chinese internet. In P. Link, R. P. Madsen, & P. G. Pickowicz (Eds.), *Restless China*. Lanham, MD: Rowman & Littlefield.

Liu, F. (2011). *Urban youth in China: Modernity, the internet and the self.* New York, NY: Routledge.

Luckman, S. (2015). *Craft and the creative economy.* London: Palgrave Macmillan.

Nye, J. (2005). The rise of China's soft power. *Wall Street Journal Asia.* Retrieved December 29 from http://belfercenter.hks.harvard.edu/publication/1499/rise_of_chinas_soft_power.html

O'Connor, J. (2010). *The cultural and creative industries: A literature review* (2nd ed.). Newcastle upon Tyne: Creativity, Culture and Education.

O'Connor, J., & Gu, X. (2006). A new modernity? The arrival of 'creative industries' in China. *International Journal of Cultural Studies, 9*(3), 271–283.

O'Connor, J., & Gu, X. (2012). Creative industry clusters in Shanghai: A success story? *International Journal of Cultural Policy.* DOI:10.1080/10286632.2012.740025.

O'Connor, J., & Gu, X. (2016). Creative clusters in Shanghai: Transnational intermediaries and the creative economy. In J. Wang (Ed.), *Making cultural cities in Asia.* London: Routledge.

Pei, C. (2006, January). Attracting foreign direct investment and restructualization of traditional industries in China—Strategies in utilizing foreign direct investment in the era of the 11th Five Year Plan. *Chinese Industry and Economy.*

Peters, M. A., & Bulut, E. (Eds.), (2011). Cognitive capitalism, education and digital labor. New York, NY: Peter Lang.

Piore, M. J., & Sabel, C. F. (1984). *The second industrial divide: Possibilities for prosperity.* New York, NY: Basic Books.

Powell, A. (2012). Democratizing production through open source knowledge: From open software to open hardware. *Media, Culture & Society, 34*(6), 691–708.

Qiu, J. L. (2009). *Working-class network society: Communication technology and the information have-less in urban China.* Cambridge, MA: MIT Press.

Rofel, L. (2007). Desiring China: Experiments in neoliberalism, sexuality, and public culture. Durham, NC: Duke University Press.

SCA (State Council Announcement). (2016). State Council Announcement No. 28. Made in China 2025 [zhongguo zhizao 2025]. Retrieved from May 1, 2016 from http://www.gov.cn/zhengce/content/2015-05/19/content_9784.htm

Tanenbaum, J., Williams, A. M., Desjardins, A., & Tanenbaum, K. (2013). Democratizing technology: Pleasure, utility and expressiveness in DIY and maker practice. *Proceedings of SIGCHI conference on Human Factors in Computing Systems* (pp. 2603–2612). New York, NY: ACM.

Tierney, J. (2015). The dilemmas of maker culture—thinking through the consequences of the proliferation of powerful tools and technologies. *The Atlantic.* Retrieved

April 20, 2015 from http://www.theatlantic.com/technology/archive/2015/04/the-dilemmas-of-maker-culture/390891/

Turner, F. (2006). *From counterculture to cyberculture: Stewart brand, the whole earth network, and the rise of digital utopianism*. Chicago: University of Chicago Press.

Voight, J. (2014). Which big brands are courting the maker movement, and why—from Levi's to Home Depot. *Adweek*. Retrieved March 17, 2014 from http://www.adweek.com/news/advertising-branding/which-big-brands-are-courting-maker-movement-and-why-156315

Wallis, C. (2013). Technomobility in China: Young migrant women and mobile phones. New York, NY: New York University Press.

Wei, Y. (2008, December 4). The abuse of 'shanzhai spirit'. *Beijing News*.

Yang, G. (2011). *The power of the internet in China: Citizen activism online*. New York, NY: Columbia University Press.

Yu, H. (2012). *China in ten words*. New York City: Vintage Books.

Zhang, L., & Fung, A. (2013). They myths of 'shanzhai' culture and the paradox of digital democracy in China. *Inter-Asia Cultural Studies, 14*(3). pp. 401–416

Yu, F.-L., & Yan, H.-D. (2015). *Handbook of East Asian entrepreneurship*. London: Routledge.

Zhu, S., & Shi, Y. (2010). Shanzhai manufacturing—an alternative innovation phenomenon in China: Its value chain and implications for Chinese science and technology policies. *Journal of Science and Technology Policy in China, 1*(1), 29–49.

16. *Our Community Hacks: Exploring Hive Toronto's Open Infrastructures*

Karen Louise Smith
Brock University

Introduction

In Toronto, Canada, the Hive Toronto community is made up of educators from over 60 youth serving organizations. Hive Toronto members have joined this Mozilla stewarded network, because they seek to create educational opportunities for young people related to digital literacy. Key features of the Hive Toronto network include monthly professional development meetups for educators and catalytic funding for collaborative community projects. The Hive Toronto community also leverages and contributes to Mozilla's open educational resources (OERs) including practices, curriculum and software.[1] Within the Hive Toronto network, *hacking* and *making* are terms which circulate amongst the members—Hack Jams, Maker Parties, webmaking, toy hacking, and eduhacking stand out as some examples of ongoing community engagement with hacker/maker culture.

The presence of hacker/maker culture in the Hive Toronto community is explained in part, by Mozilla's role as the steward of the network. Mozilla is known globally as an advocate for the open web and for the development of open source software products, such as the Firefox browser. Scholars have described the importance of open source software projects as a domain where hackers can safely view the code base, and tinker with it (Coleman, 2004; Kogut & Metiu, 2001). As part of its mandate as an organization committed to the open web, Mozilla actively supports *webmaking* as a form of constructivist and connected learning informed strategy within Hive Toronto and other communities (Belshaw, Smith and the Mozilla Community, 2014; and see also Ito et al., 2009, 2013).

Mozilla's educational work is informed by hacking and making because it sees the necessity of teaching people to read, write and participate on the web as part of its goal to protect the open internet now and in the future (Belshaw et al., 2014). Coleman (2013) examines how hackers have "extend[ed] as well as reformulate[d] key liberal ideals such as access, free speech, transparency, equal opportunity, publicity, and meritocracy" (p. 3), but work remains to be done to consider what hacking/making practices mean in the digital literacy realm with educators at both local and global levels.

The editors of this volume raise the critical question of whether hacking and making have become buzzwords that have been taken up by organizations that are diluting the meaning of the terms. Some Hive Toronto member organizations, such as libraries, museums and after school programs, are amongst the new entrants to the hacker/maker movement (Colegrove, 2013; Dougherty, 2013; Peppler & Bender, 2013). This chapter takes the stance that informal educators are integral participants in hacker/maker culture. For hacker and maker culture to exist and carry on, beginners, novices and newcomers must find ways to join the community. I argue that the open educational resources (OERs) of Hive Toronto—including curriculum, software and practices—need to be interrogated as socio-technical infrastructures, which foster opportunities for people to read, write and participate on the web as a component of hacker/maker culture.

In many instances, Mozilla's open source culture, has influenced Hive Toronto. For example, the logo for a Hive event or learning network is made available under Creative Commons license for easy community reuse.[2] Additionally, educational software tools, such as X-Ray Goggles, Thimble, and Popcorn, have served as a technical infrastructure for Hive Toronto's educators. [3] These resources and software tools are used as a kind of infrastructure for the open web, to teach that the web is hackable and remixable.

Numerous scholars in Science and Technology Studies (STS) have explored infrastructure in relation to digitally mediated life (Bowker, Baker, Millerand, & Ribes, 2010; Bowker & Star, 1999; Edwards, Bowker, Jackson, & Williams, 2009; Knobel & Bowker, 2011; Sandvig, 2013; Star, 1999). Infrastructure is described as "relational," and consists of "the balance of action, tools, and the built …" (Star, 1999, p. 137). Star (1999) also describes that infrastructure is often built upon an installed base and invisible to us, but it can be "learned as part of membership" and achieve "reach or scope" (pp. 381–382; see also Star & Ruhleder, 1996). Sandvig (2013) identifies that "the Internet is in the process of becoming foundational" to many social endeavours (p. 91). In relation to education, Hunsinger (2011) argues that hacklabs draw upon the internet for personal learning. Infrastructures that are shaped by the value of *openness* for hackability, are somewhat unique and distinctive. Knobel and Bowker (2011) identify that often, "infrastructures reveal human values most often through counterproductivity, tension, or failure" (p. 27) instead of reflecting our collective aspirations. In the case of Hive Toronto, the value of openness serves to enhance the visibility of the unique practices and perspectives relevant to hacking/making to emerge over time, and in connection with the open web as a foundational base for education.

Research Involvement

My involvement in the Hive Toronto network commenced with the community's first Hack Jam event in February 2012, where I volunteered. I later became a post-doctoral research fellow with Hive Toronto and the Mozilla Learning Networks team and served in this role from August 2013 to June 2015.[4] My role with the Hive Toronto network was to use participatory action research strategies to build and sustain the network, while simultaneously studying it. I also had close contact with the Mozilla software developers, as well pedagogically-oriented team members. This case study shares reflective insights from an iterative research process surrounding Hive Toronto's hacking and making, which in may ways was scaffolded upon Mozilla's open source culture.

The analysis I share in this chapter was based upon reflective interviews with 25 members of the Hive Toronto community.[5] This chapter references the recollections of the Hive Toronto network members between the 2012–2015 time period. The Hive Toronto timeline (Figure 16.1) provides an overview of some of the major moments in the network's history. The next section of this chapter begins by exploring the beginnings of Hive Toronto in winter of 2012. I provide community members' reflections and my analysis of the activities of the network during the Ontario Trillium Foundation (OTF) grant period (November 2012–2014). Hive Toronto convened youth facing events, during this time period as well as completing two rounds of Collaborative Community Projects (CCPs) and a round of Remixable Open Educational Resource (ROER) creation projects.

Forming Hive Toronto and the Network's First Hack Jam

Hive Learning Networks were first established in New York City and Chicago in 2009 with support from the MacArthur Foundation and other collaborators (Mozilla, n.d.-a). In brief, a Hive Learning Network has "an operational budget and staff [to] commit to promoting innovative, open-source learning

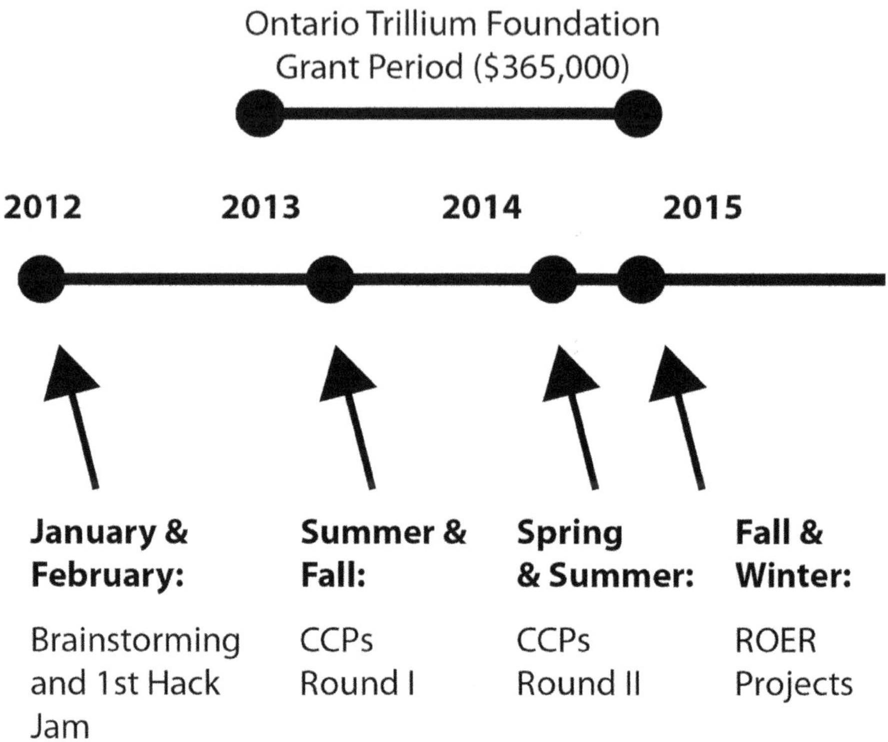

Figure 16.1. Hive Toronto timeline (not to scale).

models in partnership with a communities' civic and cultural organizations, businesses, entrepreneurs, educators and learners" (Mozilla, n.d.-b). Mozilla currently stewards Hive Learning Networks in New York City, Chicago and Toronto.[6] The Hive Toronto network first emerged in 2012, through a series of volunteer powered community events focused on hacking and making, and the network later received funding over a two-year period from the Ontario Trillium Foundation (2012–2014). This section of the chapter describes the initial process to spark hacking and making education in the Greater Toronto Area (GTA) through Hive Toronto.

To encourage collaboration amongst organizations and individuals for Hive Toronto's first event, the community's initial catalyst, Heather Payne, began by convening a conversation. "We decided to start with a brainstorming session, [to] bring together people who [were] interested in talking about how we can create more opportunities for youth in Toronto to make stuff and hack and have interesting outside-of-the-classroom learning experiences related to digital literacy" (Payne, Founder, Ladies Learning Code). Payne describes that the first meeting for the community that would become the Hive Toronto network, was "a speed geek-style event" for people to learn about digital literacy initiatives happening in Toronto and also to learn from guests from Hive New York City (NYC). One of the attendees at the first event, Juan Gonzalez, Founder of Fabspaces, now a Hive member organization, explained that he attended the first meeting, "mostly as a parent. I just wanted to know what this was all about." The first brainstorming and speed geek session culminated in plans for Hive Toronto's first community event—a youth Hack Jam, which was held in the Mozilla community space in February, 2012.

The first Hack Jam convened in Toronto contained six stations for youth. Some of the stations included a hypertext markup language (HTML) station involving a Mozilla software tool called Hackasaurus (now called X-Ray Goggles), game design with the Scratch programming site, and a paper prototyping station to create applications (apps). The registration page for the event described that the Hack Jam was "meant to be a semi-unstructured, collaborative, creativity-inspiring experience for youth" without a "strict agenda."[7] Youth were encouraged to choose which stations to try out at the event. Kathryn Meisner, Hive Toronto's Director, stated that events such as Hack Jams were "really just [the] seeds of Hive Toronto" which were planted to grow later network activity.

Hack Jams, pop-ups and Maker Parties, are each terms that Hive Toronto and the broader Mozilla community has used to signify events or gatherings that can feature station-based learning for digital literacy. Rafi Santo (2012) a researcher engaged with Hive NYC, describes that these events are part of the innovation infrastructure of Hive. Positioning events as innovation infrastructure aligns with Star's (1999) idea that members of a community need time and space to become familiarized with infrastructure as part of their learned membership in a community. At these types of events, organizations can take up the ideas of hacking and making by trying out their curriculum ideas, or providing access to new software tools. The events also provided a template for reuse and remix within organizations. Hannah,* a youth worker involved in Hive Toronto states, "We've hosted Maker parties all throughout the summer. Our participants who've had the training through Hive, they would facilitate the same workshop for younger children."

Hack Jams, Maker Parties and related events made Hive Toronto members familiar and comfortable with the types of hacking and making that would come to be associated with the network. Jennifer Chan, Founder of Exhibit Change, described that she has "… always thought of hacking as […] altering something that already existed whereas making could be altering something that already existed or coming up with an idea and then just putting your idea into physical form." Hive Toronto staff member Meisner, stated that during her directorship of the community, "the word hacking has become more of a commonly used phrase. …" Nivedita Lane, a Hive independent (or indie) member shared how her understanding of hacking has developed within the network:

> I believe my understanding and my own connection to the words hacking and remixing has definitely changed over the last couple of years. I think growing up when you hear the word hack, you think "*That is a really bad thing!*" […] Working with [educational] programming … hack now means something … very exciting, and something that we want to encourage. We want to encourage creativity.

Hive members' understandings of hacking and making are further demonstrated through the learning activities they implemented, adjacent to, or in conjunction with Mozilla's open source software tools for teaching web literacy.

Learning Activities for Making Producers Not Consumers of the Web

One of Mozilla's central interests in bringing hacker and maker culture to informal learning settings, is to encourage young people to become producers, not just consumers of the web. This goal intertwines the objectives of media literacy, the participatory culture of the web, and the ideals of hacker and maker culture. Developing the skills to *create* your our own media, which is distinct from mass media is central in many definitions of media literacy and digital literacy (Buckingham, 2006; Hoechsmann & Poyntz, 2012; Livingstone, 2004) and the web is described by some to have a participatory culture (Jenkins, Purushotma, Weigel, Clinton, & Robison, 2009). Similarly, maker and hacker culture encourage individuals to create their own media or innovations. Dougherty (2013) states that "makers are seeking an alternative to being regarded as consumers, rejecting the idea that you are defined by what you buy" (p. 8). Dery (1993) points to hacking as an inspiration for creative processes where "tales told for mass consumption" can be "reworked" (online). Countering the consumption of mass culture, and encouraging youth to create their own digital media is woven into various hacker and maker learning activities that have been featured at Hive Toronto events, and in some cases documented in OERs, such as curriculum kits.

Various learning activities implemented by Hive Toronto members, serve to demonstrate webmaking and media literacy through the hacking of mass culture to create a new artefact. In a previous whitepaper, a co-author and I described, "'webmakers' to be those participating in the culture of building the social and technical elements required for an open web" (Belshaw et al., 2014, para 15). To operationalize the goal of creating opportunities for people to become producers of the web in educational settings, educators frequently design and implement learning activities that utilize the open web, or open culture more broadly.

Hannah,* a youth worker involved in Hive Toronto, described that her organization made frequent use of Popcorn, after being introduced to it through training. Popcorn was a timeline based web editing tool, which Mozilla provided for users to create video mashups with elements including text, images, video, maps, articles and other elements. Hannah,* described, "we do a lot of workshops on […] media literacy, so the youth incorporate a lot of Popcorn into it. …" The youth in the workshops have created video mashups on topics like body image, which they use to "show what they've learned about the topic after the workshop that they participated in." Hannah* also explained how her exposure to open source led to youth opportunities:

> … it was just something that was new to me, open source technology. Most of what I've been using didn't have that kind of open sharing and that kind of culture of making […] I knew I had to go back and introduce it to my youth and it was such a good tool for them, as well. (Hannah, Youth Worker)*

Hannah* felt that open source assisted her to instil and understanding "that community of open and sharing and being able to create and not just be a consumer of the media."

Ashley Jane Lewis, a youth technology educator, who has presented at numerous Hive events, explained a similar experience. Lewis often has youth hack a movie or celebrity webpage to teach webmaking skills. She explained that it is possible to have youth install Mozilla X-Ray Goggles, a browser plug-in that allows a user to inspect and alter the code of a webpage. Using X-Ray Goggles a youth can for example, insert a cartoon character image where there was not one originally on a webpage. This makes the editable structure of the HTML, which is comprised of "tiny code-able items[,]" understandable for youth.

Another learning activity introduced was toy hacking. The registration materials for the June 2012 Hack Jam describe toy hacking as an opportunity to take apart a toy and remake it with additional parts.[8] At the event, toy hacking was taught at a station physically adjacent to webmaking tables. Susan Fohr, Education Programs Coordinator at the Textile Museum of Canada, recalled collaborating with Andy Forest, now of STEAMLabs, on the toy hacking station. Fohr stated, that this type of activity enables "traditional craft, and then [...] more [...] technical components [... to be] morphed." Andy Forest of STEAMLabs recalled that the deconstruction and reconstruction of the toys was a significant opportunity for youth to become makers and hackers. He described, "the first step after they've chosen the toy is to take it apart, and very frequently [... you hear as a facilitator], '*What, what do you mean? Are you sure? Am I allowed to do that?*'" As the toy hacking facilitator, Forest would reply *"'Yes, this is your toy. You can take this apart and turn it into something new [...] this belongs to you and the innards of it belong to you, too.'"*

Many Hive Toronto members have identified that there are negative connotations, with the term hacking. Forest's description of youths' apprehension to open up a toy, and to examine its innards, reveals much about the social fears of hacking. Kathryn Barrett Youth Programming Lead from Ladies Learning Code stated, "I have asked kids what they think hacking is [...] and they think breaking into something [... but] it can be flipped and hacking [...] can be used in a much more positive way."

Hive Toronto members have attempted to create and facilitate learning activities that encourage youth to make and hack creatively. Hive Learning Networks are experimenting with the idea that building this kind of understanding amongst youth will build a culture where the open web is valued and protected. The facilitation of these types of activities have also raised considerations relevant for equity and inclusion, and the popularity of the activities with youth have encouraged facilitators to share their curriculum and practices.

Equity and Inclusion in Hacking/Making

When convening events such as Hack Jams or Maker Parties within the Hive Toronto community, the practices surrounding equity and inclusion were considered carefully. Hive Toronto members include organizations with diverse goals ranging from increasing access to technology and media production facilities for youth, to fostering youth empowerment and addressing the needs of marginalized communities. Ensuring youth from diverse cultures, races, genders, abilities, sexual orientations and socio-economic backgrounds benefit from access to technology is of interest to many Hive members.

Despite this interest, equal access to opportunities to make, hack and otherwise create and express oneself cannot be assumed in Toronto. The Canadian Internet Registration Authority's (CIRA's) (2014) *Factbook* states, that 38% of individuals in the lowest income quartile in Canada, do not have internet access at home. MediaSmarts' (2015) survey research found that "roughly one third of students have posted their own artwork, stories or videos, posted comments on news sites or supported an activist group online […]" in Canada (p. 4). Unequal access to internet access and devices at home was an ongoing issue faced by Hive Toronto in convening youth facing events for the network. Hannah* stated that in her organization, "some participants don't own computers at all and they don't really use computers a lot unless they're in school, if they have free time in the library. So, still those technical skills are lacking." Hive Toronto events were often advertised as BYOL (bring your own laptop), but with awareness that loaner laptops needed to be provided for youth who could not bring their own equipment. Transit fares were also sometimes provided to member organizations to defray the costs for their participating youth.

The Hack Jams and maker events that Hive Toronto ran or participated in were largely held in downtown Toronto and deemed to be important but not sufficient by Neil Price, former Senior Manager at the Boys and Girls Clubs of Canada. In describing youth participation at events, "these were experiences that I thought would […] engage […] the youth who step forward. And I think [they] did […]" but it was also noted as significant to hold making events in locations, such as neighbourhood clubs or locations that youth visit regularly. Chris Penrose, Executive Director from the organization Success Beyond Limits, noted that "geography is a big barrier" to addressing the digital divide. His organization serves youth in a high school over 20 kilometres from downtown Toronto and due to issues like budget and staffing constraints "… people don't […] seem to be very interested in bringing the equipment and their expertise out to us."

The implications of the digital divide in the Greater Toronto Area extended beyond device access and the availability of expert facilitators. Hive Toronto members were also interested in who creates the infrastructure and content of the web. A staff member from Sky's The Limit stated, "the whole world is consuming the Internet, but the creators are very small elite group of […] usually straight white men, who are creating this content for almost the whole world […]" (Adele*). Ensuring that opportunities to create content and to resist dominant media narratives were raised as an important ideal by various Hive Toronto members.

One Hive organization that has been very successful in creating infrastructure and community is Regent Park Focus (RPF) Youth Media Arts Centre. Kerry Ambrose, from RPF described that a participatory media influenced ethos shapes her organization and intersects with Hive Toronto:

> Regent Park Focus was established because of mainstream media stigma around Regent Park […] as a low income Toronto Community Housing community […]. Regent Park Focus grew out of that so that young people were able to self

> represent and have the opportunity to share their learnings and talk about issues of relevance to them and to the local area.

Today, RPF is a community youth media arts centre, which broadcasts via a neighbourhood-based closed circuit television (CCTV) channel, and through an internet radio stream.

Youth from RPF have been involved in Hive Toronto in various ways. In the summer of 2013, RPF sent a youth crew out to tell the story of a Maker Party. The youth crew were responsible "to film the [Maker] Party. It was [...] young people that got to witness other young people [...] learning and our young people then got to practice [...] the training that they learn around interviewing and filming and videography [...]" and to share that with the Regent Park community via the CCTV channel and online (Ambrose). RPF also carried out a Hive Toronto Collaborative Community Project (CCP), called RadioZilla in conjunction with Facing History and Ourselves. The RadioZilla project provided youth with the opportunity to create their own social justice-oriented radio content, to be broadcast on RPF's Radio Regent stream. Digital remixes were also produced using Mozilla webmaking tools. Ambrose explains the importance of participatory media in the Regent Park community is that "it flips the script and young people are able to create their own messages, their own media, their own resources."

It remains an ongoing challenge to provide widespread opportunities for youth from communities that are stigmatized in the mass media, to "flip the script" and share their perspectives. Star (1999) described that "infrastructure has reach beyond a single event or one-site practice" (p. 381). She also describes the significance of considering who is "*not* served by a particular infrastructure" (p. 380). The open source webmaking software tools used in the RadioZilla project exemplify how this kind of reach can be achieved but further practices need to be developed to foster wider access to this type of opportunity.

X-Ray Goggles, the Mozilla browser plug-in that was described as a tool to hack websites in the previous section of this chapter, was seen as a powerful possibility to extend the reach of participatory media by one Hive Toronto member. Chris Penrose of Success Beyond Limits (SBL) works with youth in Toronto's Jane and Finch, or Blackcreek neighbourhood, which has been described in the press as the least liveable place in the city and the most problem plagued community (see for example Doolitle, 2014). After seeing the SBL neighbourhood depicted negatively in the press many times, Penrose felt that it was important to empower youth to share counter narratives and alternative perspectives. Penrose stated:

> It's from the perspective of young people and we saw this tool [X-Ray Goggles] as a really important way to not just move from being content consumers to

> content creators, which is part of the mandate around the relationship of young people to Hive, but to add that accessibility that equity lens to it that when your narrative is not just missing, but [... the media narrative is harmful towards you] it becomes even more important to have some kind of outlet and control over that narrative.

Penrose presented the idea that he believes it is significant for youth from stigmatized communities to be able to talk back to the news articles, which depict their neighbourhoods. He expressed that youth need opportunities to express their own narratives to resist the mass media ones.

In addition to creating opportunities for youth to create media, there are also practices that educators can engage in, which may promote change. Practices such as collaboration amongst educators in the informal learning sector, may foster expanded opportunities for youth. Examples of these types of practices within Hive Toronto will be discussed next in the eduhacking section.

Eduhacking Through the Hive Toronto Network

Not all reflections about hacking and making practices in Hive Toronto are focused on youth as the recipients or end-users of educational programming. During an interview, an indie Hive Toronto member, described that she thought the "the [Hive Toronto] network *itself* is a great accomplishment," which was achieved by hacking/making (Lane, emphasis mine). She focused on what informal educators and the adults created:

> I remember attending a meeting and hearing we had so many applications and that so many new organizations were coming on board and I think that is an accomplishment, in and of itself—bringing together organizations that have been working in silos and sometimes aren't even aware that another organization is even doing the same work as them. [These organizations ...] are coming together, and coming together under a common cause, with a common idea around digital literacy and digital learning.

The endeavour of creating Hive Toronto, may be conceptualized as an eduhacking endeavour, a multi-year project to expand the kinds of digital literacy educational opportunities available for youth in Toronto and the surrounding area. This type of collaborative work, is also what Bowker (1994) calls an "infrastructural inversion," where there is a "foregrounding [of] the truly backstage elements of work practice" (Star, 1999, p. 5).

Cohen and Scheinfeldt's (2013) edited collection on *Hacking the Academy*, provides some helpful insights about the broader context of hacking and education that is relevant to Hive Toronto and infrastructural inversion. The *Hacking the Academy* book project launched publicly in 2010 when the editors of this collection posted provocative questions about reforming higher education using technology, and gave their peers just 7 days to respond. The project was premised on the basis that "every aspect of scholarly infrastructure is being questioned, and even more importantly, being <em>hacked</em>." (Hacking the Academy, n.d.). The editors of this project leveraged the participatory aspects web, including crowdsourcing and social media channels, to write a book, which was their collective shared purpose.

In a similar fashion, Hive Toronto collaboration leverages the open and participatory web as a foundation upon which to build digital literacy opportunities for youth. Collaboration flows between on and offline channels and Hive Toronto members frequently describe their understanding of hacking to involve the co-creation of solutions. For example, Ioana* states, "We can actually mix two things together [...] and come across some solutions [...]" (Hive Member). Joe Wilson, Senior Strategist in the Education Cluster at the MaRS innovation hub, described hackathons as one way to elicit this kind of collaborative problem solving where, "participants get together and quickly prototype solutions to problems." Hive Toronto has utilized hackathons and related event formats to foster collaboration amongst its network members. The socio-technical infrastructures which are most readily utilized by Hive Toronto, can in many cases be described as OERs. Emphasizing the creation and usage of OERs requires working openly and collaboratively in new ways, which can be foregrounded through reflection and description.

For Quinn,* a Hive Toronto member and librarian, the culture of experimentation in hacker and maker culture allowed her to work in new ways with Hive Toronto:

> There is a sense of [...] nothing is a failure. There is a sense of, let's give it a try and see how it works. There is a really open environment for brainstorming. Use various tools and see if they work, and I think everyone feels free to share their input. I think the biggest thing for me – and what I get out of this whole hacker movement, maker movement [...] is that there's no such thing as a failure

Other descriptions of how Hive Toronto worked to create and leverage OER infrastructures will be described here in terms of getting to know the network and co-creating Remixable OERs (ROERs).

Getting to Know the Network

The bringing together of "organizations that have been working in silos" was previously described as significant by Hive member Lane. A first step in this endeavour was building relationships and rapport. In some interviews with Hive members, individuals commented on the introductions that were made as leading to open ended possibilities for collaboration. Hive staff would frequently introduce people to "each other [...] just to see where conversations would lead" (Wilson). Hive member and librarian Tracey* also commented positively on "connecting with different [...] teaching organizations across the GTA, to hear about what they're doing and seeing if there's any room for partnership opportunities or new program delivery implementation."

Specialized events for Hive member organizations were also designed to encourage collaboration over a longer time period. As one example, Hive Toronto utilized an event format called the Hivestarter, which was adapted from Hive NYC. Meisner, Director of Hive Toronto, described Hivestarter in the following terms, "it was bringing together Hive members to surface resources and needs that were present in the network, as well as [...] focus on collaboration opportunities for the collaborative community project grants that we were about to send out a call for proposals" to fund projects in the Hive Toronto network. Hive members recalled the Hivestarter event type in various ways. One Hive member shared, "we got to know the other members of the Hive who were interested in the possibility [of working together ...] I came with an idea of a project [...] then, then we, then we did sort of the mash-up idea [...] '*What do you have? What do you need?*' That was sort of the framework." (Danielle, Executive Director of a non-profit)*. Ambrose from Regent Park Focus recalled that "[y]ou got to put ... your company [business] card" on large sheets of chart paper to indicate where you could get involved. In another iteration of the event, participants created a collaborative online document. Hannah,* a youth worker and participant from the sessions described that "visually we were able to map out what every agency needed and how we can support one another and we created a spreadsheet to include and update."

The importance of introductions and these types of events was articulated by Iona.* These processes allow you to realize, other organizations, "they're not just your competitors [...]. You [...] start to see how they go through their processes, how they think, what they have to offer, and so [...] the ambiance [of Hive] is really is of learning, of helping each other. No one's, [...] hiding away their stuff and, like, keeping their resources secret" (Iona, Hive member).* In terms of resources, Remixable Open Educational Resources (ROERs) were a special category of educational materials that members came to value through the network.

ROERs

Hive members were encouraged to develop a culture of sharing and remix, particularly around curriculum resources. Lucy Harris, a Mozilla staff member who engaged in Hive events states, "people put a lot of time and energy into making [...] really elaborate teaching kits that were easy to follow and showed you how to do cool things or maybe through different activities." In many instances, Hive Toronto members have documented the learning activities and practices already shared in this chapter. For example, the toy hacking activity was documented as part of a Collaborative Community Project (CCP) and resources for facilitators wishing to replicate or remix the programming are available.[9] Meisner, Director of Hive Toronto explained that Hive is "really encouraging that factor of sharing [...] but then also remixing what is made available through the network" through remixable open educational resources (ROERs).

In the spring of 2014, Hive Toronto members were introduced to the Mozilla Thimble tool for editing templates of teaching kits and teaching activities, using hypertext markup language (HTML).[10] Hannah,* one of the participants at the teaching kit session shared:

> I was at the teaching kits [session] as well and I brought two [young adult] participants. They were very excited. They had this idea of combining some of their interests in that teaching kit. So, they had an idea of having youth who are interested in entrepreneurship present their ideas and it was kind of similar to a Dragons' Den kind of idea and they would market and promote their business plan. So, they created that in their Thimble project which looked really amazing.

Hive indie member Lane described that she brought the idea of constructing teaching kits in Thimble back to the youth serving organization she was with at the time. She stated, "I definitely demoed it [Thimble] to my staff [...] at our own organization. We were looking at ways to incorporate online lesson planning and take a lot of our paper-based activities and make them remixable." Simona Ramkisson, a Hive member who became a Mozilla staff member and the interim manager of Hive Toronto described, that remixable teaching kits enable organizations to "do something [..] with a very small budget" it enables organizations to "take really good material, very good content, and then adapt it to the needs of their youth, of their community [...]."

Hive Toronto featured a round of grants in 2014 which were focused on organizations producing teaching kits using the Mozilla Thimble software, which were remixable by other organizations.

As one example of the ROER process, Wilson from MaRS described that with his collaborators "we got a grant, a Remixable Open Education grant to put together the playbook" for how to throw a hackathon for educators. He further explained, "the playbook is designed to be a guide for teachers or facilitators who want to put together a hackathon or an idea jam at their own institutions, but are not sure what the process or the structure looks like."

Meisner described that from one of the ROER grants, the process facilitated the documentation of successful organizational practices. She described that sometimes great educator practices "live [...] in the heads of these individuals. [...] And so we saw an opportunity to provide small funds, small amounts, to have organizations document these things." The need for a network which encourages sharing in the informal learning sector was also articulated by Quinn*, a librarian involved with Hive, "I think for libraries in general, it's a great way to almost force more openness [...]" The librarian identified there is sometimes a problem that "we're all duplicating the same work ... and sometimes copyrighting the same work" and not make resources available and reusable for others (Quinn).* The impetus to work openly and share their work, was a major outcome of participation for Hive Toronto members and this idea will be expanded upon in the conclusion.

Conclusion

This chapter argued that the experiences of informal educators, such as youth workers, librarians, and non-profit staff, need to be included when assessing the broad, social implications of hacking and making. Socio-technical OER infrastructures were presented as domains where informal educators and their learners may choose to hack and make. Exploring the experiences of informal educators in hacking and making activities assists us to see how newcomers may orient themselves to the hacking and making movements. This chapter presented an account of the initial inspirations for Hive Toronto, some hacker/maker learning activities facilitated by network members, equity and inclusion considerations, and it also shared Hive Toronto's collaborative endeavours in eduhacking. This array of examples provided rich descriptions concerning how youth, as well as adult novices, may be introduced to hacking/making.

This chapter however, was not only about introducing beginners to hacking and making. Mozilla's role as the organizational steward for Hive Toronto, brings to the narrative a longer history of open source culture, and its contributions to creating a safe sandbox where hackers can tinker, as well as build the infrastructures to support the collaborative work needed to protect the open web. In addition to creating software, Mozilla also facilitates opportunities for learners and educators to align with the open web through OERs. The examples described in this chapter contribute to building an understanding of how OERs—including practices, curriculum and software—function as socio-technical infrastructures to encourage greater openness for both learners and educators through the web. Understanding how individuals gain interest and affinity for the open web through hacking/making is integral, if we wish to move away from paywalls, silos, and walled gardens online.

Hive Toronto members' use and creation of OERs demonstrates the relational aspects of infrastructure. Including learners and educators in hacker/maker culture demonstrates how the open web is a massive social project, with multiple participant groups, who shape our communication and learning infrastructures together. This chapter suggests optimistically, that further layers of open infrastructures for digital literacy can be built upon what Hive Toronto has begun to build locally. Although this chapter, shares Toronto-based examples of hacking and making, the endeavour to leverage the open internet to enhance education is truly a global project. The story of educators participating in hacker/maker culture to enhance the programming they offer to youth, their professional practices, and their community networks, is a pattern that can be replicated and remixed in other communities around the world.

Notes

1. The William and Flora Hewlett Foundation define OERs to "include full courses, course materials, modules, textbooks, streaming videos, tests, software, and any other tools, materials, or techniques used to support access to knowledge." This definition informs this chapter. See <http://www.hewlett.org/programs/education/open-educational-resources> (last accessed 14 July 2015).
2. Template logos were obtained from <https://hivelearningnetworks.org/resources/> (last accessed 14 July 2015).
3. Through most of the 2013–2015 period this research covers, Mozilla maintained a website called Webmaker, where 4 tools for webmaking (Popcorn, X-Ray Goggles, Thimble and Appmaker) were housed. Support for Popcorn and Appmaker by Mozilla is being phased out and X-Ray Goggles and Thimble are transitioning to <https://teach.mozilla.org/>. This is discussed in a Mozilla blog post <https://blog.webmaker.org/whats-next-for-webmaker-tools> (last accessed 14 July 2015).
4. The author completed a Mitacs Elevate post-doctoral fellowship and both Mitacs and Mozilla provided funding for the position. The opinions expressed in this chapter are not intended to be representative of the funders.
5. The Hive Toronto members who are quoted or described in this chapter are either attributed fully with their names and organizations, or anonymized, depending on individual preferences. All participants who have an anonymized personal and/or organizational name are denoted with an asterisk in the text of this chapter. Roles, titles and organizational affiliations of the participants in this chapter were current at the time of the interview or through contact made with the researcher on dates between 2013–2015. Roles, titles and affiliations are expected to change over time. The author thanks all participants for their contributions.

6. The Hive Toronto Learning Network is described at <www.hivetoronto.org>. For an in-depth explanation of Hive Learning Networks see <https://hivelearningnetworks.org/about/> (last accessed 14 July 2015).
7. The Hack Jam registration page is available at <http://www.eventbrite.ca/e/hive-toronto-youth-hack-jam-changing-the-world-through-technology-tickets-2813740975> (last accessed 14 July 2015)
8. The quotation describing the toy hacking station was retrieved from Eventbrite <http://www.eventbrite.ca/e/its-a-hack-jam-tickets-3616008579> (last accessed 14 July 2015).
9. The Making Makers curriculum including a toy hacking module is available from <http://ladieslearningcode.com/making-makers/> (last accessed 14 July 2015).
10. Thank you to Michelle Gay, who facilitated the teaching kit session for Hive Toronto. Her blog post on the event is at <http://hivetoronto.org/kickstarting-hive-toronto-teaching-kits/> (last accessed 14 July 2015).

References

Belshaw, D., & Smith, K. L. and the Mozilla Community. (2014). Why Mozilla cares about web literacy. Retrieved July 14, 2015 from https://mozilla.github.io/webmaker-whitepaper/> accessed July 14, 2015.

Bowker, G. C., Baker, K., Millerand, F., & Ribes, D. (2010). Toward information infrastructure studies: Ways of knowing in a networked environment. In J. Hunsinger, M. Allen, & L. Kastrup (Eds.), *International handbook of internet research* (pp. 97–117). New York, NY: Springer.

Bowker, G. C. (1994). *Science on the run: Information management and industrial geophysics at Schlumberger*, 1920–1940. Cambridge: MIT press.

Bowker, G. C., & Star, S. L. (1999). *Sorting things out: Classification and its consequences.* Cambridge, MA: MIT Press.

Buckingham, D. (2006). Defining digital literacy-what do young people need to know about digital media? *Nordic Journal of Digital Literacy*, (04). Retrieved July 14, 2015 from http://www.idunn.no/dk/2006/04/defining_digital_literacy_-_what_do_young_people_need_to_know_about_digital

CIRA. (2014). *CIRA Factbook.* Retrieved July 14, 2015 from http://cira.ca/factbook/2014/the-canadian-internet.html

Cohen, D. J., & Scheinfeldt, T. (2013). *Hacking the academy: New approaches to scholarship and teaching from digital humanities.* Ann Arbor, MI: University of Michigan Press.

Colegrove, P. (2013). Editorial board thoughts: Libraries as makerspace? *Information Technology and Libraries, 32*(1), 2–5.

Coleman, E. G. (2004). The political agnosticism of free and open source software and the inadvertent politics of contrast. *Anthropological Quarterly, 77*(3), 507–519.

Coleman, E. G. (2013). *Coding freedom: The ethics and aesthetics of hacking.* Princeton, NJ: Princeton University Press.

Dery, M. (1993). Culture Jamming: Hacking, slashing, and sniping in the empire of signs. Originally published by Open Magazine. Retrieved July 14, 2015 from http://markdery.com/?page_id=154

Doolitle, R. (2014, March 11). Losers and gainers in Toronto's new priority neighbourhoods list. *Toronto Star.* Retrieved July 14, 2015 from http://www.thestar.com/news/city_hall/2014/03/11/losers_and_gainers_in_torontos_new_priority_neighbourhoods_list.html

Dougherty, D. (2013). The maker mindset. In M. Honey & D. E. Kanter (Eds.), *Design, make, play: Growing the next generation of STEM innovators* (pp. 7–11). New York, NY: Routledge.

Edwards, P. N., Bowker, G. C., Jackson, S. J., & Williams, R. (2009). Introduction: An agenda for infrastructure studies. *Journal of the Association for Information Systems, 10*(5), 6.

Hacking the Academy. (n.d.). A book created collaboratively in one week: May 21–28 2014. Retrieved July 14, 2015 from http://hackingtheacademy.org/one-week-one-book-hacking-the-academy/

Hoechsmann, M., & Poyntz, S. R. (2012). *Media literacies: A critical introduction.* West Sussex: John Wiley & Sons.

Hunsinger, J. (2011). The social workshop as a PLE: Lessons from Hacklabs. *Proceedings of the PLE Conference,* 10–12 July, Southampton, UK.

Ito, M., Baumer, S., Bittanti, M., boyd, d., Cody, R., Stephenson, B. H., … Tripp, L. (2009). *Hanging out, messing around, and geeking out: Kids living and learning with new media.* Cambridge, MA: MIT Press.

Ito, M., Gutierrez, K., Livingstone, S., Penuel, B., Rhodes, J., Salen, K., … Watkins, C. (2013). *Connected learning: An agenda for research and design.* Digital Media and Learning Research Hub. Retrieved July 14, 2015 from http://dmlhub.net/wp-content/uploads/files/Connected_Learning_report.pdf

Jenkins, H., Purushotma, R., Weigel, M., Clinton, K., & Robison, A. J. (2009). *Confronting the challenges of participatory culture: Media education for the 21st century.* Cambridge, MA: MIT Press.

Knobel, C., & Bowker, G. C. (2011). Values in design. *Communications of the ACM, 54*(7), 26–28.

Kogut, B., & Metiu, A. (2001). Open-source software development and distributed innovation. *Oxford Review of Economic Policy, 17*(2), 248–264.

Livingstone, S. (2004). Media literacy and the challenge of new information and communication technologies. *The Communication Review, 7*(1), 3–14.

MediaSmarts. (2015). *Young Canadians in a wired world: Phase III: Trends and recommendations.* Retrieved July 14, 2015 from http://mediasmarts.ca/sites/mediasmarts/files/publication-report/full/ycwwiii_trends_recommendations_fullreport.pdf

Mozilla. (n.d.-a). Locations. Retrieved July 14, 2015 from https://hivelearningnetworks.org/locations/

Mozilla. (n.d.-b). About hive learning networks. Retrieved July 14, 2015 from https://hivelearningnetworks.org/about/

Peppler, K., & Bender, S. (2013). Maker movement spreads innovation one project at a time. *Phi Delta Kappan, 95*(3), 22–27.

Sandvig, C. (2013). The internet as an infrastructure. In W. Dutton (Ed.), *The Oxford handbook of internet studies* (pp. 86–108). Oxford: Oxford University Press.

Santo, R. (2012). Both R&D and retail: Hive NYC as infrastructure for learning innovations. New York, NY: Hive Research Lab. Retrieved July 14, 2015 from <https://empathetics.files.wordpress.com/2013/03/both-rd-and-retail-hive-nyc-as-infrastructure-for-learning-innovation-santo-2012-draft.pdf

Star, S. L. (1999). The ethnography of infrastructure. *American Behavioral Scientist, 43*(3), 377–391.

Star, S. L., & Ruhleder, K. (1996). Steps toward an ecology of infrastructure: Design and access for large information spaces. *Information Systems Research, 7*(1), 111–134.

Afterword: Hackers and Makers Are Ordinary

Andrew R. Schrock
Chapman University

Raymond Williams wrote "Culture is Ordinary" in 1958, a sensitive analysis of everyday life that was influential on the then-nascent field of cultural studies, among others. In this chapter, instead of culture, I draw on Williams' notion of "ordinary." Ordinariness—the everyday, unexceptional, and mundane—is a useful hermeneutic to view collective action and identity in the context of hacking and making. This "ordinary" framing is a response to two types of presumed exceptionality. Critical scholars praise *activism* in hacking (Maxigas, 2012), while others look to hacking and making as the key to *economic* profitability (Anderson, 2012). Missing from these discussions is a more grounded reading of how those identifying as hackers and makers come to understand shared histories, organizing, and politics—the very themes of this book.

Hacking and making are each an odd bundling of concepts that can appear contradictory. For example, Etsy provides a platform for makers to sell crafts and obtain mutual support, even as they push them into more precarious labor (Close, 2014). Our broader argument in this book is that, while contradictions sensitize us to key problematics, we should start by unpacking the cultural logics and material arrangements that drive them. For example, hackers often tout that "anyone can be a hacker." While this claim is dubious—participation is limited by technical inclinations, skills, and exclusionary practices—the sites and phenomena in this book have certainly help produce an idea of the "ordinary hacker." We can observe hacking and making's movements from subculture to mainstream, and from edgy to popular identities. In closing this book, I argue it is fruitful to return to Williams' notion of "ordinary" during this moment of popularization. I hope this serves as a necessary counterpoint to the mythologies of hacking and making that still dominate.

In the 1950s Williams was responding to Marxism's insistence on power residing primarily in production, and a frustratingly prescriptive notion of critique among its adherents. Through a rich personal history, he narrated how culture arose from action in agrarian society, which was often simply dismissed as uncultured. Williams inverted the negative connotations of "ordinary" to make everyday lived experiences of the working class noteworthy. In the past, "ordinary" referred to a legalistic definition of "persons able to act in their own right," only later migrating to a more general sense of "the expected, the regular, the customary" (Williams, 1985, p. 225). The tea house snobs he derided took culture to refer simply to "high art." This distinction rankled him because it ignored the pastiche of meaning – making that constituted everyday life. He wondered, why should we "call certain things culture and then separate them, as with a park wall, from ordinary people and ordinary work?"

A similarly constructed wall now exists between the exceptional and ordinary hacker/maker. Journalistic and academic writing defines hackers as resistant geeks capable of bringing about changes in governments and corporations. The hacker is cast as the powerful underdog, plotted against by the powers that be. Their anti-hero status is even more salient when viewing the trope of the hacker in popular media. We cheer for these exceptional hackers, as well we should. But by many readings, hacking's subversive nature is assumed to be diluted by its mainstreaming, sometimes by making. For example, Evgeny Morozov (January 13, 2014) interpreted makers as no more than commodified hackers lacking a radical edge. With a typical shrug and sigh, Morozov believes they were been duped by a fraudulent ideology. Brett Scott echoes this perspective when he writes about how the "wild and anarchic" hacker ethos was gentrified by "yuppies" (2015). Hacking was a synonym for resistance, and popularization signaled its dilution, if not outright downfall.

Lambasting people for failing to live up to revolutionary promises was what Raymond Williams fretted drew attention away from lived experience. "To try to jump the future," he wrote, "to pretend that in some way you are the future, is strictly insane." His grounded perspective suggested there are limits to seeking out an essence to hacking/making, or reading them as ideal digital citizens or producers. It may not be a particularly popular defense at the moment (over-loaded as the term has become) but I do not believe hackers/makers are on a downward spiral. Rather, their popularization worldwide reveals new questions as they seep into new identities and political subjectivities grounded in particular contexts (Coleman, 2017). For example, this volume has explored unlikely collaborations between hackers and bureaucrats, such as "civic hackers" in Los Angeles and the Chaos Computer Club (CCC) in Germany. In each case, civil society groups worked with government, even serving as proto-institutions (Lawrence, Hardy, & Phillips, 2002) for technology design and interpretation (respectively). The way power resides and is expressed is never a simple story, and our shared future might depend on such unlikely collaborations.

What we might be seeing in expectations of exceptionality is a kind of generational disappointment—a downfall narrative of hacking and making. This too is hardly new. All evidence to the contrary, many still take Steven Levy's *Hackers* to be a guide to an innate identity from the 1980s that is still cast in stone today. Even in the 1990s "golden era" hackers from academia and information security experts alike looked to writer like Levy when they were frustrated at the emergent of "script kiddies" who they saw lacking in technical prowess (Thomas, 2002). People identifying as makers are similarly attracted to origin myths when they look to the past as evidence that we have lost what we once had: an authenticity grounded in hands-on work and tinkering. Given that both hacking and making have been so rooted in white masculine models of prowess and knowledge, worrying that hacking might be "gentrified," as Scott did, brings up a further irony. I believe that hacking's expansion beyond circles of tech-savvy, affluent white males—and making's increasing recognition—demonstrates attention to diversity, not homogeny. Seeking out how histories re-emerge is one thing, while returning to apocryphal origins that deny the progressive goals that hackers and makers aspire to is quite another. For this reason, it is dangerous to seek an essence of hacking/making. I feel that the more productive perspective is hacker/maker pluralism, with distinct spaces, politics, and histories.

Ordinariness is not meant to revive historical divisions between cultural studies and European critical traditions, or downplay questions of power. Instead, it is meant to be a sensitizing concept to explore how identities are constructed, and the socio-technical situations that sustain them over time. Hackers and makers are shaped by society, not born. In *Coding Freedom*, Gabriella Coleman described how an open-source hackers' identity emerged from a fervent brew of digital connectivity, technological concepts, and shared work (Coleman, 2012). Political awareness was only connected to liberalism through shared work with open-source and code over time. Neither is my suggesting that hackers are "ordinary" meant to discard a concern with exceptional hackers/makers. We should be concerned with the Chelsea Mannings and Limor Frieds of the world, and the causes they champion. We should not, however, confuse myth for experience, or simplify complex stories to have a more appealing narrative.

Hackers and makers are ordinary: this is where we must end. Encountering her as human opens new avenues of inquiry, rather than regarding her with disappointment or as a folk hero for the information age. Taking the hacker off the museum wall may be controversial since the term is synonymous with resistance. By contrast, the maker has always been presumed to be ordinary and, as a result, often invisible. However, a return to ordinariness is hardly unusual (Gregg, 2007). Goffman paid attention to common rituals, while Malinowski famously advocated for anthropologists to consider the "imponderabilia of everyday life." Daniel Miller and Sophie Woodward (2012) used the term to refer to how people interpret their experiences with materialities. An explicit framing of ordinary—approachable, common, and attainable—enables us consider how collectives and identities are forged among everyday people in unremarkable spaces and material contexts, not simply playgrounds for the already technically-literate and affluent. Neither are everyday interactions, feelings and events foreign to studies of hackers and makers. Among others, Paul Taylor (1999) and Gabriella Coleman (2012) wrote sensitive ethnographies of hackers that unearthed their lifeworld.

What might Raymond Williams make of the ordinary hackers and makers in this book? I hope he would find its empirical grounding appealing, as he advocated for "a new equation to fit the observable facts" during a time when what was considered culture was expanding. Similar to Williams, the *patterning* of a term—how it evolved over time and was interpreted—helps us understand its social and economic impacts (Williams, 1985). Despite the boldness of our book title, ordinary hackers and makers are as much produced by collective "worlds" (Becker, 2008) as they are in control of exotic technologies that change society at large. Williams might also be intrigued to see the continued importance of hands-on work, reciprocity, and leftist politics. For their part, hackers and makers should find much in common with Williams' progressive embrace of technology paired with disdain for formalized education. After all, Williams wasn't wistful for pre-industrial society, which he saw as increasing quality of life. It gave the, "gift of power that is everything to men [sic] who have worked with their hands." Viewing hackers and makers from afar, how they might give back is only beginning to be revealed.

References

Anderson, C. (2012). *Makers: The new industrial revolution*. New York, NY: Crown Business.

Becker, H. S. (2008). *Art worlds* (25th Anniverary Edition ed.). Berkeley, CA: University of California Press.

Close, S. (2014). Crafting an ideal working world in the contemporary United States. *Anthropology Now, 6*(3), 68–79.

Coleman, G. (2012). *Coding freedom: The ethics and aesthetics of hacking*. Princeton, NJ: Princeton University Press.

Coleman, G. (2017). From internet farming to weapons of the geek. *Current Anthropology, 58*(S15), S91–S102. DOI:10.1086/688697.

Golumbia, D. (2013). Cyberlibertarians' digital deletion of the left. Retrieved from https://www.jacobinmag.com/2013/12/cyberlibertarians-digital-deletion-of-the-left/

Gregg, M. (2007). The importance of being ordinary. *International Journal of Cultural Studies, 10*(1), 95–104. DOI:10.1177/1367877907073904.

Lawrence, T. B., Hardy, C., & Phillips, N. (2002). Institutional effects of interorganizational collaboration: The emergence of proto-institutions. *Academy of Management Journal, 45*(1), 281–290. DOI:10.2307/3069297

Levy, S. (1984). *Hackers: Heroes of the computer revolution*. Garden City, NY: Anchor Press/Doubleday.

Maxigas. (2012). Hacklabs and hackerspaces – Tracing two genealogies. *Journal of Peer Production, 2*, 1–10.

Miller, D., & Woodward, S. (2012). *Blue jeans*. Berkeley, CA: University of California Press.

Morozov, E. (2014, January 13). Making it: Pick up a spot welder and join the revolution. *The New Yorker*.

Schrock, A. (2014). "Education in Disguise": Culture of a hacker and maker space. *Interactions, 10*(2), 1–25.

Scott, B. (2015, August 15). The hacker hacked. *Aeon*.

Taylor, P. A. (1999). *Hackers: Crime in the digital sublime*. London; New York, NY: Routledge.

Thomas, D. (2002). *Hacker culture*. Minneapolis, MN: University of Minnesota Press.

Williams, R. (1985). *Keywords: A vocabulary of culture and society*. New York, NY: Oxford University Press.

Digital Formations

General Editor: **Steve Jones**

Digital Formations is the best source for critical, well-written books about digital technologies and modern life. Books in the series break new ground by emphasizing multiple methodological and theoretical approaches to deeply probe the formation and reformation of lived experience as it is refracted through digital interaction. Each volume in **Digital Formations** pushes forward our understanding of the intersections, and corresponding implications, between digital technologies and everyday life. The series examines broad issues in realms such as digital culture, electronic commerce, law, politics and governance, gender, the Internet, race, art, health and medicine, and education. The series emphasizes critical studies in the context of emergent and existing digital technologies.

Other recent titles include:

Felicia Wu Song
Virtual Communities: Bowling Alone, Online Together

Edited by Sharon Kleinman
The Culture of Efficiency: Technology in Everyday Life

Edward Lee Lamoureux, Steven L. Baron, & Claire Stewart
Intellectual Property Law and Interactive Media: Free for a Fee

Edited by Adrienne Russell & Nabil Echchaibi
International Blogging: Identity, Politics and Networked Publics

Edited by Don Heider
Living Virtually: Researching New Worlds

Edited by Judith Burnett, Peter Senker & Kathy Walker
The Myths of Technology: Innovation and Inequality

Edited by Knut Lundby
Digital Storytelling, Mediatized Stories: Self-representations in New Media

Theresa M. Senft
Camgirls: Celebrity and Community in the Age of Social Networks

Edited by Chris Paterson & David Domingo
Making Online News: The Ethnography of New Media Production

www.ingramcontent.com/pod-product-compliance
Ingram Content Group UK Ltd.
Pitfield, Milton Keynes, MK11 3LW, UK
UKHW020259051026
15470UKWH00021B/1434

9 781433 160011